Communication Networks in Smart Power Grids

IEEE Press
445 Hoes Lane
Piscataway, NJ 08854

Communication Networks in Smart Power Grids

Boyang Zhou
Zhejiang Lab
Hangzhou, China

IEEE Press
WILEY

Published by John Wiley & Sons, Inc., Hoboken, New Jersey.
Published simultaneously in Canada.

Library of Congress Cataloging-in-Publication Data Applied for:

Hardback ISBN: 9781394285099

Cover Design: Wiley
Cover Image: © Pixel-Shot/Shutterstock

Set in 9.5/12.5pt STIXTwoText by Straive, Chennai, India

To my parents and teachers, whose love shaped me, and to my wife, children, friends, and all who supported me.

Contents

About the Author

Boyang Zhou is a researcher/research expert at the Research Center for High Efficiency Computing Systems at Zhejiang Lab, and an adjunct research fellow at the School of Intelligent Science and Technology, Hangzhou Institute for Advanced Study, UCAS. Prior to this, he served as a senior engineer at State Grid Corporation of China and as a postdoctoral researcher at Zhejiang University. He earned his PhD in computer science from Zhejiang University. He has authored more than 40 papers and holds 30 patents. He has received several prestigious awards, including the first prize for Technological Progress from the China Automation Society, the first prize for Scientific and Technological Progress in Zhejiang Province, and the third prize for Technological Progress in the China Electric Power Technology Progress Awards. His research focuses on computer networks, smart grid communications, data center networks, and industrial Internet security.

Preface

Over the past three decades, communication networks have undergone a transformative digitalization process, evolving electric power grids into smart grids capable of addressing diverse global demands. Modern network architectures now seamlessly integrate optical and wireless communications, utilizing redundancy designs, such as diverse routing, to ensure reliability and optimize grid operations across various regions, including China, the USA, Europe, and Japan. The critical focus on network availability, characterized by low packet loss and low latency, has intensified due to aging grid infrastructure and the escalating demand for energy. High-profile power outages, such as the 2023 Northeast event in the USA and Canada, Hurricane Harvey in 2017, and the Texas winter storm in 2021, underscore the urgency of these challenges. The relentless pursuit of innovative network development is reshaping grid architecture through next-generation solutions that emphasize centralized control flexibility and data caching capabilities, all aimed at establishing high-availability networks for enhanced power grid stability.

This book, *Communication Networks in Smart Power Grids*, serves as a comprehensive guide that delves into the complexities and demands of smart grid communication networks. It is structured to provide readers with a broad and in-depth understanding of both classical and cutting-edge solutions essential for enhancing network availability in this dynamic field. Key areas covered in this book include:

i) **Knowledge about networks**: Developments in smart grid communications and advancements aimed at improving network availability are thoroughly explored. The technical landscape is examined, focusing on strategies and solutions for achieving high network availability in modern power grid operations, alongside the role of sustainable practices in enhancing grid resilience and efficiency.
ii) **Classical approaches**: A comprehensive presentation and analysis of classical approaches that have historically ensured the resilience and reliability of power grid communications is offered. The focus is on the standardized

efforts underpinning the IEC 62439 protocols, which incorporate redundancy concepts and are widely adopted by manufacturers such as Siemens and Texas Instruments. These protocols are deployed at both end nodes and within the network, utilizing 12 distinct approaches, including packet aggregation and fast rerouting, primarily designed for wired networks, particularly in ring topologies or doubly independent routes. The principles, interactions, and applications of these protocols are thoroughly analyzed, alongside their limitations in addressing the evolving demands of modern power grids.

iii) **Challenges and landscape**: Despite the foundation established by IEC 62439, advancements in network architectures have identified five critical demands and research directions needed to tackle ongoing challenges: reliability of data delivery, path and data transport, application data completeness, and cybersecurity threats, particularly link flooding attacks (LFAs). Incidents such as the 2021 LUMA Energy attack in Puerto Rico highlight the urgent need for robust defenses against these threats. These challenges can significantly impact power grid stability by degrading monitoring accuracy and compromising control integrity, thereby increasing the risk of power outages. Addressing these issues requires technologies that span every layer of the TCP/IP stack. While the physical and link layers depend on equipment providers, the flexibility of the network, data transport, and application layers is vital for enhancing overall network availability.

iv) **Cutting-edge solutions**: Innovative solutions that supplement and enhance classical approaches to significantly improve network availability and power grid stability are presented. Key solutions include:
   - The OpenFlow-based Non-Disjoint Path Aggregation Mechanism (OPAM) facilitates reliable data delivery through enhanced redundant route exploitation and zero-time fault recovery for large-scale power networks.
   - The Traffic-Aware Fast Reroute (TA-FRR) solution improves path reliability by utilizing topology redundancy via disjoint subpaths, leading to reduced rerouting time.
   - The Disruption-Resilient Transport Protocol (DRTP) ensures data transport reliability through hop-by-hop multipath retransmissions with less fragile route constructions, thereby maintaining grid state estimation accuracy and grid observability even under significant network disruptions. This solution incorporates a concrete implementation using state transition machines and enhancements for route reliability.
   - The Adaptive RaptorQ Encoder and Decoder (ARED) enhances reliability in wireless cellular networks by employing adaptive parameter tuning based on channel quality feedback.
   - The Persistent Route Diversification Defense Mechanism (PRDD) is designed to protect against LFAs, effectively mitigating congestion states.

v) **Practice guidelines**: Practical guidelines are provided to assist readers in developing and implementing both classical and cutting-edge solutions in real-world power grid environments. The environment is used to test and assess the impact of a protocol on the power grid's monitoring and control. The content covers features, implementation practices, and unresolved concerns, equipping readers to construct high-availability networks for modern power grids.
vi) **Future prospects**: Insights into future prospects for operators, researchers, and educators are offered, encouraging strategic shifts to navigate the challenges posed by an increasingly complex energy landscape.

By synthesizing theoretical knowledge with practical applications, this book aims to equip professionals and students alike with the tools needed to navigate the multifaceted challenges of smart grid communications. Each chapter includes pedagogical questions, with answers available online, fostering active engagement and deeper understanding.

We invite you to explore the knowledge, challenges, and solutions in smart grid communications, where availability is not merely a goal but a vital aspect of sustainability, adaptability, and innovation in this dynamic field.

October 2024

Boyang Zhou
Zhejiang Lab
Hangzhou, China

Acknowledgments

This book marks the culmination of a thrilling exploration into the domain of pioneering highly-available communication networks in smart electric power grids. Throughout this captivating journey, I express my sincere gratitude for the invaluable support received from various sources.

I extend special thanks to the National Natural Science Foundation of China (62102375) and the Key Research and Development Program of Zhejiang Province of China (2020C01021) for their generous support, laying a foundation for this research endeavor. I am also grateful for the support from Zhejiang Lab, my alma mater Zhejiang University, and the State Grid Corporation of China, where I gained significant experiences that greatly contributed to the creation of this book.

I am deeply appreciative of the researchers, developers, and operators who dedicate themselves to advancing smart grid communication technologies. Your unwavering commitment has been the driving force behind the exploration of smart power grid communication networks.

A special acknowledgment goes to the collaborative efforts of industry, academia, and research communities. The synergy of these domains has played a pivotal role in fostering innovation, enhancing the resilience of communication networks, and ensuring industrial reliability in smart grids.

Heartfelt thanks to supervisors, mentors, and colleagues whose guidance and insights have shaped the trajectory of this research. Your wisdom and encouragement have been invaluable, guiding this endeavor through challenges and triumphs.

To friends and family, your unwavering support has been a constant source of inspiration, fueling the passion for unraveling the mysteries of resilient communication networks.

This book stands as a testament to the collaborative spirit that defines the pursuit of knowledge. May it serve as a valuable resource for those delving into the intricate heart of network dynamics, offering insights and inspiration for the ongoing exploration of smart grid communication systems.

Boyang Zhou

Acronyms

AC	Alternating Current
ACSI	Abstract Communication Service Interface
ADCT	AdaRQ Dynamic Coding Tunnels
ADSS	All-Dielectric Self-Supporting Cable
AFR	Active Fault Recovery
APS	Automatic Protection Switching
ARED	Adaptive RaptorQ Encoder and Decoder
ARFCN	Absolute Radio Frequency Channel Number
ARQ	Automatic Repeat reQuest
AS	Autonomous System
ASIC	Application-Specific Integrated Circuit
BCCH	Broadcast Control Channel
BEC	Binary Erasure Channel
BFS	Breadth-First Search
BGP	Border Gateway Protocol
BMv2	Behavioral Model v2
BPDU	Bridge Protocol Data Unit
BRP	Beacon Redundancy Protocol
CDF	Cumulative Probability Distribution Function
CDT	Capsule Delivery Time
CIAT	Capsule Inter-Arrival Time
CPN	Cognitive Packet Network
CPRF	Conditional Probability of Retransmission Failure Cases
CR	Confidence Ratio
CRC	Cyclic Redundancy Check
CRP	Cross-Network Redundancy Protocol
CS	Content Store
CSMA	Carrier Sense Multiple Access

DANH	Doubly Attached Node with HSR protocol
DC	Direct Current
DDoS	Distributed Denial-of-Service
DNP3	Distributed Network Protocol
DPDK	Data Plane Development Kit
DPE	Defense Policy Enforcement
DPG	Defense Policy Generation
DRP	Distributed Redundancy Protocol
DRTP	Disruption-Resilient Transport Protocol
DSP	Disjoint Subpath
DTU	Distribution Terminal Unit
EEDT	End-to-End Payload Packet Delivery Time
EEFR	End-to-End Failure Rate in Packet Delivery
EELR	End-to-End Packet Loss Rate
EPON	Ethernet Passive Optical Network
EWSSA	Electric Wireless Security Situational Awareness System
FCS	Frame Check Sequence
FEC	Forward Error Correction
FI	Fragility Index
FIB	Forwarding Information Base
FLP	Fault Link Percentage
FPGA	Field Programmable Gate Array
FPR	False Positive Rate
FRR	Fast Reroute
FTU	Feeder Terminal Unit
GOOSE	Generic Object-Oriented Substation Event
GPON	Gigabit Passive Optical Network
GPS	Global Positioning System
GSM	Global System for Mobile Communications
HARQ	Hybrid Automatic Repeat-reQuest
HRRT	Half Round-Trip Time
HSR	High-availability Seamless Redundancy
ICMP	Internet Control Message Protocol
ID	Identification
IED	Intelligent Electronic Device
IP	Internet Protocol
IT	Information Technology
ITP	Internet Transport Protocol
IoT	Internet-of-things
KSP	K Shortest Paths
LCLA	Link Congestion Level Analysis
LFA	Link-Flooding Attack

LLDP	Link Layer Discovery Protocol
LLR	Link Loss Rate in Packets
LOF	Local Outlier Factor
LQI	Link Quality Indicator
LRT	Link Recovery Time
LSA	Link State Advertisement
LSDB	Link State Database
LSR	Label Switch Router
LTE	Long-Term Evolution
LoRa	Long-Range Radio
MADSWIP	Maximally Disjoint Shortest & Widest Paths
MAR	Messages in DRTP Arrived at the Router Per Second
MMS	Manufacturing Message Specification
MPLS	Multiprotocol Label Switching
MPPS	Million Packets Per Second
MPSG	Multipath Subgraph
MPSGA	Multipath Subgraph Generation Algorithm
MPTCP	Multipath Transmission Control Protocol
MRDP	Mean Rate of Data Flows
MRP	Media Redundancy Protocol
MSR	Message Sent and Received by Each Router of the PP in Delivering a Capsule
MSTP	Multi-Service Transport Platform
MTD	Moving Target Defense
MTTFN	Mean Time to Failure of Network
MTTRN	Mean Time to Repair Network
MU	Merging Unit
NASPI	North American Synchrophasor Initiative
NDN	Named Data Network
NDPA	Non-Disjoint Path Aggregation
NF	Noise Figure
NIST	National Institute of Standards and Technology
NMN	Number of Messages Sent Per Node of PP
NTH	Non-Threatened Host
NTP	Network Time Protocol
ODPG	Optimal Disjoint Paths Generation
OF	OpenFlow
OGR	Observable Grid Ratio
OOR	Out-of-Order Rate at the Receiver Side
OPAM	OpenFlow-based Non-Disjoint Path Aggregation Mechanism
OPC-UA	Open Platform Communications Unified Architecture
OPGW	Optical Ground Wire

OSPF	Open Shortest Path First
OT	Operational Technology
OTN	Optical Transport Network
P4	Programming Protocol-Independent Packet Processors
PAFM	Packet Aggregation and Forwarding Module
PDC	Phasor Data Concentrator
PDF	Probability Density Function
PEGASE	Pan European Grid Advanced Simulation and State Estimation
PIAT	Packet Inter-Arrival Time
PIT	Pending Interest Table
PLC	Programmable Logic Controller
PMS	Production Management System
PMU	Phasor Measurement Unit
PON	Passive Optical Network
POR	Packet Out-of-Order Rate
PP	Primary Forwarding Path
PR	Persistent Route
PRDD	Persistent Route Diversification Defense
PRF	Probability of Retransmission Failure Cases
PROTAG	Proxy-based Moving Target Architecture
PRP	Parallel Redundancy Protocol
PRT	Packet Reordering Time at the Consumer
PTN	Packet Transport Network
QSDP	Qualified Source–Destination Pairs
QSRP	Quality of Service in Receiving a Packet
QoS	Quality-of-Service
QuadBox	Quadruple Port Device
R2T	Rapid and Reliable Transport Mechanism
RFR	Reroute Failure Rate
RIP	Routing Information Protocol
RNN	Recurrent Neural Network
RQ	RaptorQ
RRP	Ring-based Redundancy Protocol
RRV	Resilient Route Vector
RSDD	Resilient Sensor Data Dissemination
RSG	Resilient Subgraph
RSGA	RSG Generation Algorithm
RSGV	RSG vector
RSP	Redundant Sub-Path
RSSI	Received Signal Strength Indicator
RSTP	Rapid Spanning Tree Protocol

RTDS	Real-Time Digital Simulator
RTT	Round-Trip Time
RTU	Remote Terminal Unit
RedBox	Redundant Box
SC	South Carolina
SCADA	Supervisory Control and Data Acquisition
SCL	System Configuration Language
SCSM	Specific Communication Service Mapping
SCTP	Stream Control Transmission Protocol
SDATP	Adaptive Transmission Protocol based on Software-Defined Networking
SDH	Synchronous Digital Hierarchy
SDN	Software-Defined Networking
SDR	Software-Defined Radio
SE	State Estimation
SFH	States of Forwarding History
SGCC	State Grid Corporation of China
SONET	Synchronous Optical Networking
SPT	Shortest Path Tree
SRAM	Static Random-Access Memory
SRE	Satisfaction Ratio of EEDT
SROP	Recent Out-of-order Packets
SRRP	Success Rate of Receiving a Packet
STD	Standard Deviation
STM	State Transition Machine
STP	Spanning Tree Protocol
SV	Sampled Values
TA-FRR	Traffic-Aware Fast Reroute
TCP	Transmission Control Protocol
TH	Threatened Host
TLV	Type–Length–Value
TTL	Time-To-Live
TTU	Transformer Terminal Unit
UDP	User Datagram Protocol
VANET	Vehicle Ad Hoc Network
WAMS	Wide-Area Monitoring System
WSN	Wireless Sensor Network
WiMAX	Worldwide Interoperability for Microwave Access
iPRP	Improved Parallel Redundancy Protocol
p.u.	power unit

About the Companion Website

This book is accompanied by a companion website:

www.wiley.com/go/SmartPowerGrids1e

The website includes:

Answers to the Questions in the Chapters

1

Introduction

Powering the journey of electric energy from generation plants to end-users involves a complex web of transmission, distribution, and consumption, all accomplished by the power grids [Kirtley, 2009]. In the evolving landscape of this electrical infrastructure, smart power grids [Fang et al., 2012] represent a significant advancement over traditional power systems. They integrate modern communication networks with traditional electrical grid components, enabling a more responsive, efficient, and reliable electricity supply. As Figure 1.1 depict, the architecture of a smart power grid is composed of two categories of systems: primary systems and secondary systems, each playing a crucial role in the grid's functions.

Primary systems. Primary systems pave the foundation of the power grid [Kirtley, 2009]. These components are responsible for the direct transmission and transformation of electrical energy. They operate at high-voltage levels and include essential elements such as transformers, circuit breakers, transmission lines, and generators. Primary devices are designed to handle substantial amounts of power and are critical for maintaining the flow of electricity from generation points to end users. Their robust design ensures that the grid can transmit electricity efficiently over long distances, while also protecting against faults and ensuring system stability.

The primary systems mainly involve high-voltage fundamental devices, responsible for generating, transmitting, and distributing electricity. The electricity is usually in Alternating Current (AC), while in hybrid with Direct Current (DC) for modern and future power grids to reduce power losses. These devices generally include power transformers, power inverters, power lines, power transducers, circuit breakers, and reactive compensation devices, which is introduced below.

- The power transformers facilitate the efficient transfer of electricity across different voltage levels. During operation, transformers can step up the voltage for long-distance travel, reducing transmission losses, and step the voltage

Communication Networks in Smart Power Grids, First Edition. Boyang Zhou.

Companion Website: www.wiley.com/go/SmartPowerGrids1e

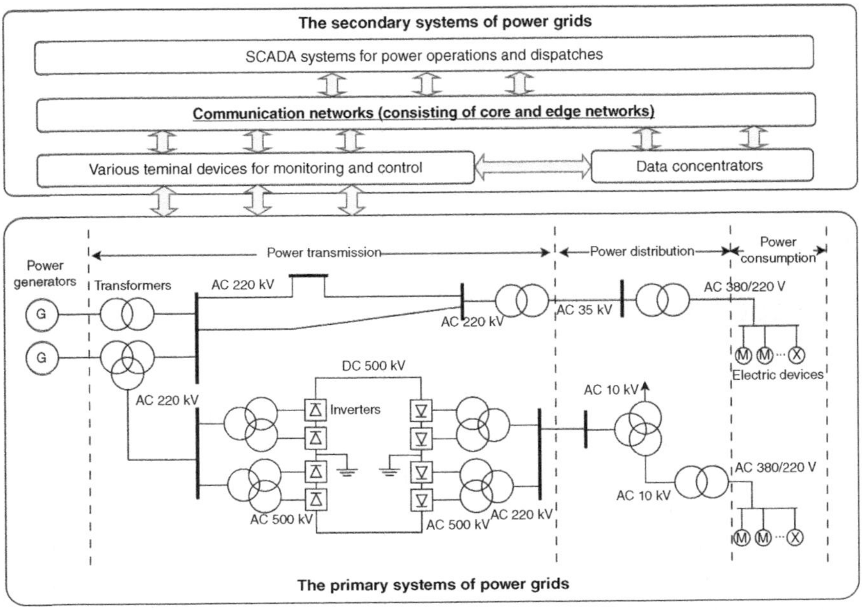

Figure 1.1 Overview of power grids.

down to safer, usable levels, as electricity approaches distribution networks and consumers.

- The power inverters can convert between DC and AC, which is useful for ultra-high-voltage (1000 kV~) and extra-high-voltage (330–750 kV) power grids, and interconnections with renewable energy sources like solar panels.
- The power transducer measures electrical quantities such as voltage, current, and frequency, providing real-time data that allows control systems to assess the grid's health and performance. Power lines ensure the physical transfer of electricity between substations and end users, forming a vast network over long distances.
- The circuit breakers can instantly disconnect a section of the grid in the event of a fault, preventing potential damage to equipment and ensuring the safety of operators.
- Reactive compensation devices, such as capacitor banks and synchronous condensers, maintain the grid's voltage stability by managing reactive power flows.

Together, these primary systems form the physical layer of the smart grid, interacting with advanced control and communication systems to ensure safe, reliable, and efficient electricity delivery.

Secondary systems. Based on the primary systems, the integration of secondary systems into the grid architecture transforms traditional electrical systems into the communication networks. The networks involve wide range of distributed power substations and central servers in large-scale geographical areas. Each substation encompasses various kinds of terminal devices and data concentrators. Meanwhile, the control servers are represented by the Supervisory Control and Data Acquisition (SCADA) systems and Phasor Data Concentrator (PDC) [Thomas, 2020]. Together, these devices of the substations, equipped with billions of sensors, meticulously measure electric parameters such as voltage magnitudes, active power, reactive power, and phase angles [Huang et al., 2012]. Meanwhile, the role of actuators comes into play as they execute commands, regulating switch gears and transformers. The interactions between the control servers and the substations enable real-time monitoring, data collection, fault detection, and automated control processes, which are vital for maintaining grid reliability and performance. The communication interaction between these devices not only enhances the functionality of the grid but also lays the foundation for future innovations in energy management and sustainability.

After evolving for more than three decades, communication networks have become an integral part of smart electric power grids [Saleem et al., 2019; NASPI, 2019]. Within this networked environment, communication network reliability takes center stage. The system must exhibit resilience, capable of navigating through multiple link faults – whether triggered by natural disasters, network glitches, or extensive cascading power line failures [Bie et al., 2017]. Additionally, frame or bit errors during transmission, stemming from signal degradation, noise, or optical impairments, pose challenges. These issues manifest as severe packet losses and an increased Link Loss Rate in Packets (LLR) [IEC, 2014]. At the transport layer, such losses can amplify the End-to-End Failure Rate in Packet Delivery (EEFR) for sensor data dissemination protocols. Ensuring robust communication networks is paramount in this dynamic landscape.

1.1 Purposes and History of Networks

1.1.1 Purposes

Over the decades, communication networks [Yan et al., 2013] have played an essential role in connecting billions of automating terminals, such as sensors and actuators, for supporting information and signal exchange during power monitoring and dispatch. Building these two kinds of Internet-of-things (IoT) networks is always the continuous goals of existing power grid operators across the world, requesting endless efforts. These goals can be summarized below:

- **Supporting comprehensive services**: This network not only encompasses the traditional functions of grid production and enterprise management but also paves the way for future comprehensive energy services and customer-centric service systems. Through shared data and resources, it integrates internal operations with external services, connecting the entire energy business chain – from cloud to network to terminal. It enables data to be collected once and applied universally, transforming business systems from vertical to horizontal structures. In doing so, it supports the development of an interconnected energy ecosystem.

- **Providing ubiquitous connectivity**: The network ensures seamless, real-time interconnection between people, machines, and devices, anytime and anywhere. It integrates traditional grid equipment, information systems, and internal staff, along with participants in source–grid–load coordination, upstream and downstream power enterprises, and external service providers. At the terminal layer, it connects all devices; at the network layer, it offers pervasive communication capabilities; and at the platform layer, it manages and controls data and devices comprehensively. The integration of the ubiquitous network with the next-generation power system has become a vital component of modern energy enterprises.

This network, evolving alongside the power grid, forms the foundation of the information and communication infrastructure. With this network, all elements can be comprehensively interconnected. The full interconnection across all aspects of the power system enables the real-time collection of data from power production, operation, and control. With the widespread adoption of intelligent terminals and mobile applications, this network supports the next-generation power system and delivers diverse energy services to a range of terminal devices. The benefit of this network focuses on bringing a standardized data and service system that spans the entire energy business chain. This ensures real-time consistency, open data sharing, and unified management across all business scopes, data types, and time dimensions, enabling seamless integration and service delivery across the energy sector. In this way, the network fosters optimized architectures, improved operational efficiency, smarter decision-making, and increased value across the power sector.

1.1.2 History

Building such networks involve evolving roadmap that initiated the prosperity of digitalization in the 1990s and accelerated in the early 21st century. Before this period, the old analog communication, popular from 1970s to 1980s, was challenged by the development of SCADA systems that allow to monitor and control grid assets remotely, such as substations and transmission lines, reducing the need for manual intervention. This old communication technology impeded the increased automation need for real-time data collection and control. Later, this technology was replaced by more advanced communication technologies that can be categorized into the two classes below:

a) By 1990s, the fiber-optic communications [Tornatore et al., 2020], such as the Synchronous Digital Hierarchy (SDH), the Optical Transport Network (OTN), the Gigabit Passive Optical Network (GPON), and the Ethernet Passive Optical Network (EPON), provided and is providing faster, more reliable data transmission, enhancing the ability of utilities to manage grid operations [NASPI, 2019]. In this period, the digital protective relays began replacing traditional electromechanical relays. These digital devices enabled faster, more reliable responses to faults in the grid and introduced programmability, which allowed for more sophisticated grid protection schemes. Additionally, the development of Phasor Measurement Units (PMUs) and their integration into Wide-Area Monitoring System (WAMSes) marked a significant leap in grid stability and situational awareness [Thomas, 2020; Xie et al., 2006]. PMUs offered real-time monitoring of the grid's electrical parameters, improving reliability.
b) From 2000s to now, wireless communications were and are being increasingly integrated into grid operations, enabling more comprehensive control and monitoring. This kind of technologies began with Worldwide Interoperability for Microwave Access (WiMAX) and cellular networks, such as 2G to 5G networks, which mark the full-scale deployment of digital devices throughout the grid. This brought about the growth of smart power meters and data concentrators. Since 2010s, wireless sensor networks, with low power usage, have facilitated more granular control over grid assets.

Development of the two kinds of networks above has now formed a hybrid optical–wireless network architecture which has been widely adopted across the world. Examples include:

a) the nation-wide ubiquitous power IoT networks built by the State Grid Corporation of China, featured with the 2G to 5G networks [Huawei Technologies Co., Ltd., 2025; Yuan et al., 2014],
b) the networks built by the South California Edison Corporation of United States, employing WiMAX, 4G Long Term Evolution (LTE), and radio-based systems [Vyas and Combs, 2011],
c) the networks of European power grids, widely using LTE or narrow-band IoT networks for connecting huge amount of smart meters [Sequans, 2023; Giordano et al., 2013],
d) the networks employed by Tokyo Electric Power Company of Japan, used 4G, 5G, and WiFi mesh networks [Weissberger, 2022], and
e) the networks built by India, utilizing WiMAX, WiFi, and ZigBee for extending smart meter connections [Kumar et al., 2022].

The hybrid architecture is an essential part of modern smart grids, offering a balance between high-performance, high-capacity fiber-optic networks, and the flexibility and coverage of wireless technologies. These communication technologies can enhance grid management, integrate renewables, and respond to the growing complexity of modern power networks.

1.2 Networks in Power Grids

Today, these communication networks in smart power grids are ubiquitously and comprehensively integrated with power grid systems on a large scale, forming large-scale and area-crossing IoT networks IEEE 2013. By incorporating new communication technologies, these networks achieve mutual penetration and deep integration with the next-generation power system. They enable real-time, online connectivity between people, machines, and devices at every stage of energy transmission, distribution, and consumption. This integration supports and streamlines operations across grid production, enterprise management, and customer services, forming the backbone of a more efficient, economical, and secure power grids at national and regional levels.

With those networks, a number of data concentrators and terminal devices, involving lots of sensors and actuators, interact with SCADA systems and Production Management System (PMSes) [Thomas, 2020]. Wherein, the terminal device directly controls the underlying electrical equipment for providing power flows for end consumers and the primary systems. This process is complicated, but we can understand their building purposes and functions, which is helpful to see why we need these networks to be highly available.

1.2.1 Network Structures

Figure 1.2 specifies the overall structure of IoT communication networks of power grids, where the network is tightly coupled with physical power lines for supporting interactions of electronic devices and SCADA systems and PMSs, and its performance can be challenged by many adversarial factors.

1.2.1.1 Network Types

As depicted in the middle of Figure 1.2, the modern networking structure can be partitioned into backbone network and edge network. These networks are specified as follows.

i) **Backbone network**: A resilient backbone network, woven with optical ground wires adhering to the IEC 60794-4-10 standard, creates a robust

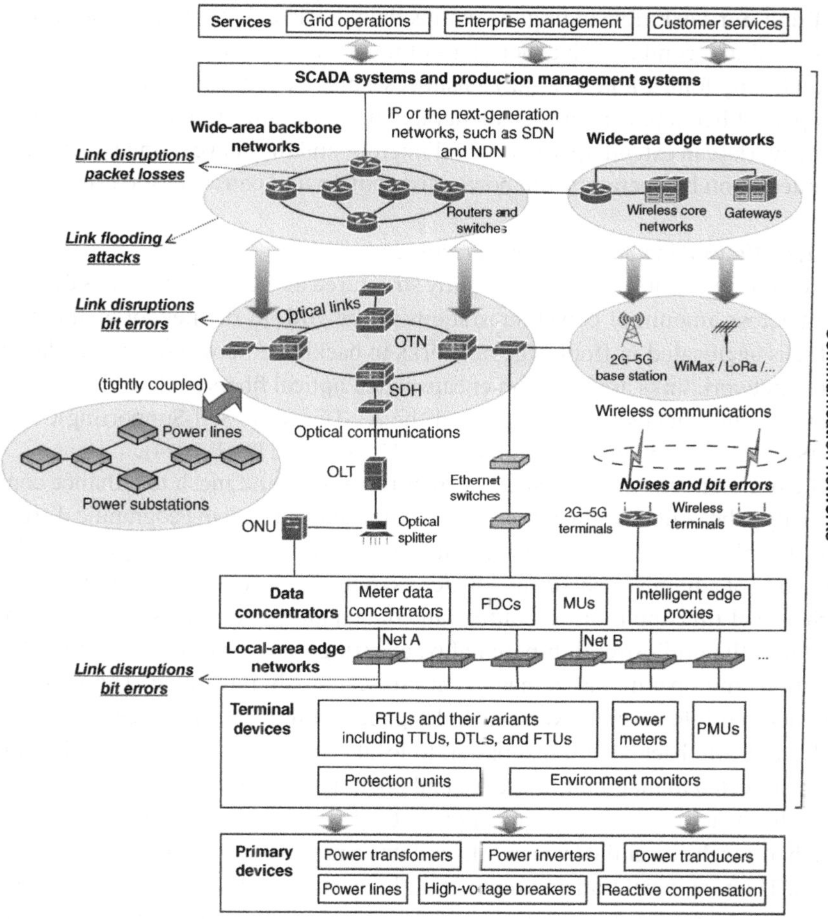

Figure 1.2 IoT communication networks in power grids.

framework of redundancy through SDH/Synchronous Optical Networking (SONET)/OTN and EPON/GPON technologies. This interconnected and ringed architecture, detailed in references [IEC, 2014], [Zhou et al., 2019], and [Zhou et al., 2022], caters to bandwidth demands ranging from 10 to 400 Gbps. Its strategic deployment is especially prominent in power transmission and distribution grids. Different types of backbone networks are deployed in scenarios where communication distances matter. The communication distances of SDH/SONET/OTN can extend to several hundred kilometers, making them suitable for power transmission grids. Meanwhile, EPON/GPON cover distances of up to 20 kilometers and 60 kilometers, respectively, making them preferred choices for power distribution networks.

ii) **Edge network**: Expanding the connectivity horizon with wireless networks, spanning second to fifth generations of mobile networks (2G to 5G), WiMAX, and Long Range Radio (LoRa), ensures seamless connectivity in areas where optical fiber deployment is challenging. The communication distance of these networks can extend up to several kilometers, and they have found widespread integration by operators into power distribution and consumption grids.

1.2.1.2 Networking Strategies and Practices

The communication networks are tightly structured along the power lines deployment across amount of power substations in the gird. In the networks, the data traffic is aggregated in from edge networks to backbone networks layer by layer. These network links are based on encapsulated optical fibers such as the Optical Ground Wire (OPGW) [IEC, 2014] and the All-Dielectric Self-Supporting Cable (ADSS) [IEEE, 2019] being compliant to ITU-T G.652 [ITU-T, 2024].

The backbone network is usually structured in ring and mesh to enhance connectivity to be fault tolerant, where the network across different geographical areas must be tolerant to no less than two fault links. The network topology forms the backbone of a highly reliable and efficient system that supports critical functions such as real-time monitoring, control, and data management. This infrastructure is vital to the grid's overall functionality, ensuring that power transmission, distribution, and operational management are seamlessly integrated. The network is structured into several layers, each designed to fulfill specific roles within the grid, from core backbone transmission to local access networks, all of which are interconnected to maintain resilience and operational efficiency.

At the heart of the communication network is the transmission backbone, which forms the essential framework for data transport across provincial and local levels. The provincial transmission network, operating with dual-plane architectures (SW-A and SW-B), ensures redundancy and disaster recovery capabilities. The SW-A plane, utilizing MSTP/SDH technology, primarily supports production control, covering high-voltage substations (220 kV and above) and ensuring that critical control signals are prioritized. The SW-B plane, on the other hand, relies on OTN technology to carry management information services, with a larger capacity to handle the growing demands of administrative data across provincial and municipal sites. These two planes work in parallel, ensuring that the grid remains operational even in the event of a failure on one side, thus enhancing overall network reliability.

The local and county transmission networks mirror the provincial structure, maintaining dual-plane designs (DW-A and DW-B) for redundancy. The DW-A plane is also based on Multi-Service Transport Platform (MSTP)/SDH [Sibley, 2020], focusing on controlling local substations (35 kV and above), while the DW-B plane, utilizing Packet Transport Network (PTN) technology [Murakami and Koike, 2014], is dedicated to carrying management information. This layered approach, with dedicated transmission channels for both control and management data, not only ensures security by isolating critical operational signals but also maximizes the use of network resources. The network's ring topology, combined with point-to-point connections at the access layer, adds another level of resilience, ensuring that even localized failures do not disrupt the broader communication system.

Optimizing these transmission networks is crucial for future scalability and performance. Network administrators continuously improve the layered aggregation structures to ensure efficient traffic management and better load distribution. This optimization enhances the network's ability to expand and increases its capacity for handling new services as the power grid evolves. In addition, ensuring high node connectivity and implementing N-1 reliability practices – where the network can withstand the failure of any single component – are key to maintaining uninterrupted operations across the grid.

Alongside the transmission network, the comprehensive data network underpins much of the grid's communication infrastructure. This network adopts advanced technologies such as IPv4/IPv6 and Multiprotocol Label Switching (MPLS) [Rosen et al., 2001] VPN to enable flexible and secure data transmission. The backbone of the data network follows a dual-star topology, with two core nodes positioned at the provincial aggregation points. These core nodes are connected by dual routers and dual 10GE links, forming a high-reliability structure that ensures seamless communication between core and backbone layers. The backbone network connects upward to the national data communication network and downward to local access networks, creating a cohesive and hierarchical communication structure that supports various power grid operations.

The access network plays an equally important role in ensuring that all levels of the grid, from provincial to local substations, have secure and efficient connectivity. It is designed with aggregation and access layers that connect the grid's different operational units. Provincial and municipal companies link through dual 10GE connections to the backbone, ensuring redundancy and robust communication. At lower levels, county companies and local substations use dual GE links and specialized technologies like MSTP [Sibley, 2020] and PTN [Murakami and Koike, 2014] to establish secure communication pathways, depending on their specific operational needs. This hierarchical structure allows for seamless integration of substations, control centers, and management systems into the broader network, ensuring that even the smallest nodes remain connected and operational.

At the terminal level, the communication access network is divided into 10 kV and 0.4 kV systems. The 10 kV network handles critical tasks such as distribution automation and power quality monitoring, serving as the primary link for control systems within the grid. The 0.4 kV network focuses on tasks like electricity information collection and load management, ensuring that even low-voltage areas of the grid remain integrated into the larger system. Both networks leverage fiber-optic technology, where possible, to ensure high-speed and reliable communication, with alternative technologies like wireless and carrier-based systems used in areas where fiber is impractical.

The backbone and access layers of the smart grid's communication network are supported by robust synchronization and management systems. The frequency synchronization network ensures that the entire grid operates on a unified time base, which is essential for coordinating grid operations. This network is anchored by a provincial clock synchronization zone, which is supported by high-precision cesium clocks and satellite systems such as Beidou or Global Positioning System (GPS) [Hofmann-Wellenhof et al., 2012]. These synchronization systems ensure that both production and management services within the grid operate in harmony, minimizing the risk of data misalignment or timing issues.

Furthermore, the communication management system serves as an integrated platform that oversees the entire communication infrastructure [Der Wielen, 2020]. It centralizes real-time monitoring, resource management, and operational controls, enabling the grid to function as a cohesive unit. This system is designed to integrate with key management tools, allowing for streamlined communication across various levels of grid management. As the grid expands, continuous improvements in standardization, intelligence, and usability within the communication management system ensure that operational efficiency keeps pace with the growing complexity of the grid's infrastructure.

The communication network also relies on a robust power supply system to maintain operation during outages or disruptions. At major substations and control centers, dual independent power systems are installed to provide redundancy, ensuring that $N-1$ reliability is maintained [IEC 62439, 2008]. These power systems are backed up by batteries capable of providing at least four hours of emergency power, guaranteeing that critical communication systems remain functional even during power disruptions.

In conclusion, the communication network topology of smart power grids is an intricate and highly resilient system. Through its layered structure, dual-plane redundancy, and advanced technologies, it supports the efficient and reliable operation of the grid. From backbone transmission networks to terminal access systems, every component of this network is designed with reliability, scalability, and security in mind, ensuring that the smart power grid can continue to meet the growing demands of modern power management.

1.2.2 Networked Systems

1.2.2.1 Terminal Devices

Terminal devices provide monitoring and control points. The terminal devices are the critical interface between the physical grid and its control and communication layers. Chief among them are the Remote Terminal Units (RTUs) [Thomas, 2020], which monitor electrical variables such as voltage, current, and phase angles, and report this data to centralized control systems. RTUs operate as the data acquisition hubs for substation automation, enabling real-time monitoring of power flows and system conditions. Specialized devices, such as Transformer Terminal Units (TTUs), Distribution Terminal Units (DTUs), and Feeder Terminal Units (FTUs) [Han et al., 2014], are deployed for specific components in the grid, enhancing the granularity of data collection. These terminal devices not only provide detailed performance data but also facilitate rapid control actions, such as opening or closing circuit breakers, to protect critical infrastructure in case of faults or overloads.

Complementing RTUs are *smart power meters* [Barai et al., 2015], which are key to the implementation of demand-side management in smart grids. These meters track energy consumption patterns at consumer endpoints in real-time and allow utilities to implement dynamic pricing models or grid load shedding during peak demand periods. For real-time system stability, PMUs measure electrical waves on an electricity grid using synchronized phasor measurements [IEEE, 2011]. These devices are crucial in monitoring grid stability, detecting grid anomalies, and enabling WAMSs to manage the dynamic behavior of the grid under various operating conditions [Lin et al., 2012; Xie et al., 2006].

Additionally, protection units and environmental monitors serve to enhance the reliability and safety of the grid [Thomas, 2020]. Protection units guard against system faults by automatically isolating problem areas, thus preventing the cascading failures that can lead to widespread outages. Environmental monitors, on the other hand, track ambient conditions like temperature, humidity, and other environmental factors that could impact grid performance. These terminal devices act as critical agents of the smart grid's real-time control systems, enabling a higher level of operational intelligence and automation in grid management.

1.2.2.2 Data Concentrators

Data concentrators enables wide-area and large-scale networked power grid control [IEEE, 2011; Thomas, 2020]. Data concentrators play a pivotal role in the communication networks of a smart power grid by gathering, processing, and forwarding large volumes of data from the terminal devices to the central control systems. These concentrators reduce the burden on communication channels by aggregating data from geographically distributed endpoints and delivering it in a consolidated form. *Meter data concentrators* collect energy usage data from numerous smart meters, compressing and transmitting this information to utilities for billing, load forecasting, and energy management [Barai et al., 2015]. The scale of data handled by these devices is significant, as millions of smart meters may be feeding information into a concentrator.

For real-time grid monitoring, Phasor Data Concentrators (PDCs) process synchronized phasor data from PMUs, merging inputs from across the grid into a unified dataset [IEEE, 2011; Thomas, 2020]. PDCs are integral to real-time situational awareness, enabling grid operators to detect instability patterns or faults across vast areas within milliseconds. This is critical for ensuring grid stability in high-demand scenarios or during extreme weather conditions.

Merging Units (MUs) combine multiple measurements from various sensors to ensure synchronized data flows from substations and other key assets in the grid [IEC, 2011d]. These units often use the IEC 61850 protocol [IEC, 2020], which standardizes communication within substations and across the broader network. Additionally, *intelligent edge proxies* enhance the smart grid's efficiency by performing pre-processing at the network edge, reducing latency and the volume of data transmitted to central servers [Feng et al., 2021]. These proxies also contribute to local decision-making capabilities, such as detecting anomalies or performing load-balancing operations before reporting back to central systems. The role of data concentrators is fundamental in transforming vast amounts of raw data into actionable intelligence for grid control, allowing for better forecasting, faster response times, and enhanced resilience.

1.2.2.3 SCADA

The SCADA system is the heart of modern grid operations and dispatches, which provide real-time monitoring, control, and automation of the power grid [Thomas, 2020]. SCADA systems form the backbone of the grid's control infrastructure, enabling operators to oversee and manage assets across vast geographical areas from centralized control centers. These systems gather data from RTUs, PMUs, and other terminal devices, offering a comprehensive view of the grid's operational status. SCADA systems are also responsible for executing control commands, such as remotely opening or closing breakers, adjusting transformer tap settings, or rerouting power flows to balance loads. Furthermore, advanced SCADA systems are integrated with predictive maintenance algorithms, enabling utilities to identify potential equipment failures before they occur and proactively manage grid health.

Alongside SCADA systems, production management systems optimize energy generation and distribution by aligning production with real-time demand. These systems coordinate the dispatch of power from various sources, including traditional generation (such as coal and natural gas) and renewable energy sources (such as wind and solar), to ensure grid reliability. In modern smart grids, production management systems must balance the intermittency of renewables with grid stability requirements. As grid operators shift toward decarbonized energy systems, these systems use sophisticated forecasting algorithms, demand response techniques, and energy storage solutions to stabilize energy supply. The coordination between SCADA and production management systems is crucial for maintaining not only operational efficiency but also the economic viability of the grid in a future where decentralized energy generation is increasingly common.

By integrating real-time data collection, control, and automation with predictive and adaptive algorithms, SCADA and production management systems form the nervous system of the smart grid, ensuring it can respond dynamically to changes in demand, generation capacity, or system health.

1.3 Evolving Networks for Smart Grids

As digitalization advances within the cyber–physical realm of smart grids, the interactions between networked components have grown increasingly diverse and complex. This evolution necessitates the development of more advanced monitoring, control, and dispatch capabilities within power grids. Key functions, such as detecting line and transformer overloads, identifying equipment faults, responding to power outages, and managing environmental changes, now rely on more sophisticated and efficient communication networks.

These networks serve as the backbone for both wide-area and local-area control systems, enabling unified management across the grid. Consequently, modern power grid networks must exhibit flexibility, security, efficiency, and reliability to meet these expanding demands. Often referred to as the "Energy Internet" [Wang et al., 2018], this concept draws parallels with the transformative impact of the traditional Internet on information exchange, underscoring the importance of robust communication infrastructures. The primary goal of these networks is to protect power grids from cascading failures and outages through resilient architectures. Designing such architectures with resilience at their core is critical to ensuring continuous grid operation.

Governments and research institutions worldwide have increasingly recognized the importance of resilient communication systems in smart grids. Agencies such as the U.S. Department of Energy [United States Department of Energy, 2014], the German Federal Government [Federation of German Industries (BDI), 2008], and the Japan Digital Grid Group [Boyd, 2013] have all prioritized advancements in smart grid communication technologies. In 2013, the Chinese government, in collaboration with the State Grid Corporation of China (SGCC), launched the Global Energy Internet project [Liu, 2015], which aimed to establish a strong, intelligent grid infrastructure supported by "Internet+" innovations. These global efforts highlight the growing significance of transitioning from traditional power grid architectures to more sophisticated, next-generation networks.

Emerging network architectures are designed to meet higher standards of reliability, performance, efficiency, flexibility, and manageability. These developments are driven by the increasing complexity of power grids, which must accommodate rising consumer demand for electricity while integrating a diverse range of energy sources. These sources include not only traditional fossil fuels such as coal and nuclear energy but also renewables like wind, solar, and hydroelectric power, distributed across vast geographic regions. This complexity in energy generation, distribution, and consumption places significant pressure on grid operations.

In addition, the evolution of power meters – now equipped with advanced computing capabilities – has enabled more complex metering and communication functions. These smart meters, when combined with the rising number of plug-in electric vehicles, exacerbate data management challenges within the grid. The growing interconnections between terminals, data concentrators, and SCADA systems [IEEE, 2013] further contribute to this complexity, requiring new and innovative communication solutions. These challenges can be broadly classified into two key areas:

a) **Diverse communication requirements**: Different services within the power grid have distinct communication requirements, particularly in terms of real-time data transfers and reliability. For instance, monitoring systems must be capable of transmitting critical grid health information rapidly and without fail. To support such varied and dynamic service needs, a new network architecture is required – one that offers granular control over data flows, ensuring high availability for essential services. *Software-Defined Networking (SDN)* [Farhady et al., 2015] is an architecture that addresses these needs by centralizing the control of network traffic. SDN enables power grids to adapt quickly to changing conditions, optimize performance, and improve resilience by isolating network faults and rerouting traffic when necessary.
b) **Hierarchical and on-demand communication patterns**: Unlike traditional Internet communication, which relies on point-to-point data transfers, power grids employ hierarchical data collection, multicasting, and on-demand data requests. For example, smart meters may periodically report data to centralized systems, while certain services, such as demand–response mechanisms, might only request data when specific conditions are met. This hierarchical and conditional communication pattern necessitates an architecture that ensures secure and efficient data transfer for mission-critical grid

services. *Named Data Network (NDN)* [Zhang et al., 2014] offers a promising approach by shifting the focus from the location of data to the content itself. By prioritizing information over its physical location, NDN enhances both the reliability and security of data transfers, which is particularly crucial in environments where timely and accurate data is essential for maintaining grid stability.

These cutting-edge architectures, SDN and NDN, represent promising solutions to the evolving communication needs of smart grids. In the following, we will delve deeper into the motivations behind these architectures, their core technical features, and the practical solutions they offer to enhance the performance, reliability, and resilience of modern power grids.

1.3.1 Embracing Software-Defined Networks

1.3.1.1 Advancements and Practices

To meet the diverse communication demands of smart grids, traditional approaches have often relied on link virtualization techniques such as VLAN and MPLS. While effective, these techniques involve complex configuration tasks for operators, where even a single configuration error can disrupt end-to-end network connectivity. This highlights the pressing need for automated network operations [Sydney et al., 2013]. Since the 1990s, significant advancements have been made, culminating in the realization of the vision of SDN [Farhady et al., 2015] through the concept of the separation routing from routers by Jeniffer Rexford et al. [2004 and Greenberg et al. [2005] et al. in 2004, and the invention of the OpenFlow protocol allowing rule-based switch programming by Nick McKeown et al. [2008 and Hu et al. [2014]. This OpenFlow protocol greatly enhances data plane programmability and introduces a centralized control paradigm.

OpenFlow-enabled SDN devices, including switches and controllers, have been commercially implemented in modern Layer 3 switches from major companies like Huawei [2025] and Cisco [2025]. These technologies have extended their applications into various sectors, including power grids [Rehmani et al., 2019; Huawei, 2025; Cisco, 2022]. The primary objectives of these commercial applications are to address the three key demands of consumers: (1) flexible control and management of isolated virtual networks, with Quality-of-Service (QoS) guarantees tailored to different service level agreements, supporting diverse applications such as power line protection, SCADA, video monitoring, and office automation; (2) dynamic and flexible routing capabilities to handle network failures and ensure seamless access for a wide range of terminal devices; (3) intelligent, centralized network monitoring and operations that enable precise and efficient fault location and prediction; and (4) simplified network control and management by abstracting device heterogeneity through a unified control protocol.

In 2014, Pat Bosshart and a team of researchers, led by Nick McKeown and Jennifer Rexford, introduced Programming Protocol-Independent Packet Processors (P4) [Bosshart et al., 2014], a programming language designed to enable deeper programmability at the switch level. Unlike previous solutions like OpenFlow, which primarily focus on configuring forwarding rules, P4 allows for defining custom packet processing behaviors, offering greater flexibility. This flexibility makes P4 particularly well suited to handling the complex and evolving network requirements of smart grids, such as supporting custom protocols like P4-based NDN [Miguel et al., 2018; Zhang et al., 2014] or meeting the real-time demands specific to power systems.

Since its introduction, P4 has gained significant traction, with numerous hardware and software platforms now offering support. P4-compatible switches, including those from Barefoot Networks (now part of Intel), deliver high-performance packet processing through programmable Application Specific Integrated Circuits (ASICs). The Tofino chip [Intel, 2025], one of the most widely adopted P4-programmable ASICs, provides line-rate packet processing with full programmability, making it ideal for both data centers and high-performance networks, including critical infrastructure like smart power grids.

Beyond switches, the P4 ecosystem includes P4 compilers and integrators such as P4Runtime [p4lang, 2025], which enable seamless integration with various network management systems. Additionally, platforms like Field Programmable Gate Arrays (FPGAs) and software-based switches such as Behavioral Model v2 (BMv2) [GitHub, Inc., 2024] offer flexible environments for both experimentation and deployment. These diverse platforms ensure that P4 can be utilized in both research and production settings, providing the scalability needed for smart grid applications.

An example illustrating the use of SDN to support communication services in a power grid, applicable to both OpenFlow and P4, is presented in Figure 1.3. SDN represents a fundamental shift in network architecture by separating the control plane from the data plane, allowing network control to be centralized within a logically centralized controller. This controller provides a comprehensive global view of the network, benefiting from significantly higher computational resources compared to traditional networking approaches. By enabling programmable and customizable behaviors in the data plane, SDN enhances both the reliability and availability of networked systems – key factors for maintaining the stability of power distribution and transmission. As power grids increasingly rely on interconnected components and real-time data flows for optimal operation, the adaptability provided by SDN becomes essential. By abstracting network management and enabling programmability, SDN empowers operators to dynamically allocate resources and adjust policies across the network with unmatched agility. Proven in various operator networks, SDN has shown its ability to significantly

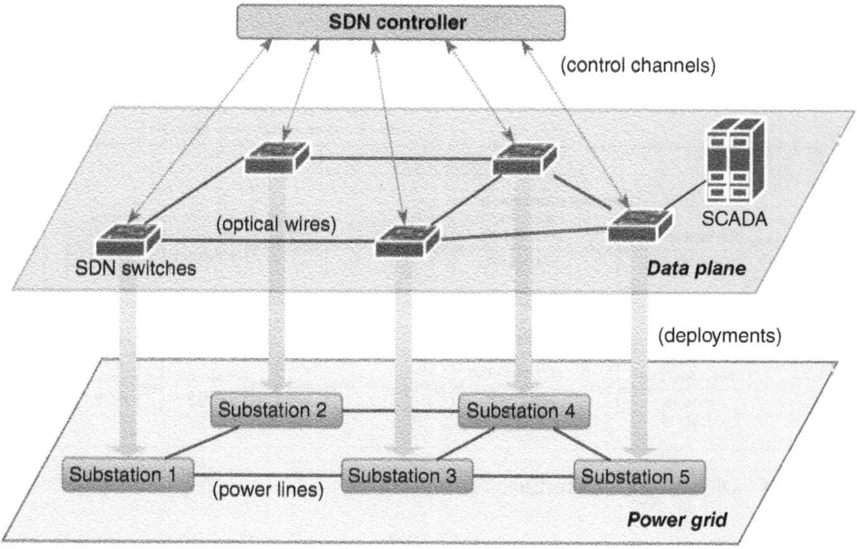

Figure 1.3 SDN network structure in power grids.

improve network flexibility and adaptability. These capabilities lead to enhanced operational efficiency and greater responsiveness in network management, particularly within the critical infrastructure of power grids.

Here, the SDN model takes shape as a graph, denoted as $G = (R, L)$, where each switch r_i finds its place within a substation or a control center [Thomas, 2020]. The SCADA system, orchestrating control over substations through various network protocols (detailed in Section 1.4). Each link $l_l(j, k)$ in this model is composed of two independent channels, seamlessly traversing both downstream and upstream paths between switches r_j and r_k within the set R. The controller establishes connections with the switches through in-band or out-of-band channels, illustrated as dashed lines. Facilitating communication among switches, a set of forwarding paths P comes into play. This model is commonly employed by the majority of SDN solutions in contemporary smart grids.

We will present the details of OpenFlow and P4-based SDN solutions in the following.

1.3.1.2 OpenFlow-Based SDN

The OpenFlow (OF) protocol [McKeown et al., 2008; Open Networking Foundation, 2012; Hu et al., 2014] is a foundational technology behind SDN, separating the control plane from the data plane. This decoupling enables centralized network control and programmability, offering flexibility for managing network traffic in a dynamic and efficient manner. As shown in Figure 1.4, OpenFlow-based SDN

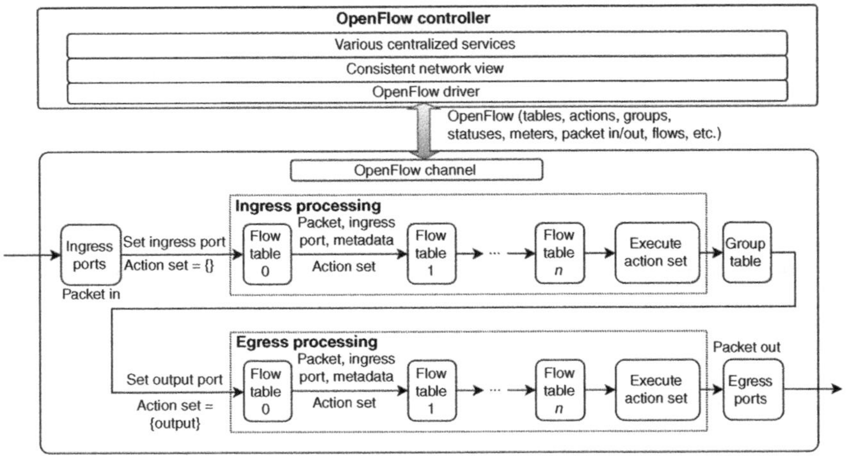

Figure 1.4 OF network architecture.

consists of several core components: the SDN controller, OpenFlow switches, the OpenFlow protocol for control-data plane communication, northbound APIs, and the applications and services that interact with the SDN controller.

Components. In an OpenFlow-based SDN environment, the controller has a global view of the network, enabling it to make real-time decisions about network policies and traffic flows. The switches, which act as the data plane, forward packets based on flow rules provided by the controller. The architecture of an OpenFlow switch [Open Networking Foundation, 2012] is critical for understanding how packets are processed:

i) **Ingress ports**: When a packet arrives at an OpenFlow switch, it first enters through an ingress port, which represents the physical or virtual network interface. The ingress port is responsible for receiving packets and passing them into the switch pipeline for processing.
ii) **Ingress pipeline**: The ingress pipeline is where the main packet processing occurs. This stage consists of multiple flow tables, each storing rules that determine how packets should be handled. The packet processing in the ingress pipeline involves the following key steps:
   - **Flow table matching**: As the packet enters the ingress pipeline, it is sequentially evaluated through a series of flow tables. Each flow table contains match fields based on packet header fields (e.g., source/destination IP addresses, VLAN tags, Transmission Control Protocol (TCP)/User Datagram Protocol (UDP) ports). If a matching flow entry is found, the switch applies the corresponding actions, such as forwarding the packet, modifying its header fields, or dropping the packet.

- **Action set accumulation**: As the packet passes through multiple flow tables, it may accumulate an action set. The action set specifies actions to be applied at later stages of the pipeline. These actions can include forwarding to specific ports, encapsulating the packet, or modifying header fields. The action set is executed only after the packet completes its traversal through the entire ingress pipeline.
- **Flow table miss**: If the packet does not match any flow entry in a flow table, the switch either sends the packet to the controller for further instructions using a packet-in message or drops the packet, depending on the configured table-miss behavior. The controller can then install new flow entries in the switch's flow table to handle future packets of the same flow.

The ingress pipeline processes the packet through multiple flow tables until it either reaches the end of the pipeline or is explicitly forwarded to the next stage in the network. This pipeline allows for granular control over packet processing, making OpenFlow switches highly adaptable to complex network scenarios.

iii) **Group table**: In addition to flow tables, OpenFlow switches support a group table that facilitates more complex actions such as multicasting or failover mechanisms. Group entries define sets of output actions that may affect multiple ports or involve advanced forwarding logic. The key types of group entries include:

- **All**: Forwards the packet to all the output ports specified in the group entry. This is typically used for multicast or broadcast traffic, ensuring that packets reach multiple destinations simultaneously.
- **Select**: Forwards the packet to one output port from a set of ports based on a selection algorithm, such as round-robin or weighted hashing. This mechanism is often used for load balancing, distributing traffic across multiple paths or servers.
- **Indirect**: Applies a single output action to all packets matching this group. Indirect groups simplify the application of forwarding rules across multiple flow entries, similar to traditional port forwarding techniques.
- **Fast failover**: Provides a fast reroute mechanism by forwarding the packet to a backup port if the primary port fails. This ensures high availability and resilience, which are critical in maintaining smart grid communication during network disruptions.

These group types allow OpenFlow switches to implement flexible and resilient forwarding strategies, which are especially valuable for power grid networks that require high reliability and real-time traffic control.

iv) **Egress pipeline**: Once the packet completes its ingress processing, it enters the egress pipeline, where accumulated actions from the ingress stage are executed. This may include modifying the packet headers, encapsulating the packet, or determining the final output port for forwarding. The egress pipeline mirrors the structure of the ingress pipeline, ensuring consistent processing before the packet exits through the egress port.

v) **Egress ports**: After completing all pipeline stages, the packet is forwarded through the appropriate egress ports. These ports represent the physical or virtual interfaces through which the packet exits the switch. Depending on the applied actions, the packet can be forwarded to one or more destinations, broadcasted, or mirrored for monitoring.

Pipeline. A defining feature of OpenFlow switches is their multi-stage pipeline architecture, which processes packets through several flow tables in a sequential manner. Each table refines the packet processing, with actions accumulating as the packet traverses through the pipeline. The multi-table pipeline offers several advantages, especially in smart grid environments, where complex policies and traffic conditions must be enforced:

- **Table 0**: The first flow table that every packet encounters. It evaluates the packet's header fields and determines the initial actions.
- **Intermediate Tables**: The packet may pass through additional flow tables (Tables 1, 2, etc.) to further refine its processing. Match fields and actions in these tables provide more granular control over packet forwarding.
- **Final Table**: After passing through the last flow table, the packet is processed according to the accumulated actions and forwarded to the egress pipeline.

This multi-table pipeline enables OpenFlow switches to apply fine-grained, flexible processing rules, making them particularly well suited for the dynamic demands of smart grid communications, such as ensuring low-latency delivery of critical data or managing traffic under varying conditions.

Centralized flow control. The SDN controller acts as the centralized brain of the network, managing communication between the control and data planes. The controller uses protocols like Link Layer Discovery Protocol (LLDP) [IEEE, 2016] to discover the network topology and maintain a global view of the entire system. This view allows the controller to install flow rules on switches, oversee network policies, and dynamically respond to network events.

When a data flow arrives at an OpenFlow switch, the switch first attempts to match the packet's header fields against its flow tables. If a match is found, the switch forwards the packet according to the specified actions. If no match is found, the switch queries the controller to determine the best route and installs the necessary forwarding rules for future packets. This interaction ensures that the network can adapt to changing conditions, making intelligent routing decisions in real time.

The above programmability of SDN allows smart grid operators to define custom behaviors for the data plane, enabling dynamic resource allocation, traffic engineering, and fault tolerance. For instance, SDN can prioritize critical communications such as data from PMUs or SCADA systems, ensuring reliable and timely delivery even under adverse network conditions. In the event of a network failure, SDN can rapidly reroute traffic to maintain grid stability, leveraging features like fast failover to ensure continuity of service. In addition to enhancing network reliability, OpenFlow's centralized control offers opportunities for improved security. By centralizing network states, SDN enables real-time threat detection and mitigation, protecting the smart grid against cyberattacks and unauthorized access.

Overall, OpenFlow-based SDN represents a paradigm shift in power grid communication management, offering the flexibility, scalability, and resilience necessary to meet the demands of modern smart grid infrastructures.

1.3.1.3 P4-Based SDN

As communication networks evolve to meet the stringent demands of smart power grids, the need for flexibility, programmability, and customizability is becoming increasingly critical. P4 [Bosshart et al., 2014; p4lang, 2025; GitHub, Inc., 2024] emerges as a transformative technology designed to address these challenges by enabling network operators to program the behavior of network devices at the packet processing level.

At its core, P4 allows network operators to define how packets are processed at the forwarding level through a high-level programming language. The principles that govern P4 include:

- **Protocol independence**: Unlike traditional networks that rely on predefined protocol stacks, P4 allows users to define new protocols or modify existing ones, making it highly adaptable to the needs of smart grids, which may require the integration of custom or proprietary communication protocols.
- **Target independence**: P4 programs are written independently of the hardware on which they run. This means that P4 can be compiled and executed on different devices, whether programmable switches, network interfaces, or software switches, ensuring compatibility across various platforms.
- **Reconfigurability**: P4 programs can be updated in real time, allowing for dynamic changes in how packets are processed without requiring hardware replacements or firmware updates.

The architecture of P4 is designed to offer flexibility in how packets are processed within the data plane, allowing network operators to customize packet handling based on specific use cases for power grid communications. The architecture is depicted in Figure 1.5 with its components detailed below.

i) **Packet parser**: The parser is the first stage in the packet processing pipeline. It is responsible for interpreting the structure of incoming packets by extracting relevant header fields based on the protocols defined by the P4 program.

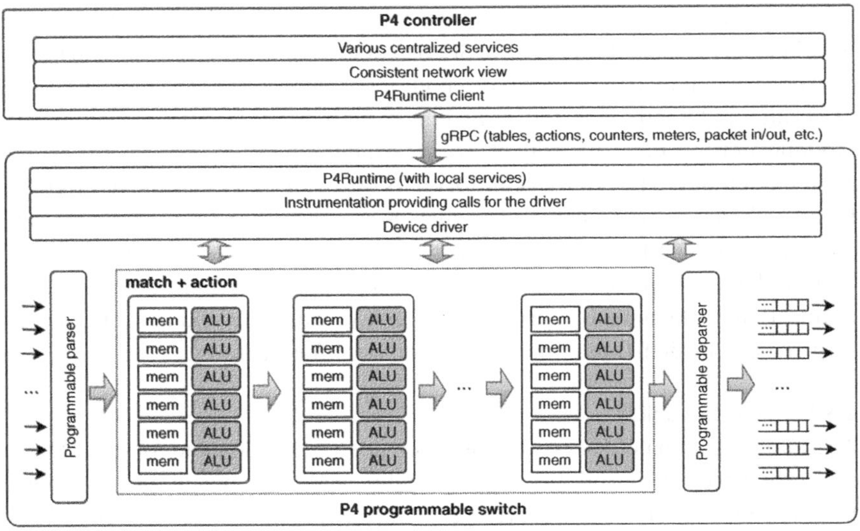

Figure 1.5 P4-based SDN network architecture.

Unlike fixed-function network devices that support only a specific set of protocols, P4 allows the developer to specify which headers to recognize, parse, and forward.

- **Header definition**: In P4, headers can be customized based on the network's needs. For example, in a smart grid communication system, specific protocols used in SCADA or IEC 61850 can be defined. The parser is capable of recognizing these protocols and their associated fields, such as timestamps, device identifiers, or measurement values.
- **Control logic**: The parser has logic that directs which header fields should be extracted in each step. It supports branching, meaning that based on the header contents, different parsing paths can be taken. This enables the recognition of complex, hierarchical, or evolving protocols common in smart grids.

ii) **Match-action pipeline**: Once the parser extracts the necessary information from the packet, it is passed to the match-action pipeline, which is the core component of P4's programmable packet processing. These match-action tables are programmable data structures that define how packets should be handled in the process below. Initially, the packet header fields are matched against pre-defined criteria. These fields can be standard fields like IP addresses or custom fields, depending on the protocols and processing logic defined in the P4 program. Once a match is found, specific actions are applied to the packet. These actions can include forwarding the packet to a specific port, modifying header fields, dropping the packet, replicating it to multiple destinations, or sending it to the control plane for further processing. The match-action paradigm is highly flexible. For example, in a smart grid communication network, a P4 switch could be programmed to prioritize control messages with low-latency requirements by forwarding them through specific paths while relegating non-critical traffic to secondary routes. Furthermore, P4 allows for the use of multiple match-action tables, each of which can be independently configured. Packets pass through these tables sequentially or conditionally, based on the results of the previous match-action stages. This enables complex decision-making processes, such as dynamic path selection in a grid network to avoid congested links or route around network failures.

iii) **Deparser**: Once the match-action pipeline finishes processing the packet, the deparser reconstructs the packet to prepare it for transmission. This stage is essential because the packet may have been modified during processing – header fields may have been altered, or entirely new headers may have been inserted. The deparser is responsible for ensuring that the modified packet is correctly reassembled before it leaves the switch. For example, in a smart grid application, the deparser might add a custom header containing diagnostic information or routing metadata, allowing other network devices to make informed decisions about how to handle the packet.

During the phases above, the control flow logic determines how packets move through the match-action pipeline and what actions are taken based on conditions defined in the P4 program. This component allows developers to implement complex packet processing flows, where different actions are applied based on the state of the network, the type of packet, or even external inputs such as data from monitoring systems. Developers can define flow control constructs, such as conditional statements and loops, to guide the packet through the appropriate tables and actions. For example, in smart grid networks, the flow logic could direct packets containing power system fault data to be processed with high priority, while routing less urgent data through less critical parts of the network. Additionally, P4 also supports stateful processing by enabling the use of counters, registers, and meters. This allows the P4 switch to track state information over time, such as the number of packets from a specific source or the volume of traffic on a particular link. In smart grid applications, this feature can be used to implement dynamic load balancing or detect anomalies in communication patterns, such as network congestion caused by power surges or equipment failures.

To enhance the P4 processing capabilities, P4 allows for the use of externs – hardware objects that perform functions outside of the P4 pipeline, such as specialized packet buffering, content storage or cryptographic operations. These externs extend the functionality of P4 programs, enabling developers to incorporate complex processing functions without having to implement them in the P4 code itself. Meanwhile, P4 supports stateful objects like registers and counters, which can be used to maintain state across multiple packets. These can be crucial in smart power grid applications, where maintaining a history of communications between devices or tracking the status of grid elements is necessary for efficient and reliable operations. For example, registers can store the last known status of a substation or track the energy consumption of different parts of the grid.

In smart grid environments, this target independence ensures that P4 programs can be deployed across different parts of the communication network, from high-performance switches in core grid networks to software switches or FPGAs in edge devices. Programmers can define custom metadata to be matched and taken actions within packets that could be used to carry status information about different grid nodes, such as voltage levels or real-time demand data. This allows the network to adapt its behavior based on real-time conditions in the power grid.

Therefore, the architecture of P4 offers a highly flexible and modular framework, from parsers to match-action tables, for programming network behavior at the packet level. This provides a platform to ensure that the network can be tailored to the specific requirements of smart power grids.

1.3.1.4 Benefits of SDN to Smart Grids

For smart power grids, the benefits of this centralized control approach are manifold. One significant advantage is the acceleration of routing convergence, which allows for swift re-routing of data in the event of link failures or network congestion. This rapid adaptability is crucial in maintaining uninterrupted power supply and enhancing grid resilience against external threats, including cyberattacks and physical disruptions. Moreover, the ability to program individual data flows within switches grants operators the flexibility to tailor network behaviors according to specific service requirements, such as prioritizing traffic for critical services like SCADA systems or integrating renewable energy sources with varying output characteristics.

By leveraging the inherent programmability of SDN, power grid operators can implement advanced monitoring and control strategies that optimize resource utilization, improve fault detection and isolation, and enhance overall system reliability. This innovative approach not only addresses the growing complexities of modern power networks but also positions them to better adapt to future energy demands and technological advancements. As the energy landscape continues to evolve, the integration of SDN into smart power grids promises to be a cornerstone for achieving a more resilient, efficient, and sustainable energy infrastructure.

These applications illustrate the vision of Software-Defined Networkings (SDNs) for power grids, where the resources of the entire grid's communication network can be managed and monitored akin to an operating system in computers. By leveraging SDN, network resilience against failures and cyberattacks, particularly against Link Flooding Attacks (LFAs), is significantly enhanced. In the event of a failure or attack, the centralized control offered by SDN facilitates the management of data flow behaviors through data plane programmability, ensuring continuity of network operations. Examples include fast routing convergence and the flexible programming of individual data flows within switches [Dorsch et al., 2014]. This capability is crucial for improving the reliability and timeliness of power grid monitoring and control, ultimately contributing to enhanced grid stability.

Furthermore, as smart grids continue to evolve, the integration of these advanced communication techniques will play a vital role in accommodating future energy demands and technological advancements. Ongoing research and development in this area will be essential for overcoming emerging challenges and optimizing the performance of power grid communication networks.

1.3.2 Unveiling Named-Data Networks

1.3.2.1 Advancements and Practices

NDN [Zhang et al., 2014] is a groundbreaking information-centric network architecture [Xylomenos et al., 2014] that leverages in-network data caching. NDN has demonstrated significant advantages in improving communication resilience, scalability, and security, making it well suited for the demanding environments of smart power grid infrastructures [Ravikumar et al., 2020]. The concept of NDN originated in 2006 when Van Jacobson, a key figure behind the development of the TCP/IP protocol stack, introduced the idea of content-centric networking [Jacobson et al., 2009]. Building on this, Lixia Zhang et al. formalized the framework of NDN in 2014 with a seminal paper [Zhang et al., 2014] that established NDN as a promising alternative to traditional host-centric networks.

Over the past decade, NDN has evolved from a theoretical concept into practical applications capable of supporting both high-speed wired and wireless networks. An open-source implementation of NDN has been made available to the community, facilitating further experimentation and development [Mastorakis et al., 2017]. Today, NDN offers a high-performance data plane, making it suitable for demanding environments such as smart grid networks. For instance, a high-speed software-based NDN router, capable of processing up to 40 Million Packets Per Second (MPPS), was developed using off-the-shelf commercial hardware [Takemasa et al., 2017]. Additionally, hardware-based NDN routers have been implemented using Intel Tofino chips, leveraging the P4 programming language (see Section 1.3.1.3) for enhanced packet processing capabilities [Intel, 2025; Bosshart et al., 2014; Miguel et al., 2018].

In 2020, a software-based NDN router was realized using the Data Plane Development Kit (DPDK) [DPDK, 2024], achieving 100 Gbps on commodity hardware [Shi et al., 2020]. This milestone was followed by the development of a hardware-based NDN router in 2021, offering scalable and cost-effective content caching solutions [Liu et al., 2021]. NDN's performance is further boosted by advanced naming lookup and content storage technologies. For example, a hierarchical aligned transition array implemented on an FPGA achieved a naming lookup rate of 125 MPPS with an ultra-low latency of 0.12 microseconds [Li et al., 2014]. In terms of content storage, the NameTrie software implementation supports 3.56 million insertions, 3.72 million lookups, and 3.25 million removals per second [Ghasemi et al., 2018].

These platforms, equipped with high-performance forwarding engines, are ideally suited for power grid networks that require high availability, scalability, and responsiveness. By leveraging the inherent strengths of NDN in flexible communication patterns, these advancements ensure that smart grid communication networks can meet the stringent demands of modern energy infrastructures. For instance, iCASM introduced the first NDN-based PMU communication system, enabling efficient broadcasting to egress faces over nondisjoint paths, significantly improving message reliability and transmission efficiency [Ravikumar et al., 2020]. Additionally, Resilient Sensor Data Dissemination (RSDD), built on NDN, introduced a robust hop-by-hop retransmission mechanism using Redundant Sub-Paths (RSPs), coupled with a recursive collaboration process guided by arrival timeouts [Zhou et al., 2023]. RSDD demonstrated a 2.44% EEFR and an average End-to-End Payload Packet Delivery Time (EEDT) of 2.7 ms during full path disruption with a 20% packet loss rate. Moreover, RSDD excels in enabling State Estimation (SE) to preserve grid observability and accuracy even under adverse conditions [Zhou et al., 2023]. These innovations highlight the transformative potential of NDN in enhancing the stability and reliability of power grids.

1.3.2.2 NDN

Figure 1.6 illustrates the NDN architecture [Zhang et al., 2014], which fundamentally departs from traditional IP-based networking by adopting a data-centric approach. Rather than routing packets based on destination IP addresses, NDN forwards packets based on content names. This shift mirrors how users typically interact with the Internet today, where the focus is on "what" data is needed rather than "where" it is located. The architecture is particularly advantageous in scenarios requiring high resilience, security, and adaptability, such as smart grids, due to its decentralized nature and built-in caching capabilities.

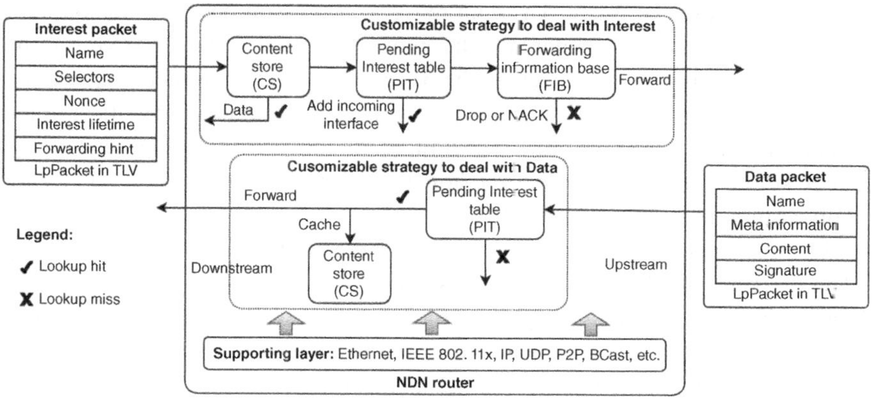

Figure 1.6 NDN architecture.

NDN is also highly versatile in its deployment, operating over various underlying network technologies. It can be implemented over Ethernet, IEEE 802.11x, IP and UDP layers, peer-to-peer (P2P) networks, or broadcasting networks, making it compatible with both legacy and modern infrastructures.

NDN messages. NDN messages use the *LpPacket* format, organized using a highly flexible Type-Length-Value (TLV) encoding scheme [Named Data Networking, 2022]. This format is employed for both Interest and Data packets and provides several key benefits:

1) **Simplicity**: The TLV format is easy to parse, with each field clearly identified by its type and length, allowing for efficient processing by both hardware and software.
2) **Extensibility**: New types of information can be introduced without disrupting existing operations, making TLV adaptable to future requirements and use cases.
3) **Compactness**: By eliminating unnecessary overhead, TLV ensures concise encoding of information, optimizing bandwidth usage.

As shown in Figure 1.6, both Interest and Data messages carry a content name similar to a URL, such as `/powergrid/RTU/voltage`, which could represent a voltage reading from an Remote Terminal Unit (RTU) in a power grid [Ravikumar et al., 2020; Zhou et al., 2023]. The Data packet contains a payload (e.g., sensor data), while the Interest packet includes a nonce for uniqueness and a lifetime to indicate its validity. Each Data packet is signed by the producer, ensuring data integrity and authenticity, regardless of how the data was transmitted or cached. This approach to securing the data itself, as opposed to securing the communication channels, is particularly effective for critical infrastructures like smart grids, where the security of individual data items is paramount.

Data retrieval in NDN. The data retrieval process in NDN, as illustrated in Figure 1.6, operates fundamentally differently from traditional IP-based routing. Instead of establishing connections between endpoints, consumers request data by sending Interest packets. These packets are routed toward the producer or any router that has cached the requested content. The routers along the way leverage three key components: the Content Store (CS), Pending Interest Table (PIT), and Forwarding Information Base (FIB).

The CS stores Data packets temporarily to serve future requests more efficiently. The PIT keeps track of pending Interest packets, ensuring that multiple requests for the same content are aggregated rather than forwarded redundantly. The FIB, akin to a routing table, maps name prefixes to next hops and guides the forwarding of Interest packets based on longest-prefix matching.

The retrieval process follows these steps:

1) A consumer sends an Interest packet with a content name. Routers forward the Interest toward the producer, creating an entry in their PIT to track the request.
2) If a router or the producer has the requested data, it responds with a Data packet that follows the reverse path of the Interest, passing through the routers that hold the corresponding PIT entries.
3) Each router along the path caches the Data packet in its CS to serve future requests, reducing redundant transmissions from the producer.

NDN's caching and forwarding strategies can be customized to optimize network performance based on specific application needs. For example, the timeout for PIT entries, the decision to cache data, and the forwarding strategy for Interests can be tailored according to network conditions or specific traffic patterns.

NDN in power grids. In the context of power grid communications, NDN can be modeled as an undirected graph $G = (R, L)$, where R is the set of routers deployed in substations and L represents the links between them [Ravikumar et al., 2020; Zhou et al., 2023]. These links are typically high-speed optical connections (e.g., 1Gbps to 100Gbps), using transmission systems like synchronous digital hierarchy or optical transport networks [IEC, 2014]. The network delay over these links is determined by the physical distance, queuing delays, and the processing time at ingress and egress routers.

In a typical scenario (Figure 1.7), a consumer sends an Interest packet to retrieve data such as voltage readings from substations. Routers forward the Interest along predefined paths, caching the retrieved Data packets to serve future requests efficiently. Flexible path selection ensures that communication can adapt to changing network conditions, enhancing resilience and performance in the power grid's communication infrastructure.

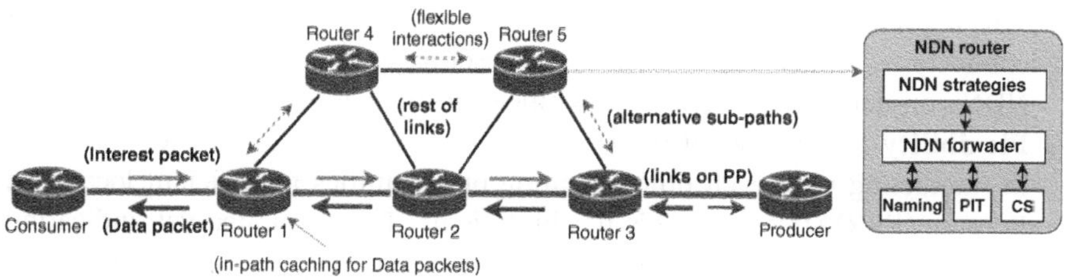

Figure 1.7 A typical communication example in NDN for the transmission grid.

1.3.2.3 Benefits of NDN to Smart Grids

As smart grids evolve into increasingly complex hierarchical structures, the implementation of NDN becomes critical. These structures necessitate hierarchical data collection, multicasting, and on-demand data requests, all of which must be exceptionally reliable to ensure grid stability in adversarial environments. This complexity fundamentally differentiates grid communications from traditional Internet architecture. The advantages of NDN in this context are outlined below:

- **Enhanced reliability for hierarchical data collection**: Smart grids often require frequent data transmissions, sometimes exceeding hundreds of updates per second. For example, communications between PMUs and PDCs, as well as between SCADA systems and RTUs, illustrate this need [IEEE, 2011; Thomas, 2020]. Traditional TCP/IP protocols rely on end-to-end transport control, which can struggle with serious packet losses common in large-scale power grids. In contrast, NDN allows intermediate nodes to cache data and manage transport control, significantly enhancing reliability and efficiency in data delivery.
- **Support for multicasting**: NDN natively supports publish/subscribe communication mechanisms, which are integral to IEC 61850-enabled devices [Brunner, 2008]. In these systems, multiple RTUs and power meters collaborate within substation networks. Unlike TCP/IP's host-to-host connections, NDN facilitates direct interaction between publishers and subscribers, improving communication efficiency. While UDP can offer multicast capabilities, it sacrifices reliability, leading to the necessity for redundancy mechanisms as outlined in IEC 62439. By caching data at intermediate nodes, NDN reduces message forwarding costs, bolstering reliability and enhancing overall data exchange efficiency. Operators can thus focus more on data analysis rather than on reliability concerns.
- **On-Demand Data Requests**: In a smart grid context, many devices do not need to maintain continuous connections; instead, they respond only when specific data requests are made. For instance, power distribution terminals, including power meters, PMUs, and environmental sensors, report their statuses upon demand [Thomas, 2020; Brunner, 2008; IEC, 2006; IEEE, 2011].

This approach contrasts with TCP/IP's requirement for persistent end-to-end connections, which can waste device resources when connections are idle. NDN's on-demand model increases efficiency and conserves device energy.

- **Flexibility in router interactions**: NDN routers exhibit greater flexibility in interacting with each other, creating opportunities for efficient local Fast Reroute (FRR) strategies. This flexibility is not fully exploited yet, indicating a significant area for further research and development to enhance local FRR mechanisms while considering global routing states.
- **Content privacy and authentication**: NDN provides robust assurances of content privacy, authentication, and integrity. This is crucial in preventing falsified data injection attacks, which pose significant threats to grid security. For more details on this aspect, interested readers can refer to the study by Haftu et al. [Reda et al., 2022].

These benefits demonstrate that the NDN architecture significantly enhances the reliability and efficiency of communications for monitoring and controlling grid states, which is vital for maintaining power grid stability. With the advancement of high-speed NDN routers and related technologies, the performance of grid communications can be greatly improved, leading to more resilient and responsive smart grid infrastructures.

1.4 Network Protocols in Power Grids

Within the communication networks, devices, control centers, and field equipment of smart power grids rely on a variety of communication protocols to ensure efficient, secure, and reliable data exchange. These protocols are vital for real-time monitoring, automation, and control within the grid. Below, we explore the most commonly used protocols in power grids, detailing their design goals, application scenarios, functions, and message interactions. These protocols include IEC 61850 [Brunner, 2008], IEEE C37.118.2 [IEEE, 2011], IEC 60870-5-104 [IEC, 2006], as well as other protocols such as Distributed Network Protocol (DNP3) [IEEE, 2010], Open Platform Communications Unified Architecture (OPC-UA) [OPC Foundation, 2023], and Modbus [Modbus Organization, 2012].

1.4.1 IEC 61850 Protocols

IEC 61850 is a global standard designed to facilitate seamless, interoperable communication in substation automation systems and the broader smart grid [IEC, 2020]. Developed to address the growing complexity and need for real-time, reliable communication in power systems, the standard provides a unified framework for managing control, protection, and monitoring devices in substations.

By defining a comprehensive set of communication protocols, data models, and configuration tools [IEC, 2011a], IEC 61850 ensures that devices from different manufacturers can operate harmoniously, enhancing both the efficiency and reliability of power grids [Brunner, 2008].

1.4.1.1 Design Concept of IEC 61850

Figure 1.8 presents the stack of IEC 61850 protocols [Brunner, 2008] and an example of interactions between client and server. The protocols mainly include Manufacturing Message Specification (MMS) based on TCP/IP [IEC, 2011c] for both inter-substation and intra-substation grid controls, as well as Generic Object-Oriented Substation Event (GOOSE) [IEC, 2011c] and Sampled Values (SV) [IEC, 2013] directly based on the link layer for the intra-substation controls. The upper circle shows the models; the core of design philosophy is the abstraction of Intelligent Electronic Devices (IEDs) which models objects and services. Meanwhile, the Abstract Communication Service Interface (ACSI), for MMS-enabled services, defines a set of abstract services that enable interaction between IEDs and higher-level control systems [IEC, 2011b]. Additionally, the Specific Communication Service Mapping (SCSM) is an essential aspect, enabling ACSI services to be transmitted using Ethernet for high-speed, local communication or through MMS for supervisory control [IEC, 2011c,2013]. The IED and ACSI are detailed as follows.

- These IEDs [IEC, 2010a] are specialized devices in substations, such as protection relays, circuit breakers, and sensors, that perform essential automation,

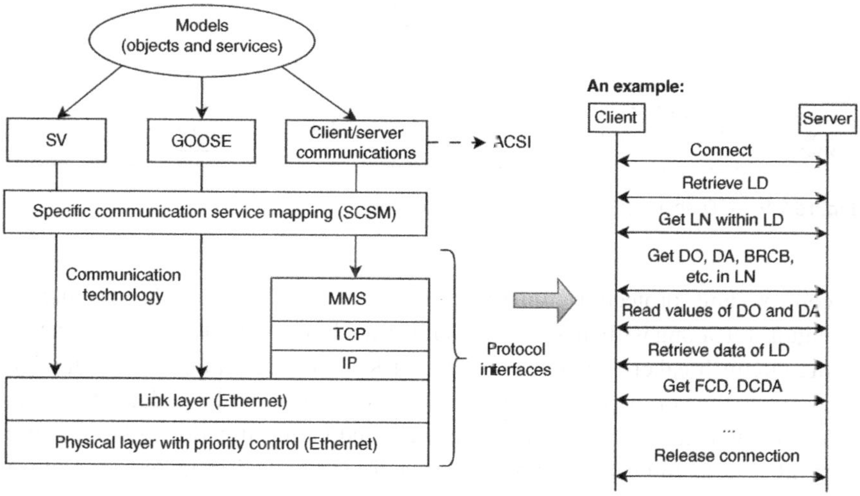

Figure 1.8 IEC 61850 protocols and interactions.

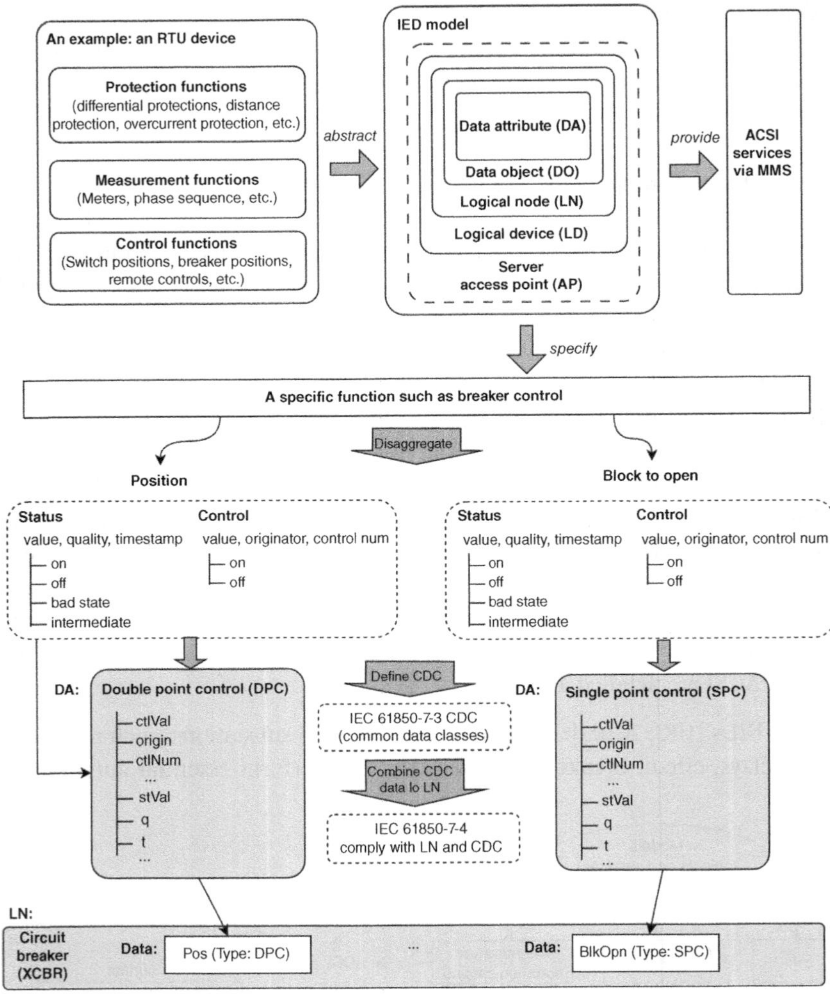

Figure 1.9 IED abstraction of IEC 61850 protocols.

control, and protection functions. The concrete structure is specified in Figure 1.9, along with an example to illustrate how the IED model can abstract the circuit breaker function of an actual RTU device. In detail, the standard abstracts the complex functionality of IEDs into standardized data structures, known as logical nodes, which represent individual functions like measurements, status indicators, or protection logic. These logical nodes are grouped into logical devices, each representing a subset of an IED's functionality. This hierarchical data modeling approach allows for uniform representation and communication across different devices, facilitating interoperability regardless of the manufacturer.

- These ACSI services [IEC, 2011b], which include data access, reporting, device control, and file transfer, are independent of the underlying communication protocols. This abstraction allows IEC 61850 to be mapped onto different transport layers, such as Ethernet or TCP/IP, ensuring flexibility and adaptability to different communication environments. By focusing on functionality rather than specific technologies, ACSI future-proofs the standard, making it adaptable to evolving communication needs in the power grid.

Another question that remains is how to define such model and services for managing the architecture and communication of devices within a substation. The answer is the System Configuration Language (SCL) introduced by IEC 61850-6 [IEC, 2010b]. The SCL file contains a detailed description of the substation's topology, including IEDs, their interactions, and the communication pathways they use. By providing a standardized format for configuring devices, SCL reduces the complexity of system setup and modification, making it easier to manage large-scale substation automation systems. This configuration flexibility also allows systems to adapt quickly to new devices or changes in network architecture.

During operation process, IEC 61850 also supports plug-and-play functionality [IEC, 2010a,2010b–2011a]. This means that when a new IED is added to the system, it can automatically describe its capabilities and data model, enabling rapid integration into the existing network. This feature streamlines the configuration process and reduces downtime, particularly in dynamic substation environments where devices are frequently added or replaced.

Additionally, this standard also integrates with time synchronization and time stamping, allowing for precise coordination between devices. The use of protocols like precision time protocol ensures that all events in the system are accurately time-stamped, which is crucial for protection schemes and fault analysis. This level of precision allows systems to operate with enhanced reliability and accuracy, preventing cascading failures and minimizing damage during faults.

1.4.1.2 Protocols Defined by IEC 61850

IEC 81650 standard's focuses, namely MMS [IEC, 2011c], GOOSE [IEC, 2011c], and SV [IEC, 2013], allow for efficient, periodic, and event-driven measurement data exchanges, crucial for fast protection and fault isolation. The MMS permits the data changes across substations and the control center of power grids for realizing power operations and dispatches, which is critical to the grid stability. Meanwhile, the use of GOOSE and SV enables instantaneous communication between field devices, ensuring rapid response to critical grid events like faults or overloads. These three protocols are specified below.

i) **MMS**: MMS [IEC, 2011c] is designed to support high-level communication between supervisory control systems and field devices in a power grid. Its primary goal is to enable the exchange of real-time data for control, configuration, and monitoring, while being vendor-neutral to ensure interoperability across devices from different manufacturers.

MMS is extensively used in substation automation systems to communicate between IEDs and SCADA systems. It is ideal for exchanging non-time-critical data such as system states, configuration parameters, alarms, and event logs. For instance, in smart grids, MMS is responsible for reporting breaker status, sending control commands to relays, and managing configuration settings for grid protection.

MMS provides a range of functions, including reading and writing data from field devices, sending control commands, and monitoring device health. It supports both event-based and periodic data transmission. MMS also handles the management of data sets, which group together related data points, allowing efficient communication of multiple values in one message.

MMS follows a client–server model, where the client (e.g., SCADA or energy management system) sends requests to the server (e.g., IED or relay) for specific data or control actions. Messages are encapsulated in a complex but efficient structure, supporting the exchange of not only process data but also meta-information about the data (e.g., quality, timestamps). MMS messages can include commands to open or close breakers, query device statuses, or retrieve event logs, ensuring that the central control system has comprehensive oversight of substation operations.

ii) **GOOSE**: GOOSE [IEC, 2011c] is designed for high-speed, event-driven communication, suited for protection and control within substations. Its primary goal is to facilitate instantaneous data transfer between devices to ensure rapid response to critical events such as faults, overloads, or system imbalances. Unlike traditional protocols that rely on cyclic polling, GOOSE operates based on events, significantly reducing response times during emergencies.

GOOSE is widely applied in protection relays, circuit breakers, and fault management systems within substations. It is used for inter-device communication where immediate action is required, such as tripping a breaker upon fault detection or isolating a section of the grid to prevent cascading failures. For example, when a fault occurs in a transmission line, protection relays generate GOOSE messages that inform adjacent breakers to open, isolating the fault and preventing damage to other parts of the network.

GOOSE supports fast communication by continuously sending multicast messages across the network to all relevant devices. These messages include critical data, such as protection triggers, fault signals, and device status. GOOSE allows direct peer-to-peer communication between field devices, bypassing the need for central control, which enhances both the speed and reliability of protection schemes. In the context of fault handling, GOOSE messages help ensure that protection relays and circuit breakers operate within milliseconds, thus minimizing equipment damage and service disruptions.

GOOSE messages are multicast over an Ethernet network and are repeated periodically to ensure delivery. The messages contain information in a structured format, including event types (such as fault triggers), device statuses (such as breaker open/close status), and priority levels. The repeated transmission ensures that even in the presence of network congestion or packet loss, critical information is delivered reliably. This communication structure enables protection devices to respond autonomously to system changes, significantly reducing the time between fault detection and response.

iii) **SV**: SV [IEC, 2013] is designed to enable the real-time exchange of analog measurement data, such as current and voltage waveforms, between field devices in substations. The goal is to provide a continuous stream of digitized measurement data that can be used for protection, monitoring, and control purposes. SV is predominantly used in substations for current and voltage measurement, with the data being shared between devices like merging units, protection relays, and control systems. For instance, SV is critical in digital substations, where traditional current and voltage transformers are replaced with electronic transformers, and the analog signals are digitized at the source. The SV data is then transmitted in real time to protection relays for fast decision-making.

SV supports the digitization of analog signals from sensors or transducers and streams this data in real time across the substation network. This is essential for dynamic protection schemes, such as differential protection, where precise and time-synchronized measurements are needed to detect issues like short circuits or phase imbalances. By continuously transmitting sampled data from multiple devices, SV enables highly accurate fault detection and real-time system monitoring.

SV messages consist of digitized values of voltage and current, which are transmitted over Ethernet at high sampling rates (e.g., 4,800 samples per second per channel). The communication is carried out in a unidirectional, publisher–subscriber model, where a device (e.g., a merging unit) publishes measurement data and subscribing devices (e.g., protection relays) receive the data in real time. The strict timing requirements ensure that the data is synchronized across multiple devices, which is crucial for coordinated protection responses. Unlike GOOSE, which is event-driven, SV continuously streams data, providing protection systems with the detailed measurements they need to make precise, timely decisions.

In summary, IEC 61850 provides a holistic, future-proof framework for substation automation and smart grid communication. By abstracting IED functionality, offering flexible configuration tools, and supporting real-time, event-driven communication, the standard enhances the interoperability, scalability, and reliability of power systems. Its ability to adapt to evolving technologies and communication needs makes it a cornerstone of modern smart grid infrastructure, ensuring that power systems can operate efficiently and respond rapidly to the dynamic challenges of today's energy landscape.

1.4.2 IEEE C37.118.2 Protocol

The IEEE C37.118.2 protocol [IEEE, 2011] is a pivotal standard for synchrophasor communications, primarily focusing on the interaction between PMUs and PDCs. This protocol plays a crucial role in modern power system monitoring and control, enabling real-time data acquisition and transmission of synchronized phasor measurements. This section will detail the design goals of the protocol, outline its application scenarios, describe its key functions, and explain the message interactions that occur within this framework.

The primary design goal of the IEEE C37.118.2 protocol is to facilitate standardized, reliable, and efficient communication of synchrophasor data across diverse devices in the power system infrastructure. One of the main objectives is to ensure interoperability among devices from different manufacturers, enabling seamless integration within various operational environments. The protocol supports high data throughput while maintaining the accuracy and integrity of the transmitted information.

Moreover, the C37.118.2 protocol is designed to accommodate varying data rates, allowing for flexibility based on the specific requirements of different applications. It supports both real-time data streaming and event-based reporting, ensuring that critical information is communicated promptly and reliably. Additionally, the protocol includes provisions for secure communication, addressing the growing concerns surrounding cybersecurity in smart grid systems.

The IEEE C37.118.2 protocol finds application in various scenarios within the electric power industry, particularly in the realms of monitoring, control, and protection. One of its primary uses is in the Wide-Area Monitoring System (WAMS), where multiple PMUs are deployed across the grid to collect and transmit phasor measurements to centralized PDCs. This data enables operators to gain comprehensive insights into the real-time state of the power grid, facilitating effective decision-making and enhancing situational awareness.

In addition, the protocol is employed in dynamic security assessment applications, where real-time data from PMUs is essential for assessing the stability and security of the power system during transient conditions. The C37.118.2 protocol also supports post-event analysis, where recorded phasor data can be reviewed to understand system behavior during disturbances, thereby informing future operational strategies and enhancing grid resilience.

The IEEE C37.118.2 protocol encompasses several key functions that contribute to its effectiveness in synchrophasor communications. Firstly, it provides a standardized format for transmitting time-synchronized phasor data, which includes voltage and current measurements along with other relevant data, such as frequency and rate of change of frequency.

The protocol supports both streaming and event-driven data transmission. Streaming mode allows for continuous transmission of measurements at defined intervals, providing real-time data for ongoing monitoring. In contrast, event-driven mode transmits data only when significant events occur, such as system disturbances or threshold breaches, thus optimizing bandwidth usage.

Another critical function is the capability for configuration and control of PMUs and PDCs. The protocol allows operators to modify settings, request specific data types, and control operational parameters, thereby ensuring that the devices function optimally based on real-time grid conditions.

Message interactions within the IEEE C37.118.2 protocol occur primarily between PMUs and PDCs and follow a well-defined structure. The protocol uses a series of message types to facilitate communication, including data frames, configuration messages, and command messages.

- Data frames contain the actual phasor measurements and are typically sent at regular intervals. Each data frame includes essential information such as timestamps, measurement values, and status indicators. The timestamping mechanism is vital, as it ensures that measurements are accurately synchronized, which is crucial for effective grid monitoring and analysis.
- Configuration messages allow PDCs to configure PMUs by setting parameters such as the data sampling rate, reporting intervals, and data types. This interaction enables dynamic adaptation to changing operational requirements.
- Command messages facilitate real-time control operations, allowing operators to issue commands to PMUs for immediate action, such as changing measurement settings or initiating specific data reporting processes.

In summary, the IEEE C37.118.2 protocol stands as a fundamental standard for synchrophasor communications in the electric power industry. By establishing clear design goals, accommodating diverse application scenarios, and defining essential functions and message interactions, the protocol enhances the ability of power system operators to monitor, control, and protect the grid effectively. As smart grid technologies continue to evolve, the C37.118.2 protocol will play an increasingly critical role in enabling reliable and efficient energy management and ensuring the stability of modern power systems.

1.4.3 IEC 60870-5-104 Protocol

IEC 60870-5-104 [IEC, 2006] is an extension of the IEC 60870-5-101 protocol [IEC, 2003] and is specifically designed for remote monitoring and control of substations and other critical infrastructure over TCP/IP networks. Its primary goal is to facilitate reliable and efficient data exchange between control centers and remote devices like RTUs, IEDs, and other automation systems. The protocol allows for real-time monitoring, control, and data acquisition from geographically dispersed locations, which is essential for the operation of large-scale power grids. By leveraging modern network infrastructure, IEC 60870-5-104 enhances communication flexibility, making it suitable for both small-scale local area networks and large-scale wide area networks.

IEC 60870-5-104 is widely used in transmission and distribution networks to support SCADA systems. Its primary application lies in transmitting control commands (such as open/close breaker operations) and collecting telemetry data (such as voltage, current, and status indications) from remote substations to central control centers. It is particularly effective for long-distance communication, such as between regional transmission control centers and substations spread across large geographical areas. In smart grids, IEC 60870-5-104 plays a vital role in integrating remote automation systems and ensuring real-time visibility and control over grid assets. This makes it suitable for applications ranging from routine operations, like load balancing, to emergency responses, such as fault isolation.

IEC 60870-5-104 supports a range of essential functions that enable real-time data exchange between control centers and remote devices. Key functions include:

- **Telecontrol and command execution**: It allows control centers to issue commands to remote field devices, such as opening or closing circuit breakers, adjusting transformer settings, or controlling other grid assets.
- **Data acquisition**: Remote devices periodically send telemetry data back to control centers, providing information about the status, health, and operational metrics of the grid infrastructure. This includes measurements like voltage, current, frequency, and breaker statuses.
- **Event reporting**: IEC 60870-5-104 supports event-based communication, where critical changes in system states (e.g., fault detection, equipment failure) are immediately reported to the control center without waiting for periodic polling. This ensures prompt action in response to grid disturbances.
- **Time synchronization**: The protocol supports time-stamping of data to ensure that all events and measurements can be correlated accurately across the network. This is crucial for activities like fault analysis, where precise timing is needed to reconstruct sequences of events.

- **Cyclic and spontaneous transmission**: IEC 60870-5-104 supports both periodic data transmission (cyclic) and on-demand or event-triggered transmission (spontaneous). Cyclic transmission is used for routine monitoring, while spontaneous transmission provides immediate updates in the event of a fault or disturbance.

IEC 60870-5-104 operates using a client–server model where the control center acts as the client and the remote devices (RTUs and IEDs) function as servers. The protocol uses TCP/IP as its transport layer, ensuring reliable data transmission over IP networks. The protocol's frame structure includes start, length, control, and information elements, ensuring robust and secure data exchange.

The communication process is divided into three main types of message exchanges:

1) **Supervisory (S) frames**: These are responsible for ensuring proper flow control in communication, handling acknowledgments and retransmissions. They ensure that data packets are reliably delivered and that any lost packets are retransmitted.
2) **Information (I) frames**: These carry the actual payload, such as telemetry data or control commands. I-frames are used for the bulk of the communication, where the control center requests data or sends commands, and the remote devices respond with appropriate data or acknowledgments.
3) **Unnumbered (U) frames**: U-frames are used for administrative purposes, such as establishing or terminating connections, or for handling events like remote device resets or failure notifications.

The protocol supports asynchronous communication, meaning that devices can communicate without waiting for a polling cycle. This is critical for event-based reporting, where the remote device can immediately notify the control center of any abnormal conditions, such as fault occurrences or equipment status changes. The protocol also allows for multiple sessions to be managed simultaneously over the same network, which is essential for large systems with multiple substations.

One of the key advantages of IEC 60870-5-104 is its use of TCP/IP for data transmission, which ensures reliable communication through error checking, packet retransmission, and flow control. The use of TCP/IP also allows for integration with modern cybersecurity measures, including encryption and authentication protocols, to protect against cyberattacks. This is increasingly important as power grids become more digitized and vulnerable to network-based threats.

IEC 60870-5-104 also implements robust error detection and recovery mechanisms. The protocol uses sequence numbers and acknowledgments to confirm the delivery of messages and ensure that lost or corrupted packets are retransmitted. This level of reliability is crucial in mission-critical power grid applications, where data integrity and availability directly impact operational safety and stability.

As an extension of IEC 60870-5-101, IEC 60870-5-104 ensures backward compatibility, allowing legacy systems to be upgraded without requiring a complete overhaul of existing infrastructure. The protocol's support for both TCP/IP and serial communication enables it to be used in a wide range of environments, from older SCADA systems to modern smart grid architectures. This flexibility makes it easier for utility companies to transition toward more advanced communication systems while maintaining compatibility with their existing assets.

IEC 60870-5-104 is also interoperable with other common power grid protocols, including IEC 61850 and DNP3. This ensures that it can coexist with different communication standards within the same system, allowing utilities to integrate various automation and control technologies seamlessly.

Overall, IEC 60870-5-104 is a critical protocol for power grid communication, offering reliable, real-time data exchange over modern IP networks. Its ability to support both legacy and modern systems, combined with its robust security and error-handling features, make it an essential component in the operation of remote substations and smart grid infrastructure. Through its flexible, event-driven communication and reliable transport layer, IEC 60870-5-104 ensures that control centers have the data and control capabilities they need to manage the grid efficiently and respond to emerging challenges in a timely manner.

1.4.4 Other Protocols

1.4.4.1 DNP3

DNP3 [IEEE, 2010] was created to ensure reliable communication in harsh environments, specifically for SCADA systems in power grids. It provides high efficiency, interoperability, and time-stamped data for better grid management. DNP3 is commonly used in power distribution networks for telemetry, remote monitoring, and control. It is particularly valuable in environments requiring long-distance communication between control centers and field devices.

DNP3 supports data acquisition, device control, and event-driven reporting. It can also provide time-stamped data for event logs and offers unsolicited reporting, where devices can send updates without waiting for a master request. DNP3 operates using a master–slave communication model. The master (e.g., SCADA system) sends requests to remote devices for data or control actions, and the devices respond with the requested information. DNP3 can also send event-based updates without continuous polling, optimizing bandwidth use.

1.4.4.2 OPC-UA

OPC-UA [OPC Foundation, 2023] was designed for secure and reliable communication in industrial automation systems, including smart grids. Its goal is to provide platform-independent communication between devices and systems from different manufacturers. OPC-UA is used in energy management, advanced metering, and substation automation. It is particularly useful in integrating systems that require secure and standardized data exchange.

OPC-UA provides data modeling, real-time data access, historical data collection, and event-driven communication. It includes built-in security features such as encryption, authentication, and integrity checking. OPC-UA supports client–server interactions over TCP/IP. Clients (e.g., SCADA systems) request real-time data from servers (e.g., sensors or meters). It also supports publish-subscribe mechanisms, allowing devices to push data updates to multiple clients simultaneously.

1.4.4.3 Modbus

Modbus [Modbus Organization, 2012] was initially developed to allow simple communication over serial links between devices such as RTUs, Programmable Logic Controllers (PLCs), and sensors. It was later extended to Ethernet (Modbus TCP), allowing its use in a wider range of industrial applications. Modbus is typically used in substation automation and energy management systems for communication between control centers and field devices. It remains popular due to its simplicity and ease of use, especially in legacy systems.

Modbus follows a master–slave communication model, where the master sends requests to the slave devices to read or write data. The protocol supports simple read-and-write operations for registers and coils. Modbus messages use a request–respons pattern. The master sends a request with a specific function code (e.g., read, write) and address, and the slave responds with the requested data or an error message. It is a straightforward protocol with limited error-handling capabilities but remains effective in low-complexity environments.

In conclusion, each of these protocols plays a vital role in the seamless operation of smart power grids, ensuring reliable, real-time communication across the grid's various components. From IEC 61850's event-driven protection schemes to Modbus's straightforward master–slave interactions, these protocols enable secure and efficient communication in modern power systems.

1.5 LFAs

The convergence of Information Technology (IT) and Operational Technology (OT) networks within smart grids offers numerous operational advantages but also introduces significant security vulnerabilities. Among the most pressing threats are LFAs, which specifically target communication links by overwhelming them with malicious traffic, disrupting critical data exchanges. This section outlines the mechanics of LFAs and their implications for smart grid security.

In the context of Link Flooding Attack (LFA), attackers utilize bots to discover secret paths, termed Persistent Routes (PRs), and subsequently unleash a flood of traffic along these routes toward a set of decoys that are not directly connected to the intended target [Kang et al., 2013].

A Persistent Route (PR) is defined as a communication link that does not connect directly to the target but instead travels through various paths, starting from certain bots to ultimately reach the target. Discovering PRs typically involves two primary methods:

1) **Probing PRs:** Bots explore potential PRs by sending Internet Control Message Protocol (ICMP) packets along links connected to the target [Kang et al., 2013].
2) **Speculating vulnerable PRs:** Once a list of PRs is generated, attackers assess their potential by verifying the impact of flooding attacks on the target via each PR.

Before initiating a crossfire attack, the adversary typically operates in two stages: the initial stage and the renaissance stage. In the initial stage, the attacker probes PRs without specific knowledge of which to target. In the subsequent renaissance stage, they speculate on the most vulnerable PRs.

Unfortunately, existing defense mechanisms encounter several challenges:

1) **Detection difficulty:** Stealthy ICMP or flooding traffic sent at low rates is challenging to detect [Hou et al., 2020; Ding et al., 2020; Kang et al., 2016; Zheng et al., 2018].
2) **Topology obfuscation:** Fixed routes and the presence of decoy addresses complicate network topology obfuscation [Meier et al., 2018; Ding et al., 2020; Hou et al., 2020; Kim et al., 2022].
3) **Traffic overload:** Managing overloaded traffic on critical PRs poses significant challenges related to route mutation and traffic throttling [Gupta et al., 2014; Kang et al., 2016; Zheng et al., 2018; Belabed et al., 2018].

To illustrate the stealthy nature of LFAs, we consider the two-stage crossfire attack model depicted in Figure 1.10. The attack unfolds in two distinct stages: the initial stage (marked with dark gray labels) and the renaissance stage (marked with light gray labels).

The initial stage of this attack consists of the following actions:

1) **Exploring paths**: Each bot acts as a probe, sending ICMP packets with incrementally adjusted Time-To-Live (TTL) values, employing randomized intervals to mimic legitimate application behavior [Kang et al., 2013].
2) **Identifying PRs**: From the probed paths, the adversary analyzes the links based on their occurrence frequency (link density) and selects links that are nonadjacent to the target but have the highest density.

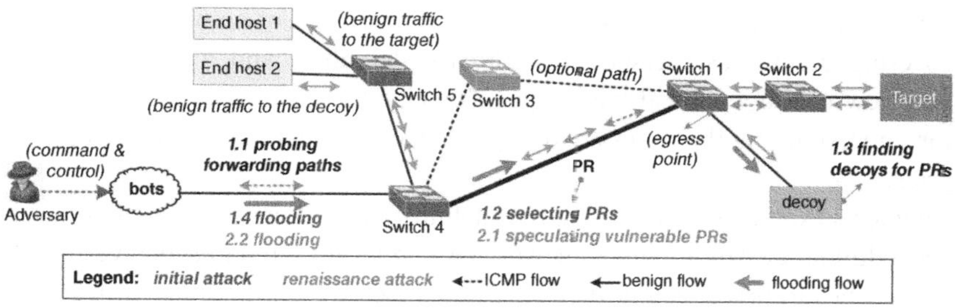

Figure 1.10 The two-stage stealthy crossfire attack model.

3) **Discovering decoys**: The attacker scans the subnet address space to identify available decoys connected to the egress points of the chosen PRs, ensuring they are nonadjacent to the target.
4) **Initiating the flood**: The final step involves orchestrating a coordinated attack, where bots flood traffic toward the decoys while maintaining valid randomized network addresses, thus creating a high-bandwidth flow (depicted by bold arrows) that disrupts related end-to-end connections while keeping the target untouched.

After completing the initial stage, **the renaissance attack stage** follows:

1) **Speculating on vulnerable PRs**: Building on insights from the initial attack, the adversary speculates on vulnerable PRs by commanding bots to flood candidate decoys gathered earlier, verifying the connectivity to the target.
2) **Executing the flood**: Upon identifying vulnerable PRs, the botnet floods traffic toward the verified decoys, replicating tactics from the initial stage. This stage has the potential to inflict disruption similar to that of the initial attack, intensifying the necessity for effective defense mechanisms.

In summary, the evolving landscape of smart grids, characterized by the integration of IT and OT networks, necessitates a proactive approach to countering the threats posed by LFAs. As these attacks exploit communication vulnerabilities, particularly targeting PRs, it becomes crucial for operators to adopt advanced detection and mitigation strategies.

1.6 Navigating the Chapters Ahead

Building on the foundational knowledge of networks and protocols, this book delves into the critical aspects of network availability, including its vision, technical landscapes, challenges, and both classical and cutting-edge solutions. Each subject will be explored thoughtfully throughout the book, with the following structure outlined:

- Chapter 2 presents the vision and technical landscape of network availability, covering definitions, perspectives, threats, demands, and a comprehensive overview of the technical landscape. This chapter establishes the framework for understanding the necessity of high network availability in modern power grids.
- Chapter 3 examines classical high-availability networking solutions that leverage redundancy. This includes an overview of relevant standards, twelve distinct approaches in practice, and their associated technical limitations, providing a historical context for the evolution of these solutions.
- Chapter 4 investigates the challenges and solutions for achieving high availability in smart power grids. It elaborates on five key research directions: non-disjoint path aggregation, fast rerouting, data transport protocols, application layer data redundancy, and defenses against LFAs. This chapter also discusses advanced technologies such as OpenFlow-based Non-Disjoint Path Aggregation Mechanism (OPAM), Traffic-Aware Fast Reroute (TA-FRR), Disruption Resilient Transport Protocol (DRTP) with its core mechanism RSDD, as well as less fragile routing solutions like HeuMPSG, Adaptive RaptorQ Encoder and Decoder (ARED), and Persistent Route Diversification Defense (PRDD). By identifying these areas, the chapter highlights the ongoing research gaps that must be addressed to improve network availability.
- Chapter 5 focuses on OPAM, addressing its challenges, providing a detailed solution, and evaluating its performance through simulations and numerical analysis. The significance of this research direction in enhancing network reliability will also be emphasized.
- Chapter 6 introduces TA-FRR, covering its routing model, problem statement, design considerations, and performance evaluation, while relating its importance to the overall context of network availability.
- Chapter 7 explores DRTP and GenMPSG, detailing the communication model, routing strategies, collaborative issues, collaboration steps, arrival timeouts, and conducting a comprehensive performance evaluation. The interconnectedness of these approaches with other solutions discussed in the book will also be noted.
- Chapter 8 discusses RSDD, presenting its specific designs in State Transition Machines (STMs), along with theoretical validation and experimental results. This chapter will clarify how these designs contribute to enhancing resilience against network failures.
- Chapter 9 analyzes HeuMPSG, addressing the fragility of Multipath Subgraph (MPSG), detailing algorithm design, complexity analysis, and evaluation results through comparative studies with GenMPSG. This will illustrate the advancements made in routing solutions.

- Chapter 10 investigates ARED, offering detailed modeling and analysis, architectural considerations, design aspects, and experimental findings. This chapter will reinforce the importance of application-layer redundancy in network availability.
- Chapter 11 examines PRDD, detailing its defensive strategies, algorithms, and evaluation outcomes, illustrating the necessity of robust defenses against link-flooding attacks.
- Chapter 12 provides practical guidelines for developing and applying both classical and cutting-edge solutions in real-world power grid environments, to enhance the reader's understanding of high-availability networks in smart power grids.
- Chapter 13 concludes this book and presents future prospects in achieving heightened availability and reliability.

The content of this book thoroughly explores network availability for power grids, addressing diverse aspects. As you read, try to connect these ideas with real-world applications. Additionally, access the online source codes for Chapters 6, 8, and 9 to reference designs and enhance systematic improvements. When using this information, please ensure accurate citations for this book and related papers to actively contribute to ongoing discussions in the field.

Questions

1.1 What key considerations drive the design goals of communication networks within the domain of power grids?

1.2 Elaborate on why the availability of communication networks holds paramount importance in the seamless functioning of smart grid operations.

1.3 Discuss the importance of network structures in the implementation of smart power grids. What factors must be considered when designing these networks?

1.4 Explain the concept of next-generation networks and their relevance to smart power grid communications. How do they differ from traditional network architectures?

1.5 What are the primary design goals of the IEC 61850 protocol, and how do these goals enhance the communication in smart power grid systems?

1.6 Explain the role of sampled measured values in the IEC 61850 protocol. How do these sampled values contribute to the performance of protection and control applications?

1.7 Describe the communication model of IEC 60870-5-104. How does it differ from other protocols used in smart grid communications?

1.8 Discuss the key features of the DNP3 protocol and explain how it supports reliable communication in SCADA systems for smart power grids. What advantages does it offer over traditional communication protocols?

1.9 Explain the operational model of OPC-UA in the context of smart grids. How does its architecture facilitate secure and reliable communication between devices from different manufacturers?

1.10 What are the main functions of Modbus in smart power grids? Discuss how its simplicity impacts its adoption in legacy systems.

1.11 Identify and analyze the stages involved in a LFA targeting communication networks in smart grids. How do these stages compromise network availability?

Bibliography

Gouri R. Barai, Sridhar Krishnan, and Bala Venkatesh. Smart metering and functionalities of smart meters in smart grid - a review. In *IEEE EPEC*, pages 138–145, 2015.

Dallal Belabed, Mathieu Bouet, and Vania Conan. Centralized defense using smart routing against link-flooding attacks. In *Cyber Security in Networking Conference*, pages 1–8, 2018.

Zhaohong Bie, Yanling Lin, Gengfeng Li, and Furong Li. Battling the extreme: A study on the power system resilience. *Proceedings of the IEEE*, 105(7): 1253–1266, 2017.

GitHub, Inc. Behavioral model version 2 (bmv2) –the reference P4 switch, 2024. URL https://github.com/p4lang/behavioral-model.

Pat Bosshart, Dan Daly, Glen Gibb, Martin Izzard, Nick McKeown, Jennifer Rexford, Cole Schlesinger, Dan Talayco, Amin Vahdat, George Varghese, and David Walker. P4: Programming protocol-independent packet processors. *ACM SIGCOMM Computer Communication Review*, 44(3):87–95, 2014.

John Boyd. An internet-inspired electricity grid. *IEEE Spectrum*, 50(1):12–14, 2013.

Christoph Brunner. IEC 61850 for power system communication. In *IEEE/PES T & D*, pages 1–6, 2008.

Cisco. Cisco –powering the new digital utility, 2022. URL https://www.cisco.com/c/dam/en_us/solutions/industries/docs/energy/digital-utilities-white-paper.pdf.

Cisco. Cisco –software-defined networking, 2025. URL https://www.cisco.com/c/en/us/solutions/software-defined-networking/overview.html.

Van Der Wielen. *Engineering and Management of Communication Networks in Powergrid's Full Digital Substations*. CIGRE, 2020.

Xuyang Ding, Feng Xiao, Man Zhou, and Zhibo Wang. Active link obfuscation to thwart link-flooding attacks for Internet of Things. In *IEEE TrustCom*, pages 217–224, 2020.

Nils Dorsch, Fabian Kurtz, Hanno Georg, Christian Hägerling, and Christian Wietfeld. Software-defined networking for smart grid communications: Applications, challenges and advantages. In *IEEE SmartGridComm*, pages 422–427, 2014.

DPDK. Data plane development kit (DPDK), 2024. URL https://www.dpdk.org/.

Xi Fang, Satyajayant Misra, Guoliang Xue, and Dejun Yang. Smart grid —the new and improved power grid: A survey. *IEEE Communications Surveys & Tutorials*, 14(4):944–980, 2012.

Hamid Farhady, HyunYong Lee, and Akihiro Nakao. Software-defined networking: A survey. *Elsevier Computer Networks*, 81(Supplement C):79–95, 2015.

Federation of German Industries (BDI). Internet of energy: ICT for energy markets of the future. Technical report, 2008.

Cheng Feng, Yi Wang, Qixin Chen, Yi Ding, Goran Strbac, and Chongqing Kang. Smart grid encounters edge computing: Opportunities and applications. *Advances in Applied Energy*, 1:100006, 2021.

Chavoosh Ghasemi, Hamed Yousefi, Kang G. Shin, and Beichuan Zhang. A fast and memory-efficient trie structure for name-based packet forwarding. In *IEEE ICNP*, pages 302–312, 2018.

Vincenzo Giordano, Alexis Meletiou, Catalin Felix Covrig, Anna Mengolini, Mircea Ardelean, Gianluca Fulli, Manuel Sánchez Jiménez, and Constantina Filiou. Smart grid projects in Europe: Lessons learned and current developments. Technical report, European Commission, 2013.

Albert Greenberg, Gisli Hjalmtysson, David A. Maltz, Andy Myers, Jennifer Rexford, Geoffrey Xie, Hong Yan, Jibin Zhan, and Hui Zhang. A clean slate 4D approach to network control and management. *ACM SIGCOMM Computer Communication Review*, 35(5):41–54, 2005.

Arpit Gupta, Laurent Vanbever, Muhammad Shahbaz, Sean P. Donovan, Brandon Schlinker, Nick Feamster, Jennifer Rexford, Scott Shenker, Russ Clark, and Ethan Katz-Bassett. SDX: A software defined internet exchange. *ACM SIGCOMM Computer Communication Review*, 44(4):579–580, 2014.

Guozheng Han, Bingyin Xu, Kaijun Fan, and Guangxian Lv. An open communication architecture for distribution automation based on IEC 61850. *International Journal of Electrical Power & Energy Systems*, 54:315–324, 2014.

Bernhard Hofmann-Wellenhof, Herbert Lichtenegger, and James Collins. *Global Positioning System: Theory and Practice*. Springer Science & Business Media, 2012.

Tao Hou, Zhe Qu, Tao Wang, Zhuo Lu, and Yao Liu. ProTO: Proactive topology obfuscation against adversarial network topology inference. In *IEEE INFOCOM*, pages 1598–1607, 2020.

Fei Hu, Qi Hao, and Ke Bao. A survey on Software-Defined Network and OpenFlow: from concept to implementation. *IEEE Communications Surveys & Tutorials*, 16(4):2181–2206, 2014.

Yih-Fang Huang, Stefan Werner, Jing Huang, Neelabh Kashyap, and Vijay Gupta. State estimation in electric power grids: Meeting new challenges presented by the requirements of the future grid. *IEEE Signal Processing Magazine*, 29(5):33–43, 2012.

Huawei. Huawei –building a grand blueprint for electric power digital twins, 2025. URL https://e.huawei.com/cn/industries/grid/power-grid.

Huawei Technologies Co., Ltd. China's super grid: Huawei 5G enables a world-leading power grid, 2025. URL https://carrier.huawei.com/en/success-stories/Industries-5G/Electric-power.

IEC. Telecontrol equipment and systems - Part 5-101: Transmission protocols - companion standard for basic telecontrol tasks, 2003.

IEC. Communication networks and systems for power utility automation - Part 1: Introduction and overview, 2010a.

IEC. Communication networks and systems for power utility automation - Part 6: Configuration language for communication in electrical substations related to IEDs, 2010b.

IEC. Communication networks and systems for power utility automation - Part 7-1: Basic communication structure - principles and models, 2011a.

IEC. Communication networks and systems for power utility automation - Part 7-2: Basic communication structure - abstract communication service interface (ACSI), 2011b.

IEC. Communication networks and systems for power utility automation - Part 8-1: Specific communication service mapping (SCSM) - mapping of MMS onto TCP/IP, 2011c.

IEC. Communication networks and systems for power utility automation - Part 9-2: Specific communication service mapping (SCSM) - sampled values over ISO/IEC 8802-3, 2011d.

IEC. Communication networks and systems for power utility automation - Part 9-1: Specific communication service mapping (SCSM) - sampled values, 2013.

IEC. IEC 61850 based substation automation system: A survey. *International Journal of Electrical Power & Energy Systems*, 120:106008, 2020.

IEC 60794-4-10. Optical fibre cables - Part 4-10: Family specification - optical ground wires (OPGW) along electrical power lines, 2014.

IEC 60870-5-104. Telecontrol equipment and systems - Part 5-104: Transmission protocols, 2006.

IEC 62439. High-availability automation networks, 2008.

IEEE. IEEE Vision for Smart Grid Communications: 2030 and Beyond, 2013.

IEEE. IEEE standard for local and metropolitan area networks - station and media access control connectivity discovery, 2016.

IEEE. IEEE standard for testing and performance for all-dielectric self-supporting (ADSS) fiber optic cable for use on electric utility power lines, 2019.

IEEE 1815. IEEE standard for electric power systems communications –distributed network protocol (DNP3), 2010.

IEEE C37.118.2. IEEE standard for synchrophasor data transfer for power systems, 2011.

Intel. Intel®intelligent fabric processors, 2025. URL https://www.intel.com/content/www/us/en/products/network-io/programmable-ethernet-switch.html.

ITU-T. Characteristics of a single-mode optical fibre and cable, 2024.

Van Jacobson, Diana K. Smetters, James D. Thornton, Michael F. Plass, Nicholas H. Briggs, and Rebecca L. Braynard. Networking named content. In *ACM CoNEXT*, pages 1–12, 2009.

M. S. Kang, S. B. Lee, and V. D. Gligor. The crossfire attack. In *IEEE S & P*, pages 127–141, 2013.

Min Suk Kang, Virgil D. Gligor, and Vyas Sekar. SPIFFY: Inducing cost-detectability tradeoffs for persistent link-flooding attacks. In *USENIX NDSS*, 2016.

Jinwoo Kim, Eduard Marin, Mauro Conti, and Seungwon Shin. EqualNet: A secure and practical defense for long-term network topology obfuscation. In *USENIX NDSS*, 2022.

James L. Kirtley. *Electric Power Principles: Sources, Conversion, Distribution and Use.* Wiley Press, 2009.

Swetha Ravi Kumar, Jose Pablo Chaves, Pradyumna Bhagwat, Rafael Cossent, Suddhasatta Kundu, Swetha Ravi Kumar, Tomás Gómez, Reji Kumar Pillai, Balasubramanyam Karnam, Jorge Moreno Barrio, Néstor Rodríguez Pérez, Matteo Troncia, and Orlando Valarezo. Smart grid replication handbook for India. Technical report, European University Institute, 2022.

Yanbiao Li, Dafang Zhang, Xian Yu, Wei Liang, Jing Long, and Hong Qiao. Accelerate NDN name lookup using FPGA: Challenges and a scalable approach. In *IEEE FPL*, pages 1–4, 2014.

Hua Lin, Santosh Sambamoorthy, Sandeep Shukla, James Thorp, and Lamine Mili. A study of communication and power system infrastructure interdependence on PMU-based wide area monitoring and protection. In *IEEE PES GM*, pages 1–7, 2012.

Z. Liu. *Global Energy Internet.* China Electric Power Press, 2015.

Bin Liu, Huichen Dai, Wenquan Xu, Tong Yun, and Ji Miao. Scalable hardware content router: Architecture, modeling and performance. In *IEEE/ACM IWQOS*, pages 1–7, 2021.

Spyridon Mastorakis, Alexander Afanasyev, and Lixia Zhang. On the evolution of ndnSIM: An open-source simulator for NDN experimentation. *ACM SIGCOMM Computer Communications Review*, 47(3):19–33, 2017.

Nick McKeown, Tom Anderson, Hari Balakrishnan, Guru Parulkar, Larry Peterson, Jennifer Rexford, Scott Shenker, and Jonathan Turner. OpenFlow: Enabling innovation in campus networks. *ACM SIGCOMM Computer Communication Review*, 38(2):69–74, 2008.

Roland Meier, Petar Tsankov, Vincent Lenders, Laurent Vanbever, and Martin Vechev. NetHide: Secure and practical network topology obfuscation. In *USENIX Security*, pages 693–709, 2018.

Rui Miguel, Salvatore Signorello, and Fernando M. V. Ramos. Named data networking with programmable switches. In *IEEE ICNP*, pages 400–405, 2018.

Modbus Organization. Modbus application protocol specification, 2012. URL https://modbus.org/docs/Modbus_Application_Protocol_V1_1b3.pdf.

Makoto Murakami and Yoshinori Koike. Highly reliable and large-capacity packet transport networks: Technologies, perspectives, and standardization. *Journal of Lightwave Technology*, 32(4):805–816, 2014.

Named Data Networking. NDN packet format specification, 2022. URL https://named-data.net/doc/NDN-packet-spec/current/.

NASPI. NASPInet 2.0 architecture guidance version 1.19, 2019. URL https://www.naspi.org/node/746.

OPC Foundation. OPC-UA specification, 2023. URL https://reference.opcfoundation.org/.

Open Networking Foundation. OpenFlow specification v1.3, 2012.

p4lang. P4Runtime, 2025. URL https://github.com/p4lang/p4runtime.

Gelli Ravikumar, Dan Ameme, Satyajayant Misra, Sukumar Brahma, and Reza Tourani. iCASM: An information-centric network architecture for wide area measurement systems. *IEEE Transactions on Smart Grid*, 11(4):3418–3427, 2020.

Haftu Tasew Reda, Adnan Anwar, and Abdun Mahmood. Comprehensive survey and taxonomies of false data injection attacks in smart grids: Attack models, targets, and impacts. *Renewable and Sustainable Energy Reviews*, 163:112423, 2022.

Mubashir Husain Rehmani, Alan Davy, Brendan Jennings, and Chadi Assi. Software defined networks-based smart grid communication: A comprehensive survey. *IEEE Communications Surveys & Tutorials*, 21(3):2637–2670, 2019.

Jennifer Rexford, Albert Greenberg, Gisli Hjalmtysson, David A. Maltz, Andy Myers, Geoffrey Xie, Jibin Zhan, and Hui Zhang. Network-wide decision making: Toward a wafer-thin control plane. In *ACM SIGCOMM HotNets Workshop*, 2004.

E. Rosen, A. Viswanathan, and R. Callon. MPLS architecture, 2001.

Yasir Saleem, Noel Crespi, Mubashir Husain Rehmani, and Rebecca Copeland. Internet of Things-aided smart grid: Technologies, architectures, applications, prototypes, and future research directions. *IEEE Access*, 7:62962–63003, 2019.

Sequans. Sequans at Enlit Europe 2023, 2023. URL https://sequans.com/sequans-at-enlit-europe-2023/.

Junxiao Shi, Davide Pesavento, and Lotfi Benmohamed. NDN-DPDK: NDN forwarding at 100 Gbps on commodity hardware. In *ACM ICN*, pages 30–40, 2020.

Martin Sibley. *Optical communications: Components and Systems (third edition).* Springer, 2020.

Ali Sydney, James Nutaro, Caterina Scoglio, Don Gruenbacher, and Noel Schulz. Simulative comparison of multiprotocol label switching and openflow network technologies for transmission operations. *IEEE Transactions on Smart Grid*, 4(2):763–770, 2013.

Junji Takemasa, Yuki Koizumi, and Toru Hasegawa. Toward an ideal NDN router on a commercial off-the-shelf computer. In *ACM ICN*, pages 43–53, 2017.

Mini S. Thomas. Power System SCADA and Smart Grids. CRC Press, 2020.

Massimo Tornatore, Peter Winzer, Yongli Zhao, Biswanath Mukherjee, and Ioannis Tomkos. *Springer Handbook of Optical Networks*. Springer Press, 2020.

United States Department of Energy. Smart grid system report. Technical report, 2014.

Kris G. Vyas and Janet S. Combs. Application of Southern California Edison Company (U-338-E) for approval of its smart grid deployment plan. Technical report, Southern California Edison Company, 2011.

Kun Wang, Jun Yu, Yan Yu, Yirou Qian, Deze Zeng, Song Guo, Yong Xiang, and Jinsong Wu. A survey on energy internet: Architecture, approach, and emerging technologies. *IEEE Systems Journal*, 12(3):2403–2416, 2018.

Alan Weissberger. Rakuten Mobile in joint venture with Tokyo Electric Power Company (TEPCO) to expand 4G/5G network, 2022. URL https://techblog.comsoc.org/2022/07/04/rakuten-mobile-in-joint-venture-with-tokyo-electric-power-company-tepco-to-expand-4g-5g-network/.

Xiaorong Xie, Yaozhong Xin, Jinyu Xiao, Jingtao Wu, and Yingdao Han. WAMS applications in Chinese power systems. *IEEE Power and Energy Magazine*, 4 (1):54–63, 2006.

George Xylomenos, Christopher N. Ververidis, Vasilios A. Siris, Nikos Fotiou, Christos Tsilopoulos, Xenofon Vasilakos, Konstantinos V. Katsaros, and George C. Polyzos. A survey of information-centric networking research. *IEEE Communications Surveys & Tutorials*, 16(2):1024–1049, 2014.

Ye Yan, Yi Qian, Hamid Sharif, and David Tipper. A survey on smart grid communication infrastructures: Motivations, requirements and challenges. *IEEE Communications Surveys & Tutorials*, 15(1):5–20, 2013.

Jiahai Yuan, Jiakun Shen, Li Pan, Changhong Zhao, and Junjie Kang. Smart grids in China. *Renewable and Sustainable Energy Reviews*, 37:896–906, 2014.

Lixia Zhang, Alexander Afanasyev, Jeffrey Burke, Van Jacobson, Christos Papadopoulos, Kc Claffy, Lan Wang, Patrick Crowley, and Beichuan Zhang. Named data networking. *ACM SIGCOMM Computer Communication Review*, 44(3): 66–73, 2014.

Jing Zheng, Qi Li, Guofei Gu, Jiahao Cao, David K. Y. Yau, and Jianping Wu. Realtime DDoS defense using COTS SDN switches via adaptive correlation analysis. *IEEE Transactions on Information Forensics and Security*, 13(7):1838–1853, 2018.

Boyang Zhou, Qiang Yang, Yansong Wang, and Chunming Wu. Reliable communication in transmission grids based on nondisjoint path aggregation using software-defined networking. *IEEE Transactions on Industrial Informatics*, 15(2):842–855, 2019.

Boyang Zhou, Chunming Wu, Qiang Yang, and Xiang Chen. DRTP: A disruption resilient hop-by-hop transport protocol for synchrophasors measurement in electric transmission grids. *IEEE Access*, 10:133898–133914, 2022.

Boyang Zhou, Chunming Wu, Yang Qiang, Qian Yaguan, and Nie Yinghui. Resilient sensor data dissemination to mitigate link faults in IoT networks with long-haul optical wires for power transmission grids. *IEEE Internet of Things Journal*, 2023.

2

Network Availability Vision and Technical Landscape

2.1 What Is Network Availability?

The concept of network availability in power grids has evolved significantly over the past several decades. In the 1970s, early power grid communication systems relied on analog circuits for basic control and monitoring. These systems were limited by their lack of flexibility and high vulnerability to failures. With the rise of digitalization in the 1980s, more advanced communication protocols and redundancy mechanisms were introduced, significantly improving the reliability and robustness of grid operations. The 1990s saw a major shift toward IP-based networks, which brought greater scalability and flexibility, enabling more complex and interconnected power grids. This evolution continued into the 21st century, with the formalization of advanced redundancy protocols in standards such as IEC 62439 [2008]. The IEC 62439 standards [IEC, 2016] introduced mechanisms like the Parallel Redundancy Protocol (PRP) [IEC, 2021] and High-availability Seamless Redundancy (HSR) [IEC, 2021], which dramatically improved fault tolerance and minimized downtime in modern power grid networks.

2.1.1 Network Availability Definition

Communication network availability in power grids is a critical metric that quantifies the time a network is accessible for sensor measurements and actuator control. It represents the percentage of time the network operates reliably compared to the total operational time. According to IEC 62439 [2016 and IEC 62439 [2008], the availability is mathematically defined as the fraction of time that a network remains functional throughout its lifecycle, as expressed in Eq. (2.1). This metric is particularly significant for repairable systems, where the focus

Communication Networks in Smart Power Grids, First Edition. Boyang Zhou.

Companion Website: www.wiley.com/go/SmartPowerGrids1e

extends beyond mere resilience to failures and encompasses the ability to recover swiftly and maintain operational continuity.

$$A_N = \frac{\text{MTTFN}}{\text{MTTFN} + \text{MTTRN}} \tag{2.1}$$

In this equation, Mean Time to Failure of Network (MTTFN) represents the average failure-free operational time of the network. Mean Time to Repair Network (MTTRN), on the other hand, quantifies the average time required to repair a network failure, thus indicating the duration during which the network is non-operational.

To put this into a practical context, consider a power grid communication network that supports critical infrastructure such as Supervisory Control and Data Acquisition (SCADA) systems [Thomas, 2020]. If the network experiences a failure and takes a significant amount of time to recover, it could severely impact grid dispatches, state estimation, and other essential functions. High network availability, therefore, is crucial to ensure consistent operation, low End-to-End Failure Rate in Packet Delivery (EEFR), and low End-to-End Payload Packet Delivery Time (EEDT), all of which contribute to maintaining high-quality power grid management for grid stability.

2.1.2 Criticality of Network Availability

2.1.2.1 Viewpoints of Network Availability

The network availability of power grids is critical, particularly in the context of the United States, where the grid is a complex and extensive system comprising over 3100 electric utilities operating more than 10,000 power plants [Tsoukalas and Gao, 2008; Aggarwal et al., 2010]. This vast infrastructure serves approximately 131 million customers, consuming over 3500 billion kilowatt-hours of electricity daily [Tsoukalas and Gao, 2008; Aggarwal et al., 2010]. With such high stakes in terms of energy consumption and the need for reliable service, ensuring continuous network availability becomes essential for maintaining operational efficiency and meeting growing demand.

One of the primary challenges facing the power grid is its aging infrastructure. For example, the average age of transmission lines of USA is between 50 and 60 years, leading to concerns about reliability and potential failures [Gungor et al., 2010]. This aging infrastructure, coupled with the fact that electricity demand increased by 30% between 1988 and 1998 while only 15% of new transmission capacity was added [Gellings, 2009], highlights the critical need for a resilient communication network that can effectively support the operations of these utilities. With a network that is not only stretched thin but also at risk of failures, ensuring high availability becomes paramount to prevent widespread outages and maintain service continuity.

Moreover, the current state of deregulation and inefficiencies within the power grid further complicates the optimization of resource use and environmental impact. The existing operational framework must be capable of responding to unexpected disruptions swiftly, thereby minimizing downtime and ensuring the reliable delivery of electricity to consumers. In this environment, network availability directly correlates with the ability to manage resources effectively and sustain the environmental objectives of energy production.

A failure in network availability can have cascading effects on grid operations, affecting everything from real-time monitoring to the control of power generation and distribution. The increasing integration of renewable energy sources and advanced technologies, such as smart grids, necessitates even greater reliability from communication networks to ensure seamless interactions among various components. Consequently, investing in technologies and protocols that enhance network availability is not merely a technical consideration but a fundamental requirement for the efficient and sustainable operation of the grid.

In conclusion, the criticality of network availability in power grids cannot be overstated. As the power grid faces the dual challenges of aging infrastructure, growing electricity demand and expanding architecture, ensuring robust, reliable communication networks is essential for maintaining operational integrity, optimizing resource use, and fostering a sustainable energy future. The interconnected nature of today's power systems underscores the necessity for high availability, making it a priority for utilities aiming to meet the needs of consumers while safeguarding the environment.

2.1.3 Network Availability Analysis

To truly understand the importance of network availability, we must consider the severe consequences that arise from its failure.

One critical metric that directly affects the performance of smart grid communication is the EEFR. When EEFR is excessively high on a server, such as in SCADA or Phasor Data Concentrators (PDCs), the result is incomplete data reception. This can have detrimental impacts, especially on grid-State Estimation (SE), leading to inaccuracies that compromise the stability and reliability of the power grid [Chenine and Nordström, 2010]. Ensuring the accurate reception of data enables SCADA's SE to monitor the grid with high precision, even under severe Link Loss Rate in Packetss (LLRs) [Huang et al., 2012], and ensures reliable delivery of control commands to the grid's actuators.

To highlight the gravity of this issue, consider the example of Phasor Measurement Units (PMUs), which transmit data at 50 or 60 Hz to PDCs according to the IEEE C37.118.2 protocol [IEEE, 2011]. In real-world deployments, network impairments have led to serious EEFR issues. The North American Synchrophasor Initiative (NASPI), funded by the U.S. Department of Energy, reported in 2011 that only 29% of PMUs in the eastern U.S. power grid could deliver more than 95% of their data due to network impairments [NASPI, 2011]. Similarly, Brazil experienced a 14% data loss from PMUs [Decker et al., 2011]. These significant data losses can prevent grid systems from achieving topological observability [Gonzalez-Perez and Wollenberg, 2001], which is essential for accurate grid dispatch. The economic consequences of such failures are staggering, with estimated annual losses in the U.S. ranging from $25 to $180 billion due to network-induced errors [Kabalci, 2016].

The demand for low EEFR is not limited to PMUs but extends to other communication protocols used for sensor data dissemination across Remote Terminal Units (RTUs), such as IEC 61850 [IEC, 2020], DNP3 [IEEE, 2010], and OPC-UA [OPC Foundation, 2023]. Each protocol is tailored to the needs of specific grid components, and maintaining a low EEFR is crucial across all of them to ensure grid stability. NASPI has further studied the stringent requirements for PMU data transmission, concluding that for SE to maintain accuracy, an EEFR of no more than 0.33% is required [NASPI, 2018]. Any violation of this threshold could lead to inaccurate grid-state estimation, jeopardizing the grid's overall observability and safety.

On the other hand, timely packet delivery is equally critical to preventing network failures from escalating into broader power grid outages. Network link failures, which cause packet loss and out-of-order packet delivery, can trigger Transmission Control Protocol (TCP) congestion control mechanisms, leading to packet retransmissions and a substantial increase in Packet Inter-Arrival Time (PIAT) [Leung et al., 2007]. This delay can disrupt real-time grid operations and, in severe cases, contribute to cascading failures across the grid [Liu et al., 2015].

Recent advancements in link fault resilience have aimed to address these challenges by reducing packet loss and minimizing recovery times across both the network and transport layers. These innovations employ a variety of techniques to quickly restore network functionality after failures, mitigating the risk of prolonged outages. In Section 2.3, we will explore these techniques in greater detail, highlighting their role in supporting the high-availability demands of modern power grid communication networks.

By ensuring low EEFR and timely data delivery, power grid networks can maintain the high availability necessary for reliable grid operation, avoiding the costly consequences of communication breakdowns.

2.1.4 Standard Efforts Made in Availability

The IEC 62439-1 standard [IEC 62439, 2008] plays a vital role in ensuring network availability by incorporating redundancy and fault-tolerant mechanisms in past decades. The PRP protocol [IEC, 2021], for example, allows data to be transmitted simultaneously across two separate networks. In the event of a failure in one network, the other network continues to function without interruption, effectively improving MTTFN by reducing the likelihood of total network failure. Similarly, HSR [IEC, 2021] offers seamless redundancy by forming a ring topology that allows data to flow in both directions. If a failure occurs at any point, data is rerouted in the opposite direction, ensuring uninterrupted communication.

The IEC 62439 protocols have been widely deployed in practice. For example, the PRP protocol has been implemented in swissgrid corporation's 380 kV substation Gösgen in 2018 for enhancing substation communication reliability [Swissgrid, 2021]. Siemens fully implemented PRP in virtual programmable logic controllers for digitalization encompassing power grids [Siemens, 2024]. Texas Instruments have implemented PRP and HSR on Texas Sitara processors, which has been widely applied to power grids [Illinois, 2016; Texas Instruments, 2023].

Emerging technologies such as automated fault detection and self-healing networks are also playing a significant role in reducing MTTRN. These technologies enable the network to detect failures and initiate corrective actions without human intervention, reducing recovery time and minimizing service disruption. As these techniques become more widely implemented, they will further enhance the resilience and availability of power grid networks.

In summary, network availability is not merely a technical metric; it is a vital indicator of the operational integrity and reliability of communication networks in power grids. By adhering to standards like IEC 62439-1 and adopting advanced redundancy and fault-tolerant strategies, power grid operators can ensure that their networks are highly available and resilient. This level of reliability is essential to handle the complexities of modern power grids and protect against disruptions that could compromise grid stability.

2.2 Availability Threats and Demands

Despite the great efforts made by the IEEE 62439 standard, network availability remains a critical and ongoing concern that significantly impacts the reliability and efficiency of operations within evolving smart power grids. As the demand for ubiquitous connectivity and comprehensive services continues to grow, the network architecture and external environments are undergoing substantial changes. These emerging trends not only expand the threats to network availability beyond what IEC 62439 addresses, but they also introduce new requirements for maintaining availability in increasingly complex grid environments. Consequently, it is imperative to develop adaptive strategies and technologies that can effectively respond to these challenges, ensuring the resilience and operational integrity of modern power systems.

Figure 2.1 illustrates the main threats to network availability across different layers of the protocol stack. At the lower layers, the physical and link layers can experience unavoidable packet losses due to frame errors, despite the implementation of reliability measures. As we move up the stack, packet losses may worsen at the network and transport layers, particularly due to Link Flooding Attacks (LFAs) [Kang et al., 2013] or network congestion. Even when protocols such as TCP are employed, congestion control mechanisms can halt packet transmission, leading to payload packet losses. These transmission challenges present significant threats to network availability, including issues with data delivery, path reliability, data transport, application data completeness, and LFAs. This section analyzes these five categories of threats and highlights the key availability requirements that must be addressed to ensure a resilient smart grid network.

2.2.1 Data Delivery Reliability

Data delivery reliability is a critical aspect of the operational stability of smart power grids, heavily reliant on the SCADA system, which serves as the backbone of power transmission grids.

These systems depend on reliable end-to-end data delivery mechanisms to ensure seamless operations. The efficient transmission of IEC 61850 Manufacturing Message Specification (MMS) [IEC, 2011] data packets over TCP/IP protocols is vital. These MMS messages are categorized, based on their delivery time requirements, with medium-speed (Type 2) messages needing delivery within 100 ms and low-speed (Type 3) messages within 500 ms [IEC, 2013]. Meeting these timelines is essential, particularly in mission-critical applications such as terminal-state monitoring and control.

However, network link failures can significantly disrupt these delivery timelines, resulting in packet loss and out-of-order events. Such failures activate TCP congestion control and retransmission mechanisms, which can lead to an inflated average PIAT and EEDT [Leung et al., 2007]. A violation of the EEDT can substantially increase the risk of cascading power grid failures or outages, as it delays real-time decision-making and fault responses. This results in operators relying on outdated information for critical operations, undermining their ability to manage interconnected disturbances effectively and potentially leading to widespread outages [Liu et al., 2015].

There are notable incidents from the past few decades that illustrate these issues in real power grids. For example, the 2003 Northeast blackout affected approximately 55 million people across the USA and Canada. This event was exacerbated by operators delaying actions to redistribute load after overloaded transmission lines drooped into foliage, demonstrating the critical impact of

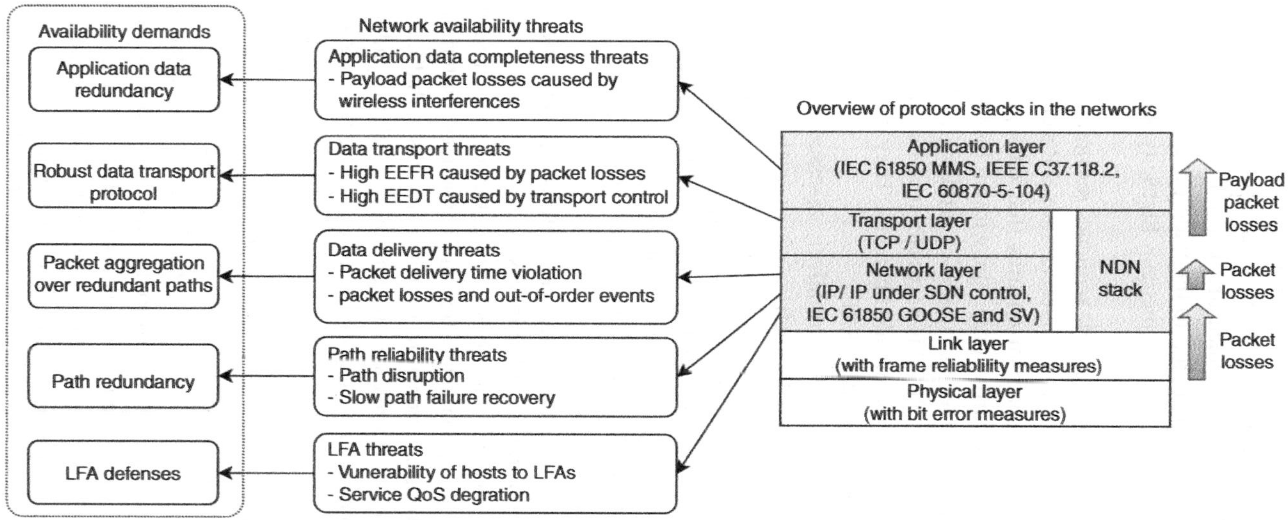

Figure 2.1 Primary threats to network availability across protocol layers.

response delays [U.S.-Canada Power System Outage Task Force, 2023]. Similarly, the 2011 Arizona–Southern California cascading outages left nearly seven million people without power in San Diego. In this case, operators faced limited visibility into grid states due to SCADA measurement errors and failed to take timely action, which ultimately contributed to the widespread outage [Federal Energy Regulatory Commission and North American Eletricity Reliability Corporation, 2011; Velay et al., 2018].

To counter these challenges, a reliable sequence restoration and packet delivery process, commonly referred to as the Success Rate of Receiving a Packet (SRRP), becomes essential. This mechanism ensures that data packets arrive at their intended destinations in the correct order and without loss. It employs techniques such as packet aggregation mechanisms over multiple paths to maintain data integrity. Nonetheless, a high SRRP is necessary to avoid introducing additional delays that could counteract its intended benefits. They indicate the necessity of data delivery reliability with a high SRRP.

Reliability in data delivery is quantitatively assessed through the Qualified Source-Destination Pairs (QSDP) metric, which evaluates how well source-destination pairs meet their operational requirements. A high QSDP value indicates efficient packet delivery, minimal delays, and low rates of packet loss, all of which are crucial for the reliability of the SCADA system. For example, in instances where timely data transmission is critical, such as in the operation of protection relays, even minor delays can lead to significant operational disruptions or safety risks.

Additionally, the complicated networks can complicate the reliability of data delivery. The integration of wireless and optical networks introduces variability in performance that must be addressed. Each communication medium possesses unique characteristics and challenges – interference in wireless links, for instance, or latency in optical networks – that can significantly impact data delivery performance. To enhance reliability, adaptive mechanisms capable of dynamically responding to changing network conditions are imperative.

In summary, data delivery reliability within smart power grids is not merely a technical requirement; it is a foundational element underpinning the overall effectiveness and safety of grid operations. The reliance on timely and accurate message exchanges, coupled with the potential impacts of network failures and the necessity of robust reliability mechanisms such as SRRP, highlights the critical importance of this aspect in modern grid management. Ensuring a high QSDP is essential for achieving the operational goals of smart grids and necessitates ongoing research and development of advanced communication protocols capable of maintaining reliability in increasingly complex environments.

2.2.2 Path Reliability

Path reliability is a critical factor for ensuring consistent end-to-end connectivity, especially in mission-critical applications such as terminal-state monitoring, control systems, and substation automation [Xing et al., 2021]. In smart power grids, the failure of communication links or devices can lead to severe disruptions in the flow of essential data, potentially resulting in delayed or erroneous control signals. Such delays can have significant consequences, including incorrect load shedding, power blackouts, or failure to detect critical faults, all of which could compromise the operational integrity of the power grid.

Smart power grids operate with stringent real-time requirements, particularly for critical data such as protection relays or synchrophasor information, where even millisecond-level delays in communication can disrupt the system's ability to react to contingencies. These applications demand path reliability with near-zero tolerance for delays or packet losses, making the need for robust Fast Reroute (FRR) mechanisms non-negotiable, which requires researches and development.

There are lots of accidents mainly caused by natural disasters and power grid disruptions in past decades. For instance, in 2017 Hurricane Harvey communication infrastructure was severely impacted, which led to delays in reporting outages and coordinating response efforts. Utilities faced challenges in managing resources effectively due to communication link faults, resulting in prolonged outages and impaired recovery efforts [North American Electric Reliability Corporation, 2024]. 2021 Texas power grids suffered a major power crisis, which came about during three severe winter storms sweeping across USA. In this crisis, grid conditions deteriorated rapidly, where communication networks failed in effective power operation and dispatches, leading to blackouts, and forcing Electric Reliability Council of Texas to shed load to avoid a catastrophic failure [The University of Texas at Austin, 2021]. 2011 Arizona–Southern California outage was influenced by communication failures with SCADA measurement errors, delayed decision-making, and inadequate responses to the evolving situation [Federal Energy Regulatory Commission and North American Eletricity Reliability Corporation, 2011]. 2003 Northeast blackout massively affected around 55 million people in the U.S. and Canada; it was partly attributed to inadequate monitoring and communication failures [U.S.-Canada Power System Outage Task Force, 2023]. Operators were unable to receive timely information about transmission line statuses due to faulty SCADA communications, leading to delayed responses that exacerbated the situation. These facts demonstrate the essentiality of FRRs in dealing with grid emergence needs.

The demands on path reliability in smart power grids also extend to the need for real-time fault detection and recovery across multi-technology networks. For instance, in optical backbone using Optical Transport Network (OTN) and Synchronous Digital Hierarchy (SDH) [Sibley, 2020], the integration of redundancy protocols such as Media Redundancy Protocol (MRP) with native recovery mechanisms like Automatic Protection Switching (APS) [Sibley, 2020] is far from seamless, often leading to delays in failover that are incompatible with the strict timing requirements of smart grid applications. Wireless segments also face similar challenges. The signal interference and link degradation typical in 5G and Long Range Radio (LoRa) environments require faster and more adaptive recovery techniques than those provided by existing FRR solutions.

In conclusion, the threats to path reliability in smart power grids are multifaceted, encompassing the unique characteristics of modern communication networks, from high-capacity optical links to constrained wireless networks. Effective FRR mechanisms must account for these variables, offering not only rapid recovery from faults but also scalable, efficient, and adaptable solutions that can handle the diverse and dynamic conditions of the smart grid. Therefore, while traditional FRR strategies such as Multiprotocol Label Switching (MPLS) and segment routing have provided a solid foundation for resilience, more sophisticated approaches are needed to address the heightened demands of smart power grids, where the consequences of unreliable communication can be catastrophic.

2.2.3 Data Transport Reliability

The digital infrastructure of smart grids is fundamentally dependent on the accurate and timely transport of data, particularly that generated by PMUs for electrical statistics measurements utilizing the IEEE C37.118 protocol [IEEE, 2011]. The integration of PMUs into grid management was strongly advocated by the Federal Energy Regulatory Commission in USA [NASPI, 2011, 2018, 2019], underscoring the critical need for robust data transport reliability frameworks as part of guidelines for implementing advanced grid technologies. This endorsement reflects a growing recognition of the vital role that data transport reliability plays in maintaining the stability and resilience of the power grid.

PMUs continuously sample electrical measurements from essential components, such as transformers and buses, producing high-resolution data packets that are precisely timestamped using Global Positioning System (GPS) signals. The synchrophasor data generated is transmitted in real time from a network of PMUs – potentially numbering in the hundreds or thousands at substations – to PDCs. The latency for this data transmission is often kept under 100 ms, with sampling rates ranging from 30 to 60 Hz. Such high-frequency data generation results in substantial volumes of synchrophasor data that must be transmitted with minimal delay to maintain real-time situational awareness [IEEE, 2011]. This data enables grid operators to make informed decisions within their SCADA systems, which incorporate electric statistics from power meters to provide comprehensive visibility into grid operations and maintain the health and stability of the power system. For effective performance, PMUs typically require an accuracy of $\pm 1\%$ for voltage and current measurements. Achieving this level of precision necessitates robust data transport mechanisms, such as the implementation of TCP/IP protocols with error correction features and redundancy measures to prevent packet loss and ensure correct data sequencing.

Reliable data transport in smart grids is intrinsically linked to maintaining low levels of EEFR to ensure accurate power grid-state estimation [Huang et al., 2012]. A high EEFR can lead to delays in data transmission and potential inaccuracies, posing serious risks to the operational integrity of the grid. For example, in dynamic environments characterized by rapid fluctuations in load and generation, any delays in real-time data availability can hinder timely decision-making. This, in turn, may lead to inappropriate responses to changing conditions, such as the need for load shedding or generator adjustments, ultimately jeopardizing grid stability.

Moreover, the implications of unreliable data transport extend beyond immediate operational concerns for SE accuracy. Notable historical incidents underscore the critical importance of data transport reliability in smart grids. In the 2003 Northeast blackout [U.S.-Canada Power System Outage Task Force, 2023], failures in the state estimation process were significant contributors to the outage. Operators struggled to accurately assess the grid's status due to delays and inaccuracies in data transmission, resulting in inappropriate operational responses and failure to alleviate overload conditions on transmission lines. Similarly, in the 2011 Arizona-Southern California outages [Federal Energy Regulatory Commission and North American Eletricity Reliability Corporation, 2011], operators lacked timely access to accurate PMU data due to communication failures, which led to errors in SCADA measurement data. This situation delayed necessary actions to effectively manage system disturbances, heightening the risk of cascading failures. Such examples underscore the imperative for high data transport reliability, particularly regarding PMU data transport, where real-time accuracy is essential for operational stability and outage prevention.

In conclusion, maintaining data transport reliability is a foundational requirement for the effective operation of smart power grids. Given the integral role of PMUs and the reliance on timely, accurate data for state estimation and decision-making, achieving a low EEFR is not just beneficial but essential. The challenges posed by modern grid architectures necessitate ongoing research and development of innovative communication strategies, such as integrating artificial intelligence for predictive analytics, that can adapt to the evolving demands of the smart grid landscape, ensuring resilience and stability in the face of potential disruptions.

2.2.4 Application Data Completeness

In modern smart distribution networks, ensuring the reliability and completeness of application data has become critical for maintaining grid operations. These networks increasingly rely on wireless private networks to connect terminal devices such as Distribution Terminal Units (DTUs), Feeder Terminal Units (FTUs), and Transformer Terminal Units (TTUs) to the master station [Thomas, 2020]. These devices facilitate telemetry, telecontrol, and remote signaling – collectively referred to as the "three-remote" commands – which demand high-speed, low-latency, and secure data transmission. These operations are essential to the accurate control of grid components and monitoring feedback, both of which are crucial for maintaining the stability and responsiveness of the smart grid.

The IEC 60870-5-104 protocol [IEC, 2006] is widely adopted to ensure the timely and reliable transmission of application data in these networks. This protocol mandates a correctness rate of over 99.99% for telecontrol commands, and the Round-Trip Time (RTT) for these commands should not exceed five seconds. Achieving these stringent requirements is critical to preventing data loss and ensuring that the right commands reach grid devices in time, particularly during fault events. Compromised data completeness can lead to delayed responses to critical grid issues or incorrect control actions, degrading grid stability.

However, ensuring data completeness in wireless communication environments is fraught with challenges, particularly when operating over shared or open frequency bands such as Long Term Evolution (LTE) [3GPP TS 36.401, 2016]. The inherent openness of these frequency bands exposes communications to interference from both authorized devices sharing the spectrum and unauthorized or malicious sources. For instance, nearby communication devices operating on similar frequencies or intentional jamming can introduce significant interference, resulting in data packet loss. The loss compromises the integrity of the transmitted data, potentially delaying the execution of critical commands as command retrying increases latency.

A notable real-world incident highlighting the criticality of application data completeness occurred during the 2015 cyberattack on Ukraine's power grid [Bock et al., 2017]. In this attack, adversaries intentionally disabled end terminals, so that communication channels are disrupted, causing incomplete and inaccurate grid-state information to be relayed to the SCADA systems. As a result, SCADA was unable to correctly estimate the grid's real-time operational state, leading to delays in response and incorrect decision-making. These compromised grid states directly contributed to blackouts across several regions. This incident underscores the essential role that application data completeness plays in maintaining accurate grid-state estimation, which is vital for ensuring the operational stability and security of power grids.

To mitigate these risks, the application layer must be equipped with advanced mechanisms that ensure data completeness even in unreliable or interference-prone environments. Techniques such as RaptorQ coding can help recover lost data packets and enhance the reliability of data transmitted over unreliable wireless links, ensuring minimal data loss. This allows grid operators to minimize the impact of interference on communication reliability and ensure the uninterrupted execution of "three-remote" commands.

In conclusion, the completeness of application data is essential for the reliable operation of modern smart distribution networks, which can impact the operational stability of the grid. As smart grid architectures evolve, so too must the communication frameworks support them. Ongoing development of lost packet recovery techniques is necessary to maintain this reliability.

2.2.5 LFA Defenses

In smart grids, where real-time data from PMUs, SCADA, and distributed sensors is essential for effective monitoring and control, LFAs can have catastrophic consequences. LFAs [Kang et al., 2013] exploit the interconnected nature of smart grid communications, where data flows through various routers and switches, often over congested or shared links. By flooding these links with excessive traffic, attackers can cause severe congestion, preventing legitimate control signals and telemetry data from reaching their destinations. This degradation of service jeopardizes grid stability, as delayed or missing data leads to inaccurate SE and improper execution of grid control commands, potentially resulting in widespread power outages [Huang et al., 2012; Chakrabarti et al., 2009]. The economic repercussions of such disruptions can be significant, affecting both utilities and consumers [Nash and Rundle, 2021; Meehan, 2022].

Notable real-world incidents about smart power grids illustrate the risks associated with LFAs. In 2021, LUMA Energy in Puerto Rico experienced a large-scale LFA that produced an overwhelming 2 million hits per second, severely hindering users' access to customer portal accounts during critical outages [Nash and Rundle, 2021]. Similarly, Ignitis Group in Lithuania encountered multiple LFA incidents, described as its "biggest cyber-attack in a decade," which disrupted vital digital services and rendered their websites inaccessible [Meehan, 2022]. These examples highlight the escalating threats posed by sophisticated LFA tactics to smart power grid infrastructures.

The potential impact of LFAs can be understood in their ability to degrade Quality-of-Service (QoS), leading to excessive packet loss, increased latency, and communication breakdowns within the grid's monitoring and control systems. Such conditions can cause deviations in SE, delayed fault detection, and misdirected control dispatches, severely compromising grid stability.

Recent developments in LFA techniques, including stealthy methods like the crossfire attack, target select links to cause disruptions while minimizing immediate detection [Kang et al., 2013]. These sophisticated attacks pose challenges that existing security measures may struggle to address.

As power grids increasingly integrate cloud computing servers, they face heightened risks associated with LFAs. Recent large-scale attacks against major cloud service providers—such as the 400 Gbps attack on Cloudflare in 2014 [KrebsonSecurity, 2014], the 620 Gbps assault on Dyn in 2016 [LowEndTalk, 2016], and a significant attack on Amazon in 2019 [Dark Reading, 2019] – underscore the scalability and evolving sophistication of LFAs. These incidents expose vulnerabilities within communication infrastructures, even in systems designed for high resilience, revealing the critical need for operators to adopt proactive defense strategies. Implementing advanced detection and mitigation techniques is essential to safeguard smart grid networks against the increasing threat of LFAs, ensuring the reliability and security of power delivery systems.

As LFAs evolve in sophistication, smart grid operators must adopt a multifaceted approach to security, incorporating advanced anomaly detection systems, traffic analysis tools, and diversified communication pathways. Such measures will ensure that grid communications remain secure and resilient against these pervasive and evolving threats.

In conclusion, the threats to network availability in smart power grids are diverse and require comprehensive mitigation strategies. Ensuring reliable data delivery, path resilience, data transport accuracy, application data completeness and protection against LFAs is vital to maintaining the stability and efficiency of modern power grids. By addressing these threats systematically, the smart grid's network infrastructure can be fortified, ensuring a more resilient energy future.

2.3 Technical Landscape

To address the various threats and demands related to network availability, high-availability networks must be meticulously designed to accommodate potential communication exceptions. This technical landscape, illustrated in Figure 2.2, encompasses a multi-layered architecture that ensures seamless operation and interaction among its components. The network functions can be classified into five distinct layers: the physical, data link, network, transport, and application layers. Each layer incorporates specific reliability measures tailored to enhance overall network performance and resilience.

The physical layer serves as the foundation, consisting of the hardware and transmission mediums that enable data transfer. The data link layer ensures reliable point-to-point communication, implementing error detection and correction protocols. The network layer manages data routing and forwarding, maintaining

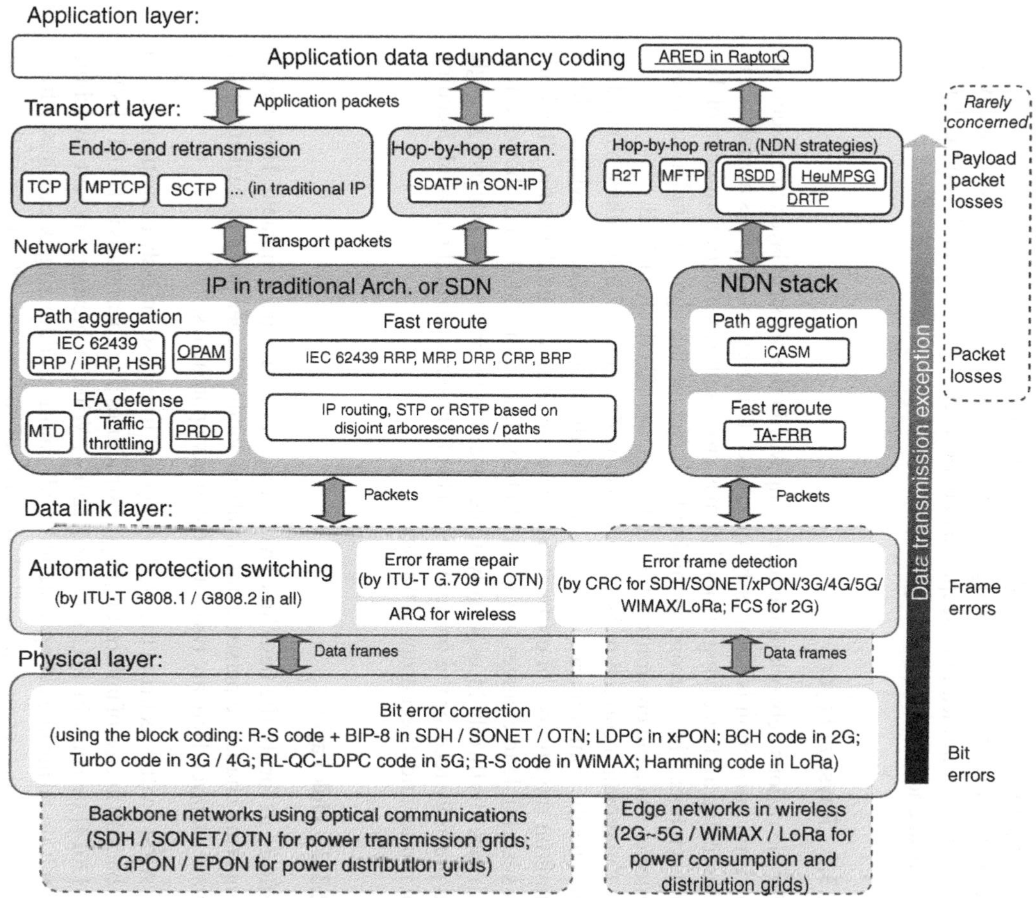

Figure 2.2 Landscape of innovative network architecture, solutions, and issues.

optimal pathways for information flow. The transport layer is responsible for end-to-end communication, providing flow control and ensuring data integrity during transmission. Finally, the application layer interfaces with end-users, delivering services that leverage the underlying network capabilities.

Together, these layers work in harmony to navigate the complexities of the power grid, ensuring high availability and reliability in the face of diverse operational challenges. By adopting a layered approach, the network can effectively manage communication exceptions, adapt to changing demands, and enhance the resilience of smart power grid operations.

2.3.1 Physical Layer

For the various network modes operating within the physical layer, the essence of bit error correction is seamlessly integrated. This technology plays a vital role in identifying and rectifying errors that may crop up during the transmission of digital symbols. These errors, stemming from factors like signal degradation, noise, interference, and environmental conditions, are effectively tackled.

The standard correction methods, illustrated in Figure 2.2, are chosen based on the specific communication mode. Each method, however, comes with a defined limit on the number of errors it can correct within a given data unit. This limit is susceptible to changes in the channel condition. Should the number of errors surpass this threshold, especially in the case of burst errors, the code may struggle to precisely recover the original data. This failure results in the invalid decoding of frames and direct drops of frames.

2.3.2 Link Layer

Efforts to fortify the data link layer's resilience have been notable. However, a lingering challenge persists: link faults inevitably trigger massive packets in the network layer. Here are two complementary approaches:

- **APS:** Swiftly securing the working optical line by seamlessly transitioning to a backup in case of failure, employing structures like linear trails in ITU-T G.808.1 [2014] or rings in ITU-T G.808.2 [2019]. While achieving efficient switching within 50 ms in existing systems, it falls short in addressing packet losses from transmission errors, network device malfunctions, and critical link faults.
- **Automatic Repeat reQuest (ARQ):** Widely adopted in wireless communications, ARQ automatically detects and requests the retransmission of corrupted or lost data during transmission. This process enhances the reliability of data transfer in the presence of errors. However, ARQ faces challenges with persistent channel errors triggered by factors such as signal interference,

signal-to-noise degradation, fading, multipath propagation, and physical obstructions. Consequences may include frame losses or prolonged frame failure notifications to the upper network layer.

- **Error frame repair:** At the receiver side, this approach involves repairing error frames. OTN achieves this by appending redundant bits to the original data payload, as specified in ITU-T G.709 [2020]. While providing limited error correction on a link, it proves inadequate in handling packet losses in an upstream link.
- **Error frame detection:** This approach is utilized to identify and handle frames that have been corrupted or compromised during transmission. Different network technologies employ various error detection techniques to ensure the integrity of transmitted frames. The specific technologies are presented in Figure 2.2, including Cyclic Redundancy Check (CRC) and Frame Check Sequence (FCS). Frames that fail to pass the check are dropped and treated as lost packets at the network layer.

2.3.3 Network Layer

Moving up to the network layer, two distinct approaches emerge:

- **IP-based approaches:** Leveraging Internet Protocol (IP) [IETF RFC 791, 1981], TCP [IETF RFC 9293, 2022], Multipath Transmission Control Protocol (MPTCP) [IETF RFC 8684, 2020], Stream Control Transmission Protocol (SCTP) [IETF RFC 4960, 2007], and Adaptive Transmission Protocol based on Software-Defined Networking (SDATP) [Chen et al., 2019, 2020], these approaches can be applied in traditional IP networks or within the Software-Defined Networking (SDN) paradigm. SDN empowers the data plane with programmability for flow-level forwarding and control.

  In these approaches, the pressing need is to enhance reliability through methods including multi-path aggregation and FRR. Available aggregation methods include IEC 62439 PRP, Improved Parallel Redundancy Protocol (iPRP), HSR, and OpenFlow-based Non-Disjoint Path Aggregation Mechanism (OPAM), where OPAM outperforms others. FRR encompasses various techniques that involve the related IEC 62439 protocols and route-based solutions. The IEC 62439 protocols are Ring-based Redundancy Protocol (RRP), MRP, Distributed Redundancy Protocol (DRP), Cross-Network Redundancy Protocol (CRP), and Beacon Redundancy Protocol (BRP), while the route-based solutions depend on IP routing, Spanning Tree Protocol (STP), or Rapid Spanning Tree Protocol (RSTP), distinguished for their route structures in either arc-disjoint arborescences or link-disjoint paths. Meanwhile, the security demand involves defending against LFAs, with Persistent Route Diversification Defense (PRDD) being one of the outstanding approaches, in comparison with other solutions such as Moving Target Defense (MTD) and traffic throttling.

2.3.4 Transport Layer

At the transport layer, the retransmission control is categorized into two ways: end-to-end and hop-by-hop. For the well-known TCP, its end-to-end retransmission aims to recover lost packets in a best-effort manner. However, TCP's performance is hindered by its congestion control, striving for min–max fairness in bandwidth sharing among multiple paths. This can lead to slowdowns or significant suppression of data sending, resulting in low reliability assurance. MPTCP, an extension of TCP, allows the simultaneous use of multiple network paths for a single TCP session, enhancing reliability and efficiency. Alternatively, SCTP offers features such as multi-streaming, multi-homing, and improved support for the transmission of real-time data. These end-to-end protocols perform retransmissions over a lengthy path on the Primary Forwarding Path (PP), making them susceptible to collapse in the face of massive packet losses caused by any intermediate link failure.

Recognizing the drawbacks of end-to-end retransmission, the appeal of hop-by-hop solutions becomes a research trend, as suggested by Chen et al. [2020 and Garcia-Luna-Aceves and Albalawi [2020]. Available solutions include SDATP based on SDN, enabling hop-by-hop retransmission control.

- **Named Data Network (NDN):** NDN treats a stream of data as named content chunks, enabling an Interest-based content retrieval pattern. In this pattern, each chunk from a producer is requested by a consumer. This unique approach in NDN mitigates threats from LFAs, as the retrieval pattern thwarts adversaries from probing paths to victim hosts. NDN emerges as a promising player in enhancing data dissemination reliability for power grids.

  The innate flexibility of NDN in multicast support, as highlighted in Ravikumar et al. [2020], strategically positions it in the pursuit of grid resilience. This advantage provides NDN with new opportunities to support various transport strategies for achieving higher resilience compared to IP. These strategies include Resilient Sensor Data Dissemination (RSDD), HeuMPSG implementing Disruption Resilient Transport Protocol (DRTP), Rapid and Reliable Transport Mechanism (R2T) [Wang et al., 2018], Internet Transport Protocol (ITP) [Garcia-Luna-Aceves and Albalawi, 2020], as well as iCASM [Ravikumar et al., 2020]. Among these, RSDD and HeuMPSG significantly outperform others.

2.3.5 Application Layer

Based on the transport layer, the application layer still needs to address the challenge of payload packet losses, a focal issue in wireless networks. These networks may be insensitive to EEDT demands, while the reliability of monitoring remains crucial. In this context, the focus shifts to redundancy codes, including

Adaptive RaptorQ Encoder and Decoder (ARED), which will be discussed in Chapter 10.

2.3.6 Discussions

As illustrated in Figure 2.2, the technical landscape across the five specified layers underscores the substantial research and development efforts aimed at mitigating various data transmission challenges, including bit errors, frame errors, packet losses, and payload packet losses. In the context of smart power grids, practitioners emphasize the network, transport, and application layers due to their capacity for customization, enabling operators to adapt solutions to diverse operational needs. The underlying communication infrastructure, primarily situated within the link and physical layers, is constructed through collaborations with telecommunications manufacturers and builders, forming the backbone of smart grid systems.

The network layer has received considerable attention, particularly regarding classical approaches outlined in the IEC 62439 standard. This standard has catalyzed the development of advanced technologies designed to enhance network reliability and resilience, such as redundancy protocols and fault-tolerant mechanisms. Simultaneously, the transport layer has evolved alongside these developments, with numerous solutions emerging to optimize data flow and ensure efficient communication based on the specific architecture of the network layer. Conversely, the application layer remains less explored, often due to a focus on the more foundational layers and their immediate operational needs.

To foster stable power grids, it is crucial to explore technologies that achieve high-availability communication networks while addressing the inherent challenges, particularly within the network, transport, and application layers. Chapters 3–11 will overview and delve into these diverse technical solutions, illustrating their roles, contributions, advantages, and shortcomings in enhancing the performance of smart power grid communication systems.

2.4 Summary

Network availability, essential for maintaining power grid stability, faces challenges in five key areas: reliable data delivery, path stability, secure data transport, application data completeness, and defense against LFA. These challenges have driven significant advancements in the network, transport, and application layers over the past decades, while the physical and link layers, due to their more rigid structures, offer less flexibility for innovation. As a result, research has increasingly focused on developing solutions in the upper layers to enhance overall reliability.

Traditional methods, such as the redundancy protocols defined in IEC 62439, have contributed to improving path stability and data delivery at the network layer. However, these approaches have not fully addressed all of the challenges in the five critical areas, leaving gaps in secure data transport and LFA defense, among others.

Chapters 3–11, we will explore how advanced solutions, including classical and state-of-art methods, aim to resolve these gaps and provide a safer, more secure communication infrastructure to support the sustainable operation of modern power grids. By addressing these challenges, these innovations will help ensure that grid communications remain resilient in the face of evolving demands and threats.

Questions

2.1 Discuss the significance of a multi-layered architecture in enhancing network availability within smart power grids. How does each layer contribute to overall network resilience?

2.2 Analyze the impact of bit error correction technologies in the physical layer on data integrity during transmission. What limitations do these technologies face, and how can they affect the performance of high-availability networks?

2.3 Compare and contrast the efficacy of APS and ARQ in the data link layer. Under what conditions would each be preferred in a smart power grid environment?

2.4 Evaluate the role of the IEC 62439 standard in improving network reliability at the network layer. How do the various redundancy protocols contribute to fault tolerance in power grid communication systems?

2.5 Critically assess the potential of NDN to revolutionize data dissemination in smart power grids. What are the key advantages of NDN over traditional IP-based approaches in the context of security and resilience?

2.6 Examine the challenges associated with payload packet losses in the application layer of smart grid communication networks. What strategies can be employed to mitigate these challenges, and what are their potential trade-offs?

2.7 Investigate how advancements in the transport layer, specifically MPTCP and SCTP, can enhance reliability in smart grid communications. Discuss the scenarios in which one might be favored over the other.

2.8 How can these challenges inform future research directions and technological developments in smart power grid communications?

Bibliography

3GPP TS 36.401. LTE; universal terrestrial IO access network (E-UTRAN); architecture description version 13.0.0 release 13, 2016.

Amit Aggarwal, Swathi Kunta, and Pramode K. Verma. A proposed communications infrastructure for the smart grid. In *ISGT*, pages 1–5, 2010.

Patrice Bock, Jean-Pierre Hauet, Romain Françoise, and Robert Foley. Ukrainian power grids cyberattack, 2017. URL https://www.isa.org/intech-home/2017/march-april/features/ukrainian-power-grids-cyberattack.

Saikat Chakrabarti, Elias Kyriakides, Tianshu Bi, Deyu Cai, and Vladimir Terzija. Measurements get together. *IEEE Power and Energy Magazine*, 7(1):41–49, 2009.

Jiayin Chen, Si Yan, Qiang Ye, Wei Quan, Phu Thinh Do, Weihua Zhuang, Xuemin Sherman Shen, Xu Li, and Jaya Rao. An SDN-based transmission protocol with in-path packet caching and retransmission. In *IEEE ICC*, pages 1–6, 2019.

Jiayin Chen, Qiang Ye, Wei Quan, Si Yan, Phu Thinh Do, Weihua Zhuang, Xuemin Sherman Shen, Xu Li, and Jaya Rao. SDATP: An SDN-based adaptive transmission protocol for time-critical services. *IEEE Network*, 34(3):154–162, 2020.

Moustafa Chenine and Lars Nordström. Performance considerations in wide area monitoring and control systems. *Water and Energy International*, 67: 1–8. 2010.

Dark Reading. Eight-hour DDoS attack struck AWS customers, 2019. URL https://www.darkreading.com/cloud/eight-hour-ddos-attack-struck-aws-customers/d/d-id/1336165.

Ildemar C. Decker, Aguinaldo S. e Silva, Marcelo N. Agostini, Fernando B. Prioste, Breno T. Mayer, and Daniel Dotta. Experience and applications of phasor measurements to the Brazilian interconnected power system. *European Transactions on Electrical Power*, 21(4):1557–1573, 2011.

Federal Energy Regulatory Commission and North American Eletricity Reliability Corporation. Arizon-Southern California outages on September 8, 2011, Causes and recommendations. URL https://www.nerc.com/pa/rrm/ea/September

%202011%20Southwest%20Blackout%20Event%20Document%20L/AZOutage_Report_01MAY12.pdf.

J. J. Garcia-Luna-Aceves and Abdulazaz Ali Albalawi. Connection-free reliable and efficient transport services in the IP Internet. In *IEEE CNSM*, pages 1–7, 2020.

C. W. Gellings. *The Smart Grid: Enabling Energy Efficiency and Demand Response.* Fairmont Press, 2009.

C. Gonzalez-Perez and B. F. Wollenberg. Analysis of massive measurement loss in large-scale power system state estimation. *IEEE Transactions on Power Systems*, 16(4):825–832, 2001.

Vehbi C. Gungor, Bin Lu, and Gerhard P. Hancke. Opportunities and challenges of wireless sensor networks in smart grid. *IEEE Transactions on Industrial Electronics*, 57(10):3557–3564, 2010.

Yih-Fang Huang, Stefan Werner, Jing Huang, Neelabh Kashyap, and Vijay Gupta. State estimation in electric power grids: Meeting new challenges presented by the requirements of the future grid. *IEEE Signal Processing Magazine*, 29(5):33–43, 2012.

IEC. Communication networks and systems for power utility automation - Part 8-1: Specific communication service mapping (SCSM) - mapping of MMS onto TCP/IP, 2011.

IEC. Communication networks and systems for power utility automation - Part 5: Communication requirements for functions and devices, 2013

IEC. Industrial communication networks - high availability automation networks - Part 1: General concepts and calculation methods, 2016.

IEC. IEC 61850 based substation automation system: A survey. *International Journal of Electrical Power & Energy Systems*, 120:106008, 2020.

IEC. Industrial communication networks - high availability automation networks - Part 3: Parallel redundancy protocol (PRP) and high-availability seamless redundancy (HSR), 2021.

IEC 60870-5-104. Telecontrol equipment and systems - Part 5-104: Transmission protocols, 2006.

IEC 62439. High-availability automation networks, 2008.

IEEE 1815. IEEE standard for electric power systems communications –distributed network protocol (DNP3), 2010.

IEEE C37.118.2. IEEE standard for synchrophasor data transfer for power systems, 2011.

IETF RFC 4960. Stream control transmission protocol, 2007.

IETF RFC 791. Internet protocol, 1981.

IETF RFC 8684. TCP extensions for multipath operation with multiple addresses, 2020.

IETF RFC 9293. Transmission control protocol (TCP), 2022.

Illinois. Texas bus network dataset, 2016. URL http://icseg.iti.illinois.edu/power-cases/.

ITU-T G.709. Interfaces for the optical transport network, 2020.

ITU-T G.808.1. Generic protection switching –linear trail and subnetwork protection, 2014.

ITU-T G.808.2. Generic protection switching –ring protection, 2019.

Yasin Kabalci. A survey on smart metering and smart grid communication. *Renewable and Sustainable Energy Reviews*, 57:302–318, 2016.

M. S. Kang, S. B. Lee, and V. D. Gligor. The crossfire attack. In *IEEE S & P*, pages 127–141, 2013.

KrebsonSecurity. The new normal: 200-400 Gbps DDoS attacks, 2014. URL https://krebsonsecurity.com/2014/02/the-new-normal-200-400-gbps-ddos-attacks/.

K. C. Leung, V. O. K. Li, and D. Yang. An overview of packet reordering in transmission control protocol (TCP): Problems, solutions, and challenges. *IEEE Transactions on Parallel and Distributed Systems*, 18(4):522–535, 2007.

Ren Liu, Ceeman Vellaithurai, Saugata S. Biswas, Thoshitha T. Gamage, and Anurag K. Srivastava. Analyzing the cyber-physical impact of cyber events on the power grid. *IEEE Transactions on Smart Grid*, 6(5):2444–2453, 2015.

LowEndTalk. Dyn.com Attack Analysis (21st October), 2016. URL https://lowendtalk.com/discussion/95204/dyn-com-attack-analysis-21st-october.

Alex Meehan. Lithuanian energy firm disrupted by DDOS attack, 2022 URL https://www.infosecurity-magazine.com/news/lithuanian-energy-ddos-attack/.

Kim S. Nash and James Rundle. Puerto Rico's power distributor suffered a cyberattack hours before a devastating fire, 2021. URL https://www.wsj.com/articles/puerto-ricos-power-distributor-suffered-a-cyberattack-hours-before-a-devastating-fire-11623453388.

NASPI. Update and technology roadmap presented to the NERC planning and operating committee, 2011. URL https://www.naspi.org/node/404.

NASPI. Categorizing phasor measurement units by application data requirements, 2018. URL https://www.naspi.org/node/736.

NASPI. NASPInet 2.0 architecture guidance version 1.19, 2019. URL https://www.naspi.org/node/746.

North American Electric Reliability Corporation Hurricane Harvey event analysis report, 2024. URL https://www.nerc.com/pa/rrm/ea/Hurricane_Harvey_EAR_DL/NERC_Hurricane_Harvey_EAR_20180309.pdf.

OPC Foundation. OPC-UA specification, 2023. URL https://reference.opcfoundation.org/.

Gelli Ravikumar, Dan Ameme, Satyajayant Misra, Sukumar Brahma, and Reza Tourani. iCASM: An information-centric network architecture for wide area measurement systems. *IEEE Transactions on Smart Grid*, 11(4):3418–3427, 2020.

Martin Sibley. Optical Communications: Components and Systems *(third edition)*. Springer, 2020.

Siemens. Network concept for redundant connections of SIMATIC virtual PLCs, 2024. URL https://cache.industry.siemens.com/dl/files/831/109955831/att_1277654/v2/109955831_Netzwerkkonzept_vPLC_DOC_V1_0_en.pdf.

Swissgrid. Comprehensive retpofit for substation –modern secondary system for the substation Gösgen, 2021. URL https://www.sprecher-automation.com/fileadmin/references/Swissgrid_UWgoesgen/2020_Swissgrid_UW_Goesgen_12002120enA.pdf.

Texas Instruments. Analog interfacing for grid infrastructure with Sitara processors, 2023. URL https://www.ti.com/lit/ta/sszt793/sszt793.pdf?ts=1728628204805\LY1\textbackslash&ref_url=https%253A%252F%252Fwww.google.com.hk%252F.

The University of Texas at Austin. The timeline and events of the February 2021, Texas electric grid blackouts, 2021. URL https://energy.utexas.edu/sites/default/files/UTAustin%20%282021%29%20EventsFebruary2021TexasBlackout%2020210714.pdf.

Mini S. Thomas. *Power System SCADA and Smart Grids*. CRC Press, 2020.

L. H. Tsoukalas and R. Gao. From smart grids to an energy internet: Assumptions, architectures and requirements. In *DRPT*, pages 94–98, 2008.

U.S.-Canada Power System Outage Task Force. Interim report: Causes of the August 14th blackout in the United States and Canada, 2023. URL https://www.ferc.gov/sites/default/files/2020-05/blackout-report.pdf.

Maxime Velay, Meritxell Vinyals, Yvon Besanger, and Nicolas Retiere. An analysis of large-scale transmission power blackouts from 2005 to 2016. In *UPEC*, pages 1–6, 2018.

Zhaoxu Wang, Hongbin Luo, Huachun Zhou, and Jiawei Li. R2T: A rapid and reliable hop-by-hop transport mechanism for information-centric networking. *IEEE Access*, 6:15311–15325, 2018.

Jiarong Xing, Wenqing Wu, and Ang Chen. Ripple: A programmable, decentralized link-flooding defense against adaptive adversaries. In *USENIX Security Symposium*, 2021.

3

Classical High-Available Networking Solutions

3.1 High-Availability Standards

For the technical landscape mentioned in Section 2.3, the classical solutions for high-available networks are the protocols defined in the IEC 62439 standard [IEC, 2017a], which has been widely adopted in today's power grids in practice.

IEC 62439 standards fully realize the concept of redundancy as a key strategy for improving communication network availability in smart power grids. Redundancy involves incorporating additional resources that remain unused under normal operating conditions but serve as critical backups during system failures. In the context of smart power grids, this may manifest as extra communication links or alternative pathways that can be automatically activated when a primary channel fails. This fault-tolerant approach enables systems to integrate redundancy seamlessly, allowing for automatic reinsertion of components once repairs are completed, thereby ensuring continuous operation and enhanced reliability.

Moreover, a comprehensive understanding of availability necessitates considering all transitions within the system – not solely the aftermath of a failure. This includes the dynamics of how systems switch between operational states, the activation of redundant resources, and the reestablishment of functionality post-repair. By analyzing these transitions, we can better grasp the resilience of communication networks in smart power grids, ensuring that they meet the demanding requirements of modern energy management.

To safeguard the communication network availability of power grids, many industrial organizations have begun to work on standards to promote the availability of communication networks in power grids, over past 20 years. Representatives include a complete series of IEC 62439, defined by IEC technical committee 57, focused on the networks among power substations [IEC, 2017a]. IEC 62439 is a key standard for defining network redundancy protocols tailored

Communication Networks in Smart Power Grids, First Edition. Boyang Zhou.

Companion Website: www.wiley.com/go/SmartPowerGrids1e

to industrial communication systems, including those used in power grids, which has been widely adopted by most of power grid operators to date.

To maximize the availability, this standard introduces an important concept of redundancy, ensuring high availability and fault tolerance, which are crucial for maintaining uninterrupted communication in critical applications such as power grid control and monitoring systems. The standard addresses not only traditional failure recovery methods but also introduces innovative solutions that significantly reduce network downtime during failures, ensuring seamless communication even in the most demanding environments.

Specifically, in IEC 62439, while the Parallel Redundancy Protocol (PRP) [IEC, 2021] and the High-availability Seamless Redundancy (HSR) [IEC, 2021] are perhaps the most well-known redundancy mechanisms, the standard also includes other protocols designed for specific network topologies and performance requirements. These additional protocols – the Media Redundancy Protocol (MRP) [IEC, 2017b], the Cross-Network Redundancy Protocol (CRP) [IEC, 2010a], the Beacon Redundancy Protocol (BRP) [IEC, 2018], the Distributed Redundancy Protocol (DRP) [IEC, 2010b], and the Ring-based Redundancy Protocol (RRP) [IEC, 2011] – complement PRP and HSR, offering tailored redundancy solutions based on the specific needs of a given communication network. Table 3.1 compares the key features of the traditional IP incorporated with the Spanning Tree Protocol (STP) [IEEE, 2004] or Rapid Spanning Tree Protocol (RSTP), employing IEEE 802.1D [1998], with the IEC 62439 standards using various technologies. These technologies are tailored to different networking scenarios, which is specified below.

3.1.1 Analysis on Fast Reroutes (FRRs)

Traditional FRR techniques, such as Multiprotocol Label Switching (MPLS) [IETF RFC 4090, 2005] and segment routing [Litkowski et al., 2022], provide essential tools for rerouting traffic during failures, but their limitations in rapid recovery and scalability become apparent when applied to the dynamic environments of smart grids. The high degree of decentralization and heterogeneity within smart grids, encompassing both wired and wireless segments, compounds these challenges. Failures may occur across optical backbone networks, wireless Internet-of-things (IoT) devices, or localized 5G networks, each requiring distinct recovery approaches and timings.

Moreover, the topology and complexity of smart grid communication networks demand FRR mechanisms that can adapt quickly and intelligently to failures. In large-scale smart grids, where communication links span vast geographical regions, network faults could occur anywhere from substations to control centers, affecting critical power system operations. For instance, a single-point

Table 3.1 Comparison of redundancy protocols in IEC 62439.

Protocol	Solution	Frame loss	Redundancy protocol	Node attachment	Network topology	Failure recovery time
PRP	IEC 62439-3	No	End node	Double	Doubly meshed, independent	0 s
HSR	IEC 62439-3	No	End node	Double	Ring, meshed	0 s
RRP	IEC 62439-7	Yes	End node	Double	Single ring	8 ms for 100BASE-X, 4 ms for 1000BASE-X
CRP	IEC 62439-4	Yes	End node	Single and double	Doubly meshed, cross-connected	Worst case for 512 endpoints: 1 s
BRP	IEC 62439-5	Yes	End node	Double	Doubly meshed, connected	Worst case for 100 endpoints: 8.88 ms
IP	IP routing	Yes	Network	Single	Single meshed	> 30 s
STP	IEEE 802.1d	Yes	Network	Single	Single meshed	> 20 s
RSTP	IEEE 802.1w	Yes	Network	Single	Single meshed, ring	Determined by rule 8
DRP	IEC 62439-6	Yes	Network	Single and double	Single or double rings	Worst case: 100 ms
MRP	IEC 62439-2	Yes	Network	Single	Single or double rings	Worst case: 500 ms, 200 ms, 30 ms, or 10 ms

failure in a substation communication link could propagate across the network, delaying state estimation or grid stability assessments, which are essential for decision-making in load balancing or fault isolation.

IEC 62439-based redundancy mechanisms [IEC, 2017a] such as PRP, MRP, CRP, and DRP provide foundational FRR solutions, but these protocols alone are insufficient for addressing the unique demands of smart grids. These redundancy mechanisms often assume deterministic, low-latency environments typical of industrial Ethernet. However, in smart grids, communication spans various layers, from high-capacity optical networks such as Optical Transport Network (OTN), to resource-constrained wireless segments (e.g., IoT and Long Range Radio (LoRa)), to public networks like 5G. Each layer presents distinct challenges that exceed the design scope of these traditional protocols.

Instead of employing FRR, PRP, and HSR, while delivering zero-recovery time by duplicating packets, introduce excessive bandwidth overhead, particularly in bandwidth-limited environments such as wireless networks. These protocols strain the already-limited resources in wireless segments, where low-power devices communicate over narrow-bandwidth links. Furthermore, ring-based protocols like CRP and DRP, though efficient in simple ring topologies, struggle with the mixed and hybrid topologies common in smart grids. The increased propagation delays in wireless and geographically dispersed networks further exacerbate the inadequacies of these protocols in ensuring rapid and reliable failover.

3.2 Redundancy Provision at End Nodes

3.2.1 PRP and HSR

PRP and HSR [IEC, 2021] are designed for applications requiring zero-recovery time in the event of a network failure. PRP provides the way for reliable data transmission between different terminal entities within power grids, as illustrated in Figure 3.1. It works by duplicating data packets and sending them simultaneously over two or more independent networks, such as LANs A and B. If one network fails, the other continues to deliver data seamlessly, without delay or packet loss. This redundant-network approach ensures maximum availability, making it a useful tool for power grids where communication disruptions can lead to severe operational failures.

On the other hand, HSR is optimized for ring and mesh topologies, where devices are interconnected in a loop, as depicted in Figure 3.2. Each node is equipped with two network ports operated in parallel, named the Doubly Attached Node with HSR protocol (DANH), where these parallel ports form a Redundant Box (RedBox). Using RedBox, data is sent in both directions around the ring, and if a failure occurs on one segment, the data continues to flow uninterrupted through the other path. Moreover, multiple rings can be required if traffic flows exceeds the capability of a single ring. To connect two different rings, we need to adopt a Quadruple Port Device (QuadBox), consisting of two RedBoxs. This mechanism provides self-healing capabilities, ensuring that communication continues with no perceptible downtime.

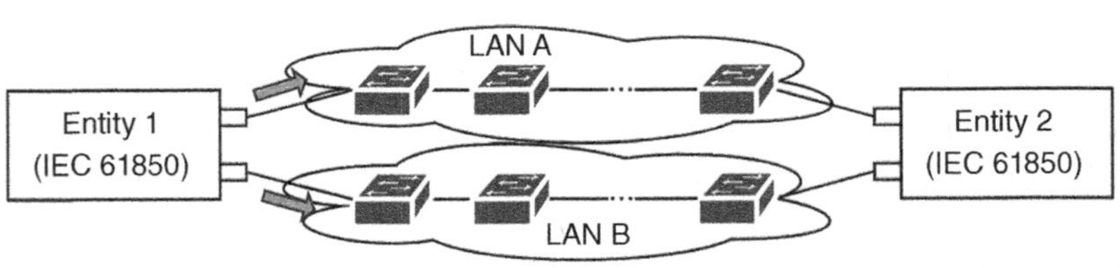

Figure 3.1 PRP illustration.

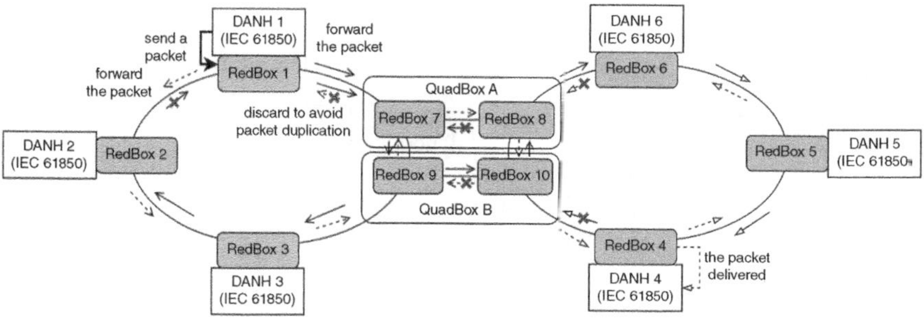

Figure 3.2 HSR illustration.

3.2.2 CRP

CRP [IEC, 2010a] is Honeywell's legacy protocol for Fieldbus Foundation, using doubly attached nodes, specifically designed to enhance network reliability in industrial settings. As depicted in Figure 3.3, CRP addresses the need for seamless redundancy across multiple networks, ensuring that data transmission remains uninterrupted even in the event of network failures. This protocol is particularly relevant in environments where downtime can lead to significant operational disruptions, such as smart power grids and automation systems.

CRP operates by creating multiple paths for data transmission, effectively coupling different networks to work together as a cohesive unit. This redundancy not only improves fault tolerance but also optimizes network performance by enabling load balancing across the available paths. The protocol intelligently manages the routing of data packets, allowing for rapid failover and recovery, thereby maintaining continuous communication and data integrity.

One of the standout features of CRP is its ability to function in diverse network topologies, making it versatile for various applications. It supports both ring and tree structures, adapting to the specific requirements of the network architecture

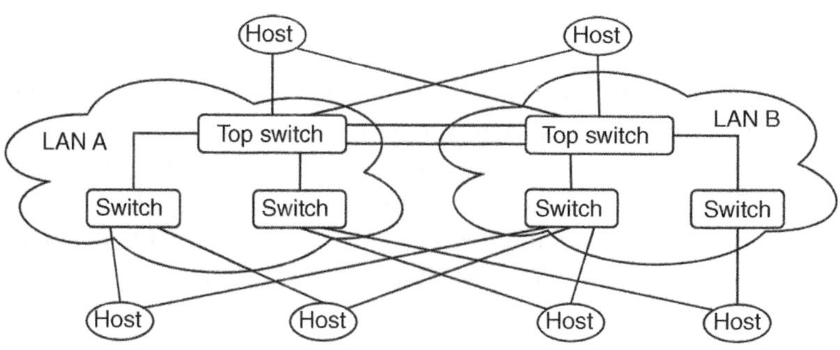

Figure 3.3 CRP illustration.

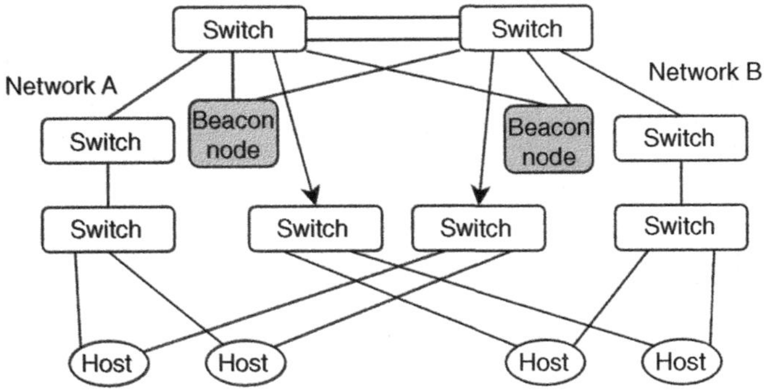

Figure 3.4 BRP illustration.

in use. By leveraging CRP, operators can achieve a higher level of network reliability, which is essential for mission-critical applications.

3.2.3 BRP

BRP [IEC, 2018] operates by using a beacon signal that constantly monitors the status of the network, where the network can be constructed in line, star, or ring. As depicted in Figure 3.4, beacons and nodes are connected to the top switches. If a failure is detected, the protocol rapidly switches to an alternative communication path. BRP is particularly effective in networks where constant monitoring is needed to ensure minimal disruption. It provides a relatively low-latency recovery mechanism, making it suitable for networks that require quick failover but not necessarily zero-recovery time. In power grid communication networks, BRP can be used in systems where continuous health checks are vital for ensuring network connectivity and operational safety and security.

3.2.4 RRP

RRP [IEC, 2011] is specifically tailored for ring topologies, offering efficient and quick recovery from network failures. Similar to HSR and MRP, RRP leverages the inherent redundancy of a ring structure to maintain communication even when one link or node in the ring fails. However, RRP focuses more on simpler, less complex ring networks, offering a cost-effective solution for environments that do not require the high performance or zero-recovery time of HSR. RRP is often used in power grid communication systems where reliability is important, but the network topology and communication needs are relatively straightforward.

As depicted in Figure 3.5, upon the event of a node or link failure, RRP facilitates automatic rerouting of data through the alternate path by turning the ring topology to a line topology to bypass the failed links or nodes.

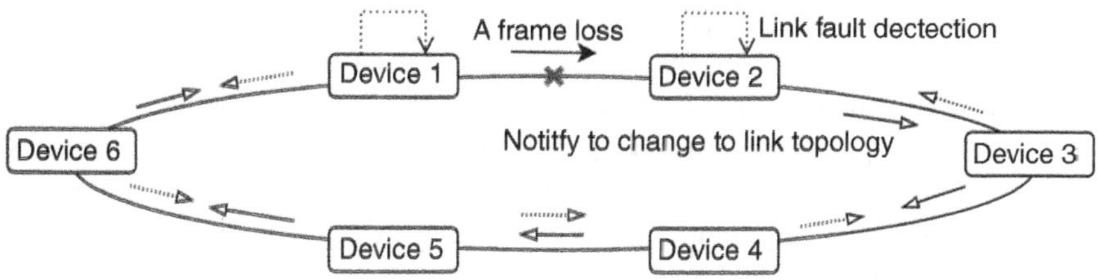

Figure 3.5 RRP illustration.

By this rerouting behavior, RRP can minimize downtime and maintaining data integrity. This dynamic rerouting capability is essential for mission-critical applications, such as those found in smart grids and automated manufacturing systems.

One of the standout features of RRP is its ability to implement a distributed control mechanism. Each node actively participates in monitoring the health of the network, detecting failures, and managing data flow. This decentralized approach reduces the reliance on a single point of failure, enhancing the overall robustness of the network. Additionally, RRP supports rapid recovery from faults, allowing networks to maintain high availability and reliability in the face of potential disruptions.

3.3 Redundancy Provision in Networks

3.3.1 Internet Protocol (IP) Routing

In smart power grids, IP routing is essential for ensuring reliable and efficient communication among various devices and systems among different substations and the control center. These routes are either established through static routing configuration or dynamic routes. The static routing is often the preferred method for ensuring reliable, secure, and predictable data transfer. While dynamic routing protocols offer flexibility and adaptability, many power grid operators choose static routing due to the highly deterministic and stable requirements of grid communication networks.

3.3.2 Static Routing

3.3.2.1 Features

One of the key advantages of static routing is its predictability. Power grids require communication systems that are stable and can be relied upon for the consistent transfer of critical data, such as protection signals, control commands, and monitoring information. With static routing, the routes between devices are manually configured, ensuring that data always follows a predetermined path. This predictability is essential in environments like substations, where minimizing disruptions and ensuring low-latency communication are paramount.

In addition to predictability, simplicity is another reason why static routing is favored. Many substations and local communication networks within the power grid consist of a relatively small number of devices, meaning that static routes can be configured with minimal effort. This reduces the complexity that comes with implementing dynamic protocols, which require constant updates and recalculations based on network topology changes. Static routing, in contrast, provides a straightforward solution where paths are fixed and do not change unless manually reconfigured.

Another important factor is security. Power grids are critical infrastructure, and any disruptions or vulnerabilities in communication could have serious consequences. Static routing enhances security because routing information is not advertised across the network, making it less vulnerable to attacks such as route hijacking or spoofing. Dynamic protocols, on the other hand, often involve continuous exchange of routing data, potentially introducing security risks if misconfigured or compromised.

Reduced overhead is a further benefit of static routing. Dynamic routing protocols require constant communication between routers to maintain up-to-date routing tables, consuming bandwidth and processing power. In the context of a power grid, where communication networks are typically stable, static routing eliminates this unnecessary overhead, leading to more efficient use of network resources and reduced operational complexity.

3.3.2.2 Routing Failover

For failover and redundancy, static routing allows operators to manually configure backup routes. This approach gives grid operators full control over how traffic is rerouted in the event of a link or node failure, ensuring that failover processes are deliberate and well understood. While dynamic protocols offer automatic rerouting capabilities, static routing offers the advantage of more predictable and intentional behavior, which is often a priority in critical systems like power grids.

Although dynamic protocols are sometimes used in large or highly interconnected parts of modern smart grids, static routing remains the method of choice for many operators. Its reliability, simplicity, and security align with the operational demands of power grid communications, where predictability and stability are crucial. However, as smart grids evolve and expand, dynamic routing may become more relevant in handling the increased complexity of wide-area networks, but static routing continues to serve as a foundational approach in ensuring the resilience of today's power grid infrastructures.

3.3.3 Dynamic Routing

Key protocols such as Open Shortest Path First (OSPF) [Moy, 1998], Routing Information Protocol (RIP) [Malkin, 1998], Border Gateway Protocol (BGP) [Rekhter et al., 2006], and MPLS [Rosen et al., 2001] have been specifically designed to meet the unique demands of power grid environments. These protocols not only facilitate seamless communication between substations and control centers but also play a critical role in maintaining data integrity in the face of network disruptions.

3.3.3.1 OSPF

OSPF [Moy, 1998] excels in environments that require rapid adaptation to network changes, making it ideal for substation automation. When a link or node fails, OSPF quickly detects the change and generates a new Link State Advertisement (LSA) to inform neighboring routers. These routers then update their Link State Databases and recalculate the shortest paths using Dijkstra's algorithm, allowing for immediate rerouting of traffic through alternative paths. This fast convergence ensures that critical communication remains uninterrupted, supporting vital functions like protective relaying and control.

OSPF deals with link or node failures by means of the rerouting process-based link state convergence in the following steps.

i) Each OSPF router initializes by discovering its neighboring routers and forming adjacencies. They exchange LSAs to share information about their links.
ii) Each router creates a Link State Database (LSDB) that contains the LSAs from all routers in the area. This database provides a complete view of the network topology.
iii) When a link or node fails, the affected router detects this change through its regular hello packets or by the lack of acknowledgment from a neighboring router.
iv) The router that detects the failure generates a new LSA indicating the change in its state (e.g., the link is down) and floods this LSA to all OSPF neighbors.
v) Upon receiving the new LSA, all neighboring routers update their LSDBs to reflect the new network topology.
vi) Each router recalculates the shortest path to all destinations using Dijkstra's algorithm based on the updated LSDB.
vii) The routing table is updated with new paths, and traffic is immediately rerouted along the best available routes.
viii) Data packets are now sent using the newly computed paths, ensuring minimal disruption to communication.

3.3.3.2 RIP

RIP [Malkin, 1998], while simpler, is effective in smaller power grid networks where basic routing capabilities suffice. Upon detecting a link failure, RIP invalidates the affected route and triggers an update to its neighbors, allowing them to adjust their routing tables accordingly. Although its slower convergence compared to OSPF can be a drawback, RIP's straightforward configuration makes it suitable for specific contexts within the IEC 60870-5-104 framework, where reliable communication is still achievable despite potential delays.

RIP performs the rerouting actions in the main steps below:

i) RIP routers send periodic updates of their routing tables to their neighbors every 30 seconds.
ii) If a router fails to receive updates from a neighbor for a certain period (default is 180 seconds), it assumes the link is down.
iii) The router marks the route to the downed neighbor as invalid and sets its metric to infinity (16 hops), signaling the unavailability of that route.
iv) The router then sends a triggered update to inform its neighbors about the invalidated route.
v) Neighbors receiving the update adjust their routing tables accordingly, potentially selecting alternative routes based on the remaining valid paths.
vi) Data packets are redirected through available routes, although this process may be slower than OSPF due to RIP's periodic nature.

3.3.3.3 BGP

BGP [Rekhter et al., 2006] plays a crucial role in managing routing between multiple autonomous systems within the smart grid. When a link or node failure occurs, BGP quickly withdraws the affected routes and communicates these changes to its peers. This protocol's ability to implement policy-based routing decisions enables it to maintain robust connectivity, even in complex network environments. By allowing for reevaluation of alternative paths based on various metrics, BGP ensures that data flows remain efficient and secure across wide-area networks.

BGP provides rerouting features that work in the primary steps below:

i) BGP routers establish sessions with peers using Transmission Control Protocol (TCP), exchanging routing information and maintaining connections.
ii) Each BGP router advertises its routes to its peers, including paths, next-hop addresses, and AS-path information.
iii) If a BGP router detects a failure in a link or node (typically through a loss of the TCP connection with a peer), it removes the affected routes from its routing table.

iv) The router sends a BGP UPDATE message to its peers, withdrawing the affected route.
v) Upon receiving the withdrawal, peers reevaluate their routing tables and may choose alternative paths based on BGP path selection rules (e.g., shortest Autonomous System (AS)-path, lowest origin type).
vi) The peers then update their routing tables and reroute traffic according to the new best paths.

3.3.3.4 MPLS

MPLS [Rosen et al., 2001] stands out for its ability to prioritize critical traffic and optimize bandwidth utilization. In the event of a link failure, MPLS can quickly activate backup paths established previously, enabling rapid rerouting with minimal latency. By leveraging label-switching techniques, MPLS ensures that data packets continue their journey along the most efficient routes, which is vital for real-time applications in power grid operations. Furthermore, its capacity to support virtual private networks enhances security, safeguarding sensitive communications within substations.

MPLS performs the rerouting process in the following major steps:

i) MPLS routers – Label Switch Routers (LSRs) – distribute labels to data packets upon entering the MPLS network based on the forwarding equivalence class.
ii) As packets traverse the MPLS network, each LSR reads the incoming label, swaps it for a new label, and forwards the packet accordingly.
iii) If an LSR detects a link or node failure (via routing protocol mechanisms or direct monitoring), it immediately initiates rerouting.
iv) If a backup path was previously established (using mechanisms like FRR), the LSR can quickly switch to this alternative path.
v) The LSR assigns a new label for the packet based on the rerouted path and forwards it along the new route.
vi) The packet continues its journey along the new path, minimizing latency and ensuring real-time data delivery.
vii) Throughout the process, MPLS can monitor the performance of the new path, adjusting as necessary if further failures occur.

3.3.3.5 Discussions

The choice of IP routing technique plays a crucial role in ensuring network reliability, especially during failover events where link or node failures may disrupt normal operations. The two primary approaches to routing – static routing and dynamic routing protocols – must be carefully evaluated to determine their suitability for maintaining effective communication across the grid infrastructure. Their features are summarized in Table 3.2.

Table 3.2 Comparison of IP routing approaches for power grid communications.

Routing method	Configuration	Predictability	Adaptability to failures	Convergence speed	Scalability
Static routing	Manual	High	Low (manual changes)	Manual (no convergence)	Limited
OSPF	Automated	Moderate	High (automatic)	Fast	High
RIP	Automated	Moderate	Moderate (automatic)	Slow	Limited
BGP	Policy-based	High	High (automatic)	Moderate	Very high
MPLS	Label-based	High	Very high (instant reroute)	Instant	Very high

Static routing is widely used in power grid networks because it offers predictability and control. Since routes are manually configured, operators have full authority over the network paths, ensuring that communication follows a predetermined and stable route. This is particularly important in substations and other critical areas of the grid, where maintaining consistent data flow is paramount. The security benefits of static routing are also notable, as it does not require the continuous exchange of routing information, reducing exposure to potential attacks like route hijacking or manipulation. Additionally, static routing eliminates the overhead associated with dynamic protocols, as routers do not need to communicate routing updates regularly. However, static routing has limitations, particularly when it comes to scalability and adaptability. Changes to the network must be manually configured, and in the event of link failures, pre-configured backup routes must be in place for any failover process. This lack of dynamic response makes static routing less suitable for large, complex networks where topology changes are more frequent.

On the other hand, dynamic routing protocols are designed to automatically adapt to changes in the network. OSPF, for example, is known for its fast convergence, meaning it can quickly detect link failures and recalculate routes to minimize downtime. This makes it well suited for large power grid networks, especially those that span across wide-area networks. OSPF is scalable and offers route optimization by continuously calculating the shortest path for data transmission. However, the protocol introduces more overhead due to the constant exchange of link state updates, consuming bandwidth and processing power, which may be a drawback in networks with limited resources. RIP, though simpler than OSPF, shares some dynamic capabilities but suffers from slower convergence and limited scalability due to its 15-hop limit. This makes RIP more suitable for smaller, less complex networks, although its simplicity and ease of configuration are beneficial for operators looking for a straightforward routing solution.

BGP is often applied in larger smart grid systems that connect multiple autonomous networks. Its strength lies in policy-based routing, which allows for granular control over traffic flow based on specific routing policies. This makes BGP highly flexible and scalable, capable of handling large, distributed networks. However, BGP's slower convergence time and higher configuration complexity can be drawbacks in scenarios where immediate rerouting is required after a failure. MPLS, while not a traditional routing protocol, is highly effective in traffic engineering. Its ability to provide fast rerouting through pre-configured backup paths makes it particularly valuable in critical applications within smart grids, where minimizing communication delays is essential. MPLS also supports traffic prioritization, ensuring that essential grid communications, such as Supervisory Control and Data Acquisition (SCADA) data, are transmitted with high priority even during periods of congestion.

In summary, the choice of routing technique depends on the specific needs of the power grid, such as the size of the network, the criticality of data, and the need for flexibility in response to network changes.

3.3.4 STP

STP, defined in IEEE 802.1D [2004], is a network protocol used to prevent loops in Ethernet networks by creating a loop-free logical topology. STP achieves this by selecting a single active path between any two nodes in a network, while placing redundant links in a blocked state. The protocol operates by electing a root bridge, which is the switch with the lowest bridge ID. From this root, it calculates the shortest path to all other switches and determines which links to forward and which to block. This ensures that frames are only forwarded along one active path, avoiding broadcast storms and network loops.

In the routing process, each switch running STP exchanges Bridge Protocol Data Units (BPDUs) with its neighbors to share information about bridge IDs and link costs. Based on these exchanges, STP determines the root bridge, the shortest path from each switch to the root, and the designated bridges on each network segment. Once the topology is determined, STP places any redundant or non-optimal links into a blocking state to prevent data loops. Traffic is then forwarded along the active paths established by the protocol.

When a link or node failure occurs, STP initiates a reroute process. It detects the failure by the absence of BPDU messages and recalculates the topology, promoting blocked links to forwarding states if necessary. The convergence time of traditional STP can take up to 50 seconds due to timers, such as the max age and forward delay, which allow the network to stabilize before activating backup links. This convergence time is a significant consideration for real-time communications, though enhancements like RSTP improve convergence by reducing these delays.

STP is suitable for communication networks in smart power grids due to its simplicity and the ability to maintain loop-free topologies, ensuring reliable data transmission across redundant paths. In substations and distribution networks, where redundancy is critical for fault tolerance, STP offers a straightforward solution for managing network loops and reconfiguring paths in case of failures. However, the traditional STP may not meet the strict low-latency demands of time-sensitive grid operations, and alternatives such as RSTP or newer protocols might be more appropriate in those cases.

3.3.5 RSTP

RSTP, defined in IEEE 802.1w [1998], is an evolution of the original STP aimed at providing faster convergence during network topology changes. While STP relies on timers and a more gradual process to re-establish network paths, RSTP introduces improvements that allow switches to react more quickly to link or node failures, reducing the time it takes to restore communication. This is particularly useful in communication networks for smart power grids, where fast rerouting is essential to maintain reliable operations and prevent disruptions in the grid's control and monitoring systems.

The principle of RSTP builds upon the idea of creating a loop-free topology, like STP, but enhances the speed of convergence by employing edge ports and a more dynamic system of exchanging BPDUs. RSTP introduces three port roles – root port, designated port, and alternate port – that allow for rapid transitions in the event of topology changes. An edge port is a key feature, as it immediately transitions to the forwarding state, bypassing the traditional blocking and learning stages used in STP. This significantly reduces the time it takes for devices to begin forwarding traffic.

In terms of routing processes, RSTP constantly monitors the state of links in the network by exchanging BPDUs between switches. If a link or node failure is detected, RSTP does not rely on timers to decide the next step, unlike STP. Instead, it actively exchanges information with neighboring devices to determine an optimal alternative path. Alternate ports, which are pre-configured as backup paths, can be immediately activated if the primary link goes down. This proactive approach allows the network to reroute traffic within a few milliseconds, ensuring continuous data flow and minimizing downtime.

For communication networks in smart power grids, RSTP's fast reroute capabilities are highly beneficial. Power grid operations demand high-availability and low-latency communication to handle real-time control signals and monitoring data. RSTP provides the necessary resiliency by quickly adapting to link or device failures, ensuring the communication network remains operational. Moreover, the protocol's ability to minimize recovery times enhances the overall reliability of power grid infrastructure, making it a suitable choice for these critical environments.

3.3.6 MRP

MRP [IEC, 2017b] is a more traditional protocol compared to PRP and HSR, but it is still highly effective for industrial networks. MRP is designed specifically for ring topologies, where it can quickly detect and recover from network failures. As depicted in Figure 3.6, unlike HSR, which provides seamless redundancy, MRP operates with a short recovery time by activating backup paths in the event of a single failure in the ring-based topology. MRP can restore communication within a few milliseconds, which is sufficient for many industrial applications, though it does not achieve the zero-recovery time of PRP and HSR. MRP is widely used

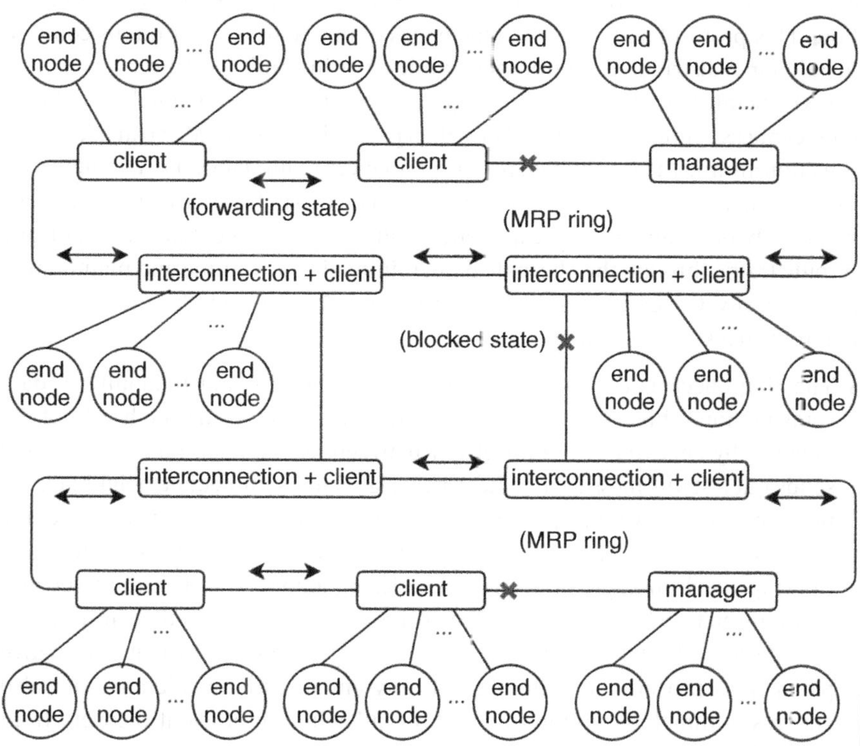

Figure 3.6 MRP illustration.

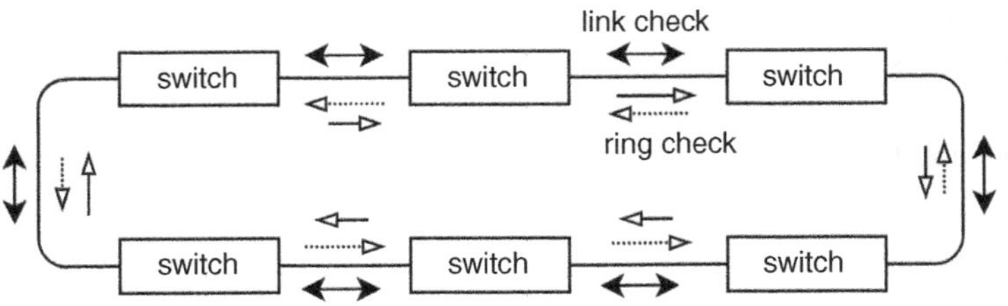

Figure 3.7 DRP illustration.

in Ethernet-based industrial communication networks and is particularly suited for networks that do not require the ultra-low recovery time that PRP and HSR offer but still need reliable redundancy. Its main advantage lies in its simplicity and ability to integrate with existing Ethernet switches without the need for specialized hardware.

3.3.7 DRP

DRP [IEC, 2010b] is specifically designed to enhance the reliability of networks configured in a ring topology that provides multiple pathways for information flow. This protocol addresses the challenges of maintaining continuous communication in a distributed environment while ensuring fault tolerance.

DRP operates by leveraging the unique characteristics of ring topology to facilitate seamless communication. When a data packet is transmitted, it can take either direction around the ring, which not only enhances redundancy but also allows for load balancing across the network. This dual-path capability is crucial in scenarios where network segments may experience failure, as the protocol can automatically reroute traffic through the alternate path, minimizing downtime and maintaining data integrity. Concretely, as depicted in Figure 3.7, DRP includes the ring check and link check mechanisms:

- The ring check is essential for ensuring the integrity and reliability of data transmission. This process involves continuously monitoring the health of the network by checking the connectivity between nodes. Each node periodically sends out a status message or a heartbeat signal to its immediate neighbors. If a node fails to receive a response within a specified timeframe, it can conclude that there is a potential failure in the network. When a failure is detected, the ring check mechanism triggers a series of actions to isolate the issue and maintain network functionality. For instance, the affected node can notify its neighbors of the failure, prompting them to reroute data through alternative paths. This proactive monitoring not only enhances fault tolerance but also ensures minimal disruption to data flow, allowing for efficient recovery from network faults.

- Link check complements the ring check by focusing specifically on the quality and performance of the connections between nodes. This mechanism assesses the status of individual links, evaluating parameters such as signal integrity, bandwidth availability, and error rates. By conducting regular assessments, the link check can identify degraded links that may not yet have failed but are at risk of causing communication issues. If a link check detects a problem with a specific connection, the protocol can automatically reroute traffic through healthier paths, thus preventing potential data loss or delays. Additionally, the link check may trigger notifications to network administrators, enabling them to take corrective action before the issue escalates.

Both ring check and link check are integral to the robust functionality of CRP in ring topologies, providing the ability to distribute control and decision-making among the network nodes. With these monitoring mechanisms, each node actively participates in monitoring the health of the network, detecting potential failures, and managing the rerouting of traffic. This decentralized approach reduces the reliance on a single point of failure. Hence, CRP ensures continuous data transmission, enhances redundancy, and maintains high availability in critical communication networks. As a result, DRP is particularly suitable for applications in industrial automation and smart grid communications, where high availability and reliability are paramount.

The range of protocols in IEC 62439 offers flexibility to network engineers, allowing them to choose the appropriate solution based on the specific technical requirements of their systems. PRP and HSR are ideal for environments where zero-recovery time is critical, such as real-time control and monitoring in power grids. MRP, while offering slightly slower recovery, is simpler to implement and can be integrated with standard Ethernet devices. CRP, BRP, DRP, and RRP provide more specialized solutions that cater to the unique needs of distributed systems, modular network segments, and ring-based topologies.

In power grid communication networks, where both high availability and low latency are essential, these redundancy protocols ensure that critical operations, such as fault detection and grid stabilization, continue without interruption. Each protocol brings a different balance of performance, complexity, and cost, allowing network designers to build robust, scalable, and resilient communication infrastructures tailored to their specific operational needs.

3.4 Limitations of Classical Approaches

IEC 62439 delivers a comprehensive suite of redundancy technologies, addressing a wide range of industrial and critical communication requirements. From the zero-recovery time solutions provided by PRP and HSR to the more traditional

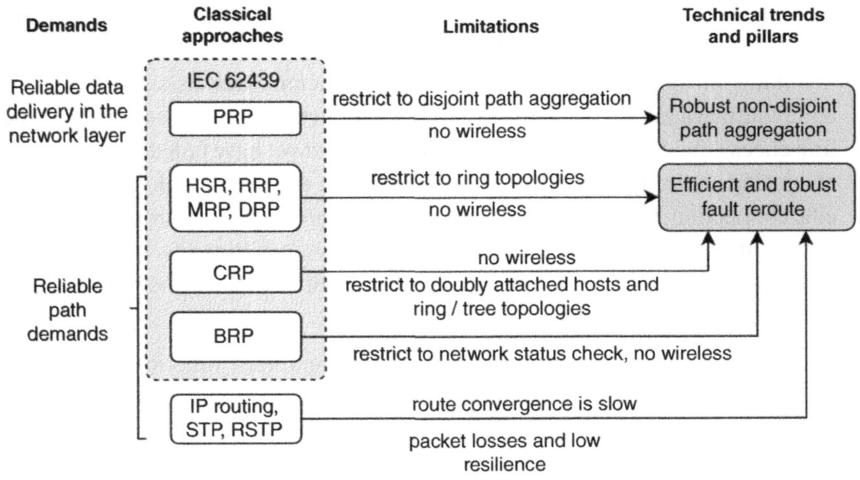

Figure 3.8 Limitations of IEC 62439 and research trends.

failover methods like MRP, the standard ensures that power grid communication networks can withstand failures and continue to function without compromising safety or operational integrity. By leveraging these protocols, power grids can enhance their resilience, ensuring that even in the face of network disruptions, critical data and commands are transmitted reliably.

3.4.1 Limitations

However, the redundancy protocols specified by IEC 62439 were designed with traditional networks in mind, and their limitations become more evident when applied to the specialized requirements of smart power grids. Smart grids involve complex, large-scale communication networks that integrate optical networks [NASPI, 2019; Sibley, 2020] and wireless technologies [Gungor et al., 2010]. These modern networks require more advanced redundancy and recovery mechanisms to meet real-time demands, scalability, and reliability. Figure 3.8 overviews the limitations and points out the research trends. Let us dive into the specific limitations in the following aspects.

i) *PRP and HSR exhibit excessive overhead in networks.* PRP and HSR, which duplicate data transmission to ensure zero-failure recovery, place significant additional traffic load on the network. In smart grid environments that rely on low-bandwidth wireless technologies such as LoRa and IoT, this overhead can overwhelm the network, reducing efficiency and making it unsustainable for large-scale deployment.

Additionally, they incur costly optical network overhead. In OTN and Passive Optical Network (PON), where bandwidth is abundant but expensive, duplicating traffic can result in unnecessary consumption of resources. While PRP and HSR are effective for critical data, using them across all network segments, especially in high-capacity optical links, can significantly increase operational costs.

ii) *CRP and DRP have limited flexibility for diverse topologies.* CRP and DRP are optimized for specific network topologies, such as rings, which limits their applicability in the mixed topologies commonly found in smart grid networks. Smart grids utilize diverse topologies, including rings, mesh, and tree structures, depending on the geographical and infrastructural requirements.

In wireless segments (e.g., 5G, LoRa, and IoT) [Gungor et al., 2010], ring-based redundancy mechanisms are vulnerable to signal interference and link degradation, leading to delays in redundancy switching and potentially catastrophic failures in grid monitoring systems.

Additionally, both CRP and DRP are insufficient for dealing with the complex multiplexing and data aggregation techniques used in OTN and Synchronous Digital Hierarchy (SDH). These protocols do not take full advantage of the recovery mechanisms native to these transport networks, leading to inefficiencies.

iii) *IP routing, STP and RSTP exhibit limited support for real-time communications.* In smart power grids, where real-time communication between substations and control centers is critical, both STP and RSTP exhibit high convergence times. When there is a network failure, the reconfiguration time can be too long to meet the stringent time constraints required for energy control systems, especially in mission-critical services such as protection relays.

There are challenges in optical networks [Sibley, 2020], where OTN, SDH, and PON form the backbone of smart grids, carrying high-speed, low-latency data. IP routing, STP, and RSTP, designed for simpler Ethernet-based networks, are not optimized for the performance needs of these high-capacity networks. Their slow recovery times become a bottleneck in maintaining service continuity.

Additionally, they are inefficient in wireless networks. Wireless networks in smart grids, especially with 5G, LoRa, and IoT-based communication, require faster, more adaptive redundancy mechanisms due to the inherent challenges of wireless communication, such as interference, limited bandwidth, and signal degradation. IP routing, STP/RSTP are not agile enough to handle these challenges effectively.

iv) *MRP raises latency concerns in multi-technology networks.* This MRP, designed for industrial Ethernet, introduces additional latency due to its recovery mechanisms. This latency is problematic in smart grids where data transmission across multiple technologies (e.g., OTN for long-haul communication and 5G for localized distribution) must happen in near real time.

Moreover, MRP has limited optical integration. In OTN and SDH environments, MRP does not seamlessly integrate with the fault recovery mechanisms native to these technologies, such as SDH's Automatic Protection Switching (APS) or OTN's dedicated recovery paths [Sibley, 2020]. This can result in inefficiencies and longer network recovery times.

v) *BRP is inefficient for optical networks.* In optical and PON-based segments of smart grids, where data is broadcast over shared mediums, BRP's redundancy mechanism can introduce unnecessary traffic overhead. The protocol is not optimized for the high-speed, low-latency requirements of optical data transmission, leading to inefficiencies in fault recovery.

   In wireless networks such as 4G/5G and LoRa [Gungor et al., 2010], BRP may generate excessive broadcast traffic, consuming valuable spectrum and leading to congestion, particularly in low-bandwidth wireless IoT networks. This congestion can disrupt time-sensitive communication in power grid applications.

vi) *RRP has resource inefficiency problem in high-density networks.* In smart grid networks, which often span large geographic areas and employ high-density wireless links (such as 5G for urban environments), RRP's additional overhead for managing redundant paths can strain available network resources. This inefficiency is especially pronounced in the data transport layer, where high-throughput and low-latency requirements are critical.

   Additionally, RRP is difficult to integrate smoothly with modern transport network recovery techniques, such as the wavelength-based restoration in OTN or the flexible multiplexing features of PON, making it less suitable for smart grid applications that leverage these advanced optical technologies.

vii) *General challenges exist for IEC 62439 protocols in wireless networks.* First, the propagation delay can be a problem in long-distance wireless communication. In rural or remote areas where 2G and LoRa technologies are used, redundancy protocols like STP, RSTP, and CRP are inefficient in handling the higher propagation delays. Wireless links in smart grids are particularly prone to failure due to environmental factors, and these redundancy protocols often fail to recover quickly enough.

Second, interference and bandwidth can be constraints in smart grids using IoT or LoRa for device-to-device communication. Those redundancy mechanisms must be lightweight and adaptive due to limited bandwidth. Most IEC 62439 protocols, which were designed for Ethernet networks, introduce too much overhead and are not optimized for the noise-prone and low-power characteristics of these wireless technologies.

Additionally, many of these redundancy protocols are not energy-efficient, which is a critical consideration for IoT-based communication networks in smart grids, where devices often have limited power supplies. PRP, in particular, is unsustainable in such environments due to the requirement for duplicate packet transmission.

To conclude, the IEC 62439 redundancy protocols, while offering basic fault tolerance for industrial Ethernet networks, fall short in addressing the specific requirements of modern smart power grids. These grids involve diverse communication mediums – ranging from optical transport to wireless technologies – and require redundancy mechanisms that can adapt to high-speed data transport, low-latency requirements, and complex network topologies. The protocols' limitations in scalability, latency, and resource efficiency make them less suitable for the highly dynamic and heterogeneous environments of smart grid networks.

3.4.2 Future Trends

Beyond the IEC 62439 protocols, as illustrated in Figure 3.8, future trends in smart grid communications are emerging in two critical directions:

i) **Robust non-disjoint path aggregation**: This method relaxes the traditional constraint of requiring disjoint paths, allowing intermediate links on different paths to overlap. By enabling this overlap, more parallel paths can be established, which enhances the reliability of end-to-end data delivery within the network layer. This approach mitigates the risk of single-link failures disrupting communication, significantly improving network resilience. Further details will be elaborated in Section 4.1.
ii) **Efficient and robust FRR**: The next generation of FRR techniques will need to accommodate a wide variety of network topologies, including ring, mesh, and other complex structures, without relying on multiple host attachments. By supporting diverse topological designs, these FRR solutions aim to provide swift rerouting capabilities under link failures, thus ensuring minimal disruption in data transmission. This will be discussed further in Section 4.2.

While these advancements are poised to improve reliability in terms of data delivery and path availability, there remain other critical challenges beyond IEC 62439. These include the need for a more **robust data transport protocol**, **application-level data redundancy**, and **defenses against Link Flooding Attacks (LFAs)**. These aspects, crucial for achieving high-availability communication networks in smart grids, will be explored in Chapter 4.

3.5 Summary

This chapter presents details of the classical high-availability networking solutions defined by the IEC 62439 standard, emphasizing their design principles, functionalities, and the interactions between various approaches. The redundancy strategies are categorized based on their implementation location: at the end nodes or within the network. The first category includes protocols such as PRP, HSR, CRP, BRP, and RRP, while the second encompasses IP routing, static routing, dynamic routing, STP, RSTP, MRP, and DRP.

While these protocols are foundational for achieving high availability, they face limitations regarding data delivery reliability and path reliability. Many are specifically designed for ring and tree topologies, which may not adequately address the complexities of modern wireless networks or other essential factors, such as data transport reliability, application data completeness, and defenses against LFAs. Addressing these challenges is crucial for meeting the demands of contemporary smart power grids. The following chapter will delve into these challenges in detail and explore potential solutions to enhance network availability in today's power systems.

Questions

3.1 What are the primary redundancy protocols defined by IEC 62439, and what unique features do they offer to power grid communication networks?

3.2 Discuss the specific limitations of PRP and HSR in the context of smart power grid networks. How does traffic overhead impact their efficiency?

3.3 Explain the challenges posed by CRP and DRP regarding network topology in smart power grids. Why are these protocols not optimal for diverse network structures?

3.4 Analyze the convergence times of STP and RSTP in relation to real-time communication needs in smart grids. What implications do these times have for mission-critical services?

3.5 What latency issues arise from using MRP in multi-technology networks within smart grids? How does this affect data transmission across different technologies?

3.6 Discuss how BRP is inefficient in optical networks and wireless segments. What traffic implications do they have for smart grid applications?

3.7 Evaluate the resource inefficiency of RRP in high-density networks. Why is this particularly problematic in smart grid environments?

3.8 What general challenges do IEC 62439 protocols face in wireless networks, specifically regarding propagation delay and interference?

3.9 Identify two emerging trends in smart grid communication that address the limitations of IEC 62439 protocols. How do these trends enhance network reliability and resilience?

3.10 Summarize the overarching limitations of classical high-availability networking solutions in smart power grids. What are the critical areas for future research?

Bibliography

Vehbi C. Gungor, Bin Lu, and Gerhard P. Hancke. Opportunities and challenges of wireless sensor networks in smart grid. *IEEE Transactions on Industrial Electronics*, 57(10):3557–3564, 2010.

IEC. Industrial communication networks - high availability automation networks - Part 4: Cross-network redundancy protocol (CRP), 2010a.

IEC. Industrial communication networks - high availability automation networks - Part 6: Distributed redundancy protocol (DRP), 2010b.

IEC. Industrial communication networks - high availability automation networks - Part 7: Ring-based redundancy protocol (RRP), 2011.

IEC. Industrial communication networks - high availability automation networks - Part 5: Beacon redundancy protocol (BRP), 2018.

IEC. Industrial communication networks - high availability automation networks - Part 3: Parallel redundancy protocol (PRP) and high-availability seamless redundancy (HSR), 2021.

IEC. Industrial communication networks - high availability automation networks - Part 1: General concepts and calculation methods, 2017a.

IEC. Industrial communication networks - high availability automation networks - Part 2: Media redundancy protocol (MRP), 2017b.

IEEE. IEEE standard for rapid reconfiguration of spanning tree, 1998.

IEEE. IEEE standard for local and metropolitan area networks: Media access control (MAC) bridges, 2004.

IETF RFC 4090. Fast reroute extensions to RSVP-TE for LSP tunnels, 2005.

Stephane Litkowski, Ahmed Bashandy, Clarence Filsfils, Pierre Francois, Bruno Decraene, and Daniel Voyer. Topology Independent Fast Reroute using Segment Routing. Internet-draft, IETF, 2022.

G. Malkin. RIP version 2, 1998.

J. Moy. OSPF version 2, 1998.

NASPI. NASPInet 2.0 architecture guidance version 1.19, 2019. URL https://www.naspi.org/node/746.

Y. Rekhter, T. Li, and S. Hares. A border gateway protocol 4 (BGP-4), 2006.

E. Rosen, A. Viswanathan, and R. Callon. MPLS architecture, 2001.

Martin Sibley. *Optical Communications: Components and Systems (third edition)*. Springer, 2020.

4

Challenges and Solutions for High-Available Networks

As previously discussed in Chapter 2 and Section 3.4, the expanded and growing demands for high network availability, coupled with the limitations inherent in the classical IEC 62439 standard [IEC, 2016], present complex technical challenges that necessitate a comprehensive redesign of networking solutions. These emerging solutions must be capable of supporting modern network structures and protocols while incorporating robust defenses against Link Flooding Attack (LFA) threats. A key requirement is that these solutions must build upon practical network architectures – primarily traditional IP networks, Software-Defined Networkings (SDNs) [Dorsch et al., 2014], and Named Data Networks (NDNs) [Zhang et al., 2014; Ravikumar et al., 2020; Zhou et al., 2023a] – to effectively enhance network resilience and availability.

In response to these demands, five critical research directions have been identified in this section as follows: (1) *robust non-disjoint path aggregation*, which allows overlapping intermediate links to increase parallelism and reliability; (2) *efficient and resilient Fast Reroute (FRR)*, offering failover protection across diverse network topologies such as rings and meshes; (3) *a robust data transport protocol* to ensure secure, low-latency transmission; (iv) *data redundancy at the application layer*, which introduces multiple safeguards for critical applications; and (v) *fortified defenses against LFAs*, which involve advanced methods to detect and mitigate LFAs, ensuring the continuous operation of the grid's communication infrastructure. These trends have been previously summarized in Figure 2.2.

The remainder of this chapter delves into the challenges and specific solutions for each of these research areas. These technological advancements form the foundation for high-availability network systems, reinforcing their resilience against external threats and ensuring long-term network stability. Sections 4.2–4.5 will explore the motivations driving research in these domains, highlighting key innovations and valuable research directions.

Communication Networks in Smart Power Grids, First Edition. Boyang Zhou.

Companion Website: www.wiley.com/go/SmartPowerGrids1e

4.1 Non-disjoint Path Aggregation

4.1.1 Technical Challenges

As we embark on this exploration, it becomes evident that existing solutions in literature have predominantly focused on the paradigm of disjoint path redundancy. These solutions, exemplified by reliable multi-path routing designed for general networks [Iqbal and Kuipers, 2015] and the IP Parallel Redundancy Protocol (PRP) (iPRP) [Popovic et al., 2016], represent commendable strides in enhancing the IEC 62439 standard [IEC 62439, 2008] for bolstering control robustness. Yet, the intricate topological characteristics of transmission grids pose a non-trivial challenge in the generation of such disjoint paths.

In unveiling this complexity, Figure 4.1 vividly illustrates that in scenarios devoid of packet delivery time requirements, only meager percentages – 14.29%, 30.34%, and 8.21% – of source–destination pairs can be interconnected via a single path in the IEEE 14, 30, and 57-bus networks [IEEE, 2025], respectively. Consequently, alternative links outside these disjoint paths are often overlooked for packet delivery, leading to an untapped potential in harnessing network resilience, as measured by the increase of Success Rate of Receiving a Packet (SRRP) and Qualified Source-Destination Pairs (QSDP). Our journey unfolds as we explore innovative strategies to maximize the utilization of these overlooked paths and propel the frontiers of network reliability.

In the pursuit of enhancing packet delivery reliability, a novel and pragmatic solution emerges – the utilization of previously overlooked alternative links. This approach involves establishing non-disjoint loop-free paths that allow shared links between two paths to overlap. Through the implementation of these non-disjoint paths, the Quality of Service in Receiving a Packet (QSRP) can be enhanced, coupled with an increase in the SRRP at the destination node. Consequently, the Packet Inter-Arrival Time (PIAT) experiences a significant

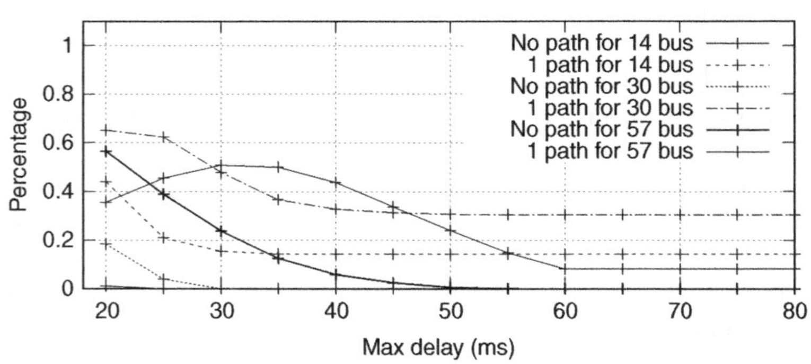

Figure 4.1 Percentage of pairs with no path and 1 path vs. max delay.

reduction during link failures, facilitated by the avoidance of unnecessary TCP retransmissions.

However, delving into the heart of this innovation reveals challenges on two fronts. First, the *Non-Disjoint Path Aggregation (NDPA) problem* presents computational complexities. The computation of non-disjoint loop-free paths, aiming to obtain the minimum number of links meeting a multiplicative constraint (SRRP) and a concave constraint (end-to-end delay), is identified with NP-Complete complexity [Wang and Hou, 2000]. This complexity is especially pertinent when reducing the number of forwarding rules is a priority. While the generation of non-disjoint loop-free paths subject to the end-to-end delay constraint has been studied, with contributions from Bernstein [2010], Hershberger et al. [2007], and Yen [1971], the exploitation of the SRRP constraint remains uncharted territory.

Second, the efficient reorganization of out-of-order packets under specific link failures poses a non-trivial challenge in terms of time and space complexity. Minimizing the Packet Out-of-Order Rate (POR) is crucial, requiring strategies to reduce routing convergence time and maximize SRRP. The study of packet reordering on multiple or time-varying paths has been explored in the context of distributed routing (e.g., [Baccelli et al., 1984; Wang and Gelenbe, 2016]). However, the existing solutions fall short in explicitly addressing the prolonged routing convergence time's impact on POR, leading to increased memory usage and Half Round-Trip Time (HRRT) in the packet reordering process.

4.1.2 Paths Aggregation Approaches

Concretely, the path aggregation approach allows intermediate routers to aggregate packets over the multiple paths either disjoint or non-disjoint in their links. This approach has seen successful adoption in real power transmission grids. The approach is more reliable than the route failover approach [Bankhamer et al., 2022; Tan and Kuo, 2022; Isyaku et al., 2021; Zhou et al., 2023c], due to no need for route re-convergence.

On one hand, researchers focus on the in-network packet aggregation, which plays crucial roles in enhancing network reliability. The aggregation depends on multiple paths in both non-disjoint and disjoint approaches.

For the disjoint approach, a representative protocol is the Parallel Redundancy Protocol (PRP) of IEC 62439 [2008], following its improved version utilizing VLAN [Popovic et al., 2016]. PRP necessitates that paths be disjoint in their links, neglecting alternative links outside these disjoint paths for packet delivery. iPRP [Popovic et al., 2016] enhances packet loss rates under link failure conditions. While effective, these solutions are not tailored for TCP protocols, and the issue of packet out-of-order remains unaddressed.

Utilizing non-disjoint paths, OpenFlow-based Non-Disjoint Path Aggregation Mechanism (OPAM) achieves the highest reliability under fault links [Zhou et al., 2019], targeting the IEEE 61850 manufacturing message specification message [IEC, 2013], by seamlessly integrating with TCP or UDP. Additionally, an architecture based on NDN, called iCASM, supports Phasor Measurement Unit (PMU) communication by enabling broadcasting for multiple egress faces over non-disjoint paths [Ravikumar et al., 2020]. UDP-based OPAM shares a similar process in aggregating payload packets with iCASM. However, these solutions that utilize non-disjoint paths, akin to Redundant Sub-Paths (RSPs), are limited to operating at the network layer and can only manage packet losses within a single transmission round.

Previously, Cognitive Packet Networks (CPNs), integrating with autonomic communications and natural computation, have garnered attention. The first issue lies in routing mechanism. CPN, epitomized by Gelenbe's work [Gelenbe, 2017; Gelenbe, 2009], employs a distributed routing architecture with autonomic features. Timotheou [2010] utilize a Random Neural Network through reinforcement learning, ensuring Quality-of-Service (QoS) by identifying egress ports for data flows. The second is packet Reordering. Wang's enhancement of CPN [Wang and Gelenbe, 2016] incorporates a packet reordering buffer for real-time traffic in industrial communication networks. The CPN, though effective, demonstrates higher HRRT and packet reordering buffer size, influenced by the context of routing architecture. The third focuses on optimizations. While CPN optimizes forwarding paths for QoS goals, certain challenges remain unaddressed, such as the potential loss of in-flight packets during path failover and the absence of explicit studies on non-disjoint path generation and packet reordering within the networks.

The above analysis underscores the need for innovative solutions in the quest for enhanced network reliability and performance. These unexplored territories pave the way for the novel OPAM presented in this book. The drawback of such approach is that they struggle to handle packet losses during failovers and critical fault links.

On the other hand, the multi-path generation is an issue concerned by existing approaches. They generally adopt the two kinds of algorithms as follows:

i) Non-disjoint path generation algorithms. These algorithms focus on generating *k* shortest loop-free paths [Zhou et al., 2019]. Notable methods encompass: (1) Yen's algorithm [Yen, 1971], (2) Hershberger's enhancement of Yen's algorithm [Hershberger and Suri, 2001], (3) the Bernstein algorithm [Bernstein, 2010] which incorporates a multiplicative approximation to the second shortest path [Frieder and Roditty, 2015], and (4) an algorithm aimed at generating non-disjoint loop-free paths while minimizing the number of links to meet constraints in both the packet success rate and end-to-end delay [Zhou et al., 2019]. These algorithms do not address the consideration above.

ii) Disjoint path generation algorithms. The algorithms are commonly employed in standards such as IEC 62439 [IEC 62439, 2008] and iPRP [Popovic et al., 2016], as reviewed in Iqbal and Kuipers [2015]. However, they predominantly deal with path disjointness, making them unsuitable for generating RSPs that are essential for MPSGs.

4.1.3 Heading Toward OPAM Solution

In the quest for enhanced reliability, this book presents OPAM [Zhou et al., 2019]. OPAM stands out by ingeniously embracing nondisjoint paths, setting it apart from alternative solutions. This distinctive approach marks a significant technical leap, strategically navigating the network layer toward heightened reliability. The key lies in leveraging the intricate meshed and ringed structures inherent in power grid networks.

OPAM is based on SDN that heuristically computes the constrained non-disjoint paths in a centralized fashion and re-organize the out-of-order packets at edge switches. The solution is evaluated through simulations of the IEEE 30-bus network scenario under 2% link failure rate, and the result confirms its effectiveness: the packet delivery success rate is significantly improved in comparison with the current non-disjoint and disjoint path algorithms. TCP throughput is improved by 121.62% with the packet reordering. Also, the memory usage for the packet reordering buffer is reduced by 76.563% compared with the distributed CPN.

This innovative solution promises to reshape the landscape of network reliability, introducing a paradigm shift that aligns seamlessly with the complexities of power grid topologies.

4.2 Efficient and Robust FRR

4.2.1 Technical Challenges

The exploration of fast rerouting in the presence of link faults on the Primary Forwarding Path (PP) using detours, crucial for minimizing Link Recovery Time (LRT), remains underexplored.

To minimize LRT, current approaches in FRR [Chiesa et al., 2021] focus on swiftly detecting and converging link states with active participation from the data plane. This approach helps bypass the slow convergence issues found in the control plane, encompassing classical distance vector and link-state algorithms, along with centralized SDN [Xia et al., 2014]-based techniques [Ganchev et al., 2020;

Liu et al., 2013]. Notably, these strategies aren't tailored for NDN [Schweiger et al., 2021; Kamisiński, 2018; Xu et al., 2021b; Shukla and Foerster, 2021]. Moreover, the challenge of minimizing distributed collaboration steps during rerouting [Xu et al., 2021b] has been overlooked. Excessive steps can induce transient states in forwarding paths, potentially amplifying delays.

Additionally, constructing linked paths for rerouting is more straightforward than traditional disjoint paths in power grids [Zhou et al., 2019; Foerster et al., 2022a], thanks to the presence of abundant meshes and rings. We refer to such linked paths as a Resilient Subgraph (RSG) [Zhou et al., 2019] for convenience. Moreover, during the rerouting process, Data packets may become out of order due to differing delays on the PP and detoured paths. This aspect needs attention to ensure optimal data flow delivery quality.

In recent research on FRR, a common approach involves a two-tier hierarchical reroute design. Essentially, when a router encounters link faults, it performs a nonoptimal local reroute and globally converges on route states. This is accomplished through either of two main types of routes: (1) multiple arc-disjoint arborescences rooted from a router [Elhourani et al., 2016; Chiesa et al., 2016; Foerster et al., 2018, 2019a, 2019b, 2022b; Bankhamer et al., 2022; Tan and Kuo, 2022], or (2) multiple end-to-end disjoint paths between two routers [Zheng et al., 2019; Jiang et al., 2020; Isyaku et al., 2021; Foerster et al., 2022b]. However, in these approaches, the reroute paths must exclusively link the two ends of the PP. Additionally, the number of available routes in the network is restricted by the minimum ingress degrees of each node, limiting reroute options [Ahlswede, 2019; Foerster et al., 2018; Bankhamer et al., 2022]. This limitation can be mitigated by exploiting Disjoint Subpaths (DSPs) between two intermediate nodes in the PP. DSPs have the potential to enhance resilience by connecting more than just the two ends. However, research on the local FRR approach [Liu et al., 2013; Ramadan et al., 2016; Shukla and Foerster, 2021] has seldom addressed multiple link failures in a distributed recovery manner.

4.2.2 FRR Approaches

The current approaches on the fast reroute are discussed in the data plane and control plane as follows.

In the data plane, researchers focus on how to enforce a reroute efficiently and effectively. Traditional IP fast reroute strategies were thoroughly explored [Kamisiński, 2018]. Liu introduced a data-driven connectivity mechanism for recovering non-partitioning failed links, lacking immediate link protection [Liu et al., 2013]. Ramadan proposed an adaptive centralized latency-constrained resilient routing without path discovery [Ramadan et al., 2016]. Shukla suggested an in-network reroute mechanism for immediate path protection from a single link failure by locally removing reroute loops [Shukla and Foerster, 2021].

However, it falls short when dealing with multiple link failures due to its agnostic nature about link states. Xu's proactive link failure recovery framework, while attempting to encapsulate backup paths in packet headers, struggles with multiple backup paths and challenges posed by multiple link failures [Xu et al., 2021b]. Li put forward a hybrid multi-failure recovery system with low forwarding rule storage overhead to optimize traffic balance [Li et al., 2022]. Miura introduced a fast failure recovery mechanism utilizing multiple routing configurations with link failure detection [Miura et al., 2022]. Unfortunately, these approaches don't focus on reducing the number of steps during rerouting and are not tailored for Named Data Networking (NDN).

In the control plane, researchers mainly study on improvement in the efficiency and effectiveness within the route failover process. Such a goal is navigated by means of a systemic design in the two hierarchies: Global reroute control and local reroute coordination. They are specified as follows.

Global reroute control mechanisms, specifically arc-disjoint arborescence routing, have been explored by Foerster. He proposed a failover path allocation method and a deterministic FRR to reduce path stretch under multiple link faults [Foerster et al., 2018, 2019a]. CASA, also introduced by Foerster, offers robust and provable FRR using backtracking techniques [Foerster et al., 2019b]. Additionally, postprocessing operations, like those proposed by Foerster et al. [2022b], optimize detour path length. Tan suggested locally and optimistically finding the shortest failover route and recomputing routes for detected loops [Tan and Kuo, 2022]. Bankhamer contributed by proposing local fast rerouting algorithms with low congestion in a complete network [Bankhamer et al., 2022]. Some solutions, however, require higher network connectivity, limiting their applicability to general networks. For end-to-end disjoint paths, Schweiger presented TREE to enhance CASA in resilience and path length quality [Schweiger et al., 2021]. Zheng proposed Sentinel, focusing on immediate local failovers and load balancing on multiple paths [Zheng et al., 2019]. Numerous studies explored traffic engineering for end-to-end fault paths [Jiang et al., 2020; Isyaku et al., 2021] or multiple ones [Foerster et al., 2022b]. Unfortunately, these approaches haven't tapped into DSP to improve resilience and aren't tailored for NDN.

In terms of local reroute coordination, Liu, Ramadan, and Shukla introduced data-driven connectivity mechanisms and adaptive centralized resilient routing for immediate protection from single link failures [Liu et al., 2013; Ramadan et al., 2016; Shukla and Foerster, 2021]. However, they lack awareness of non-neighboring network link states during rerouting. Xu's proactive link failure recovery framework, while encapsulating backup paths in packet headers, struggles with multiple backup paths [Xu et al., 2021b]. Recent studies explored local reroute with centralized routing in SDN to handle multiple link failures [Li et al., 2022; Miura et al., 2022]. Unfortunately, these reroutes aren't distributed and aren't suitable for NDN.

4.2.3 Advancing Toward Traffic-Aware Fast Reroute (TA-FRR) Solution

In tackling the formidable challenge at hand, this book unfolds the innovative concept of Traffic-Aware Fast Reroute (TA-FRR) [Zhou et al., 2023b, 2023e]. This cutting-edge approach masterfully strikes a meticulous balance between global control and local coordination, anchored in the robust foundations of NDN, custom-tailored for the unique demands of power grids. What sets TA-FRR apart is its intriguing hybrid design, artfully crafted to minimize Reroute Failure Rate (RFR) in the face of severe link failures.

TA-FRR mechanism is based on NDN that harnesses a new RSG route involving multiple DSPs. TA-FRR proactively constructs an RSG with scalable computational complexity, maximizing resilience while avoiding excessive rerouting. Within the RSG, TA-FRR efficiently computes the shortest detour path for disconnected components, actively identified based on packet arrival timeouts. Employing distributed reroute control, TA-FRR utilizes an optimized algorithm to minimize reroute steps, steering clear of unnecessary actions between connected links on the PP. Notably, the algorithm exhibits scalability in both time and space complexities. Performance evaluation in ndnSIM, using two distinct topologies, validates the efficacy of TA-FRR. The first topology mirrors a real electric transmission grid with 300 buses. Results demonstrate that TA-FRR adeptly recovers from one or multiple link faults, achieving a short LRT ranging from 0.4 to 98.9 ms, with a maximum End-to-End Packet Loss Rate (EELR) of 2.03%. Furthermore, TA-FRR excels in handling high-frequency data delivery with bounded delays. The second topology, representing synthesized Internet scenarios with varying degrees of link faults, showcases TA-FRR's efficient recovery from serious link faults in PP. The scalable RSG construction outperforms existing approaches, providing significantly higher resilience with a maximum stretch comparable to the arborescences approach.

Brace yourself for a fascinating exploration into the realms of network resilience with this groundbreaking solution.

4.3 Robust Data Transport Protocol

4.3.1 Technical Challenges

The robust data transport protocol is an urgent need, since the data link and network layers, despite considerable focus, still grapple with unresolved packet losses. Effectively recovering from these losses requires the transport layer to integrate a strong retransmission control mechanism.

The studies in the field of data transport protocol on reducing the End-to-End Failure Rate in Packet Delivery (EEFR) especially utilize the hop-by-hop retransmission control for lost packets. Such a control requires to adopt in-path caching [Su et al., 2015; Wang et al., 2018; Chen et al., 2019, 2020; Garcia-Luna-Aceves and Albalawi, 2020]. Recent studies underscore that this approach can outperform existing end-to-end transport protocols [Chen et al., 2020; Garcia-Luna-Aceves and Albalawi, 2020].

In this control process, the intermediate hops of the PP play a pivotal role by caching received packets during forwarding. Simultaneously, each hop retransmits lost packets by retrieving cached copies from its upstream hop. However, a critical enhancement has been overlooked – leveraging the potential of *RSPs* for the PP to fortify the reliability of the upstream link. These RSPs, originating from the current hop to its upstream counterpart within the PP, remain disjoint from each link between the source and the current hop. Abundant in the network due to its redundant topological structure in meshes and rings [Zhou et al., 2019], these RSPs offer a promising solution (see Section 7.1.2).

4.3.2 Hop-by-Hop Retransmission Methods

The transport layer focuses on the hop-by-hop retransmission control, which outperforms the traditional end-to-end retransmission control according to various studies [Chen et al., 2020; Garcia-Luna-Aceves and Albalawi, 2020; Zhou et al., 2022a]. Among those solutions, Disruption Resilient Transport Protocol (DRTP) stands out with the lowest EEFR, until now. We elaborate them as follows.

In simple terms, these approaches request a retransmission of lost payload packets from either the sender or receiver. In the sender-side design [Su et al., 2015; Wang et al., 2018], the current hop acknowledges its upstream hop upon receiving a message. The message falls into either of two types: (1) a forwarded packet followed by a control message (e.g., MFTP [Su et al., 2015]), or (2) simply a forwarded packet (e.g., Rapid and Reliable Transport Mechanism (R2T) [Wang et al., 2018]). Consequently, at the current hop, the received packets have not yet reached the downstream hops. This makes it challenging to utilize RSPs and simultaneously cope with the complete disruption of any PP link [Zhou et al., 2022a].

In the receiver-side design [Chen et al., 2019, 2020; Garcia-Luna-Aceves and Albalawi, 2020], the retransmission request involves first detecting the lost payload packet through one of two methods: (1) solely at the current hop based on packet disorder and inter-arrival time, using protocols like Adaptive Transmission Protocol based on Software-Defined Networking (SDATP) [Chen et al., 2019, 2020] and Internet Transport Protocol (ITP) [Garcia-Luna-Aceves and Albalawi, 2020], or (2) by considering both the recursive collaboration process and exploiting RSPs based on packet arrival timeouts, as seen in DRTP [Zhou et al., 2022a], effectively reducing the possibility of cache mismatch. Moreover, DRTP can be seamlessly integrated with both the network layer (in NDN) and the sensor data dissemination protocols thanks to NDN's extensible Type-Length-Value (TLV) message format. The analysis above indicates that DRTP is a leading approach in research focused on reducing errors within the transport layer [Zhou et al., 2022a].

Therefore, the aforementioned sender-side and receiver-side designs have rarely considered reliable hop-by-hop retransmissions by leveraging RSPs, excepting DRTP. Additionally, receiver-side designs have overlooked their distributed collaboration issue (see Section 7.2.4), potentially impacting EEFR under serious link disruptions (see Section 7.4.3).

Moreover, the hop-by-hop retransmission control design can benefit from the popularized NDN architecture [Zhang et al., 2014; Mastorakis et al., 2017]. In NDN, a consumer receives a Data packet with a prefix name stored at the producer by sending an Interest packet with the prefix along a path. In-path routers can efficiently cache received Data into their local Content Store (CS), configured with a limited size by the operator. The CS size, if appropriately configured, reduces cache mismatching, and interactions between routers can be flexibly redefined through their forwarding strategies.

4.3.3 Embarking on DRTP Solution

Directed toward enhancing reliability, let us delve into the DRTP [Zhou et al., 2022a] which will be presented in this book. This ingenious solution, firmly rooted in the NDN architecture [Zhang et al., 2014], takes shape as a receiver-side marvel, harnessing hop-by-hop caching on the PP. What propels DRTP to the forefront is its unparalleled advantage in achieving the lowest EEFR, courtesy of a meticulously orchestrated hop-by-hop retransmission control mechanism endowed with specific robustness.

Think about this: a recursive collaboration process fueled by meticulously specified arrival timeouts of payload packets. This control process dances to a rhythmic frequency during the sensor data dissemination, revolutionizing the reliability game. Brace yourself as each communication pair in the network navigates a new breed of routes, woven into the fabric of Multipath Subgraph (MPSG). Witness the reliability boost as MPSG employs disjoint RSPs, acting as stalwart backups for the PP subpath in the event of failure [Zhou et al., 2022a].

The effectiveness of DRTP was rigorously assessed in the cyber–physical environment, considering both typical and general routes derived from real-world transmission grids. The environment is built with integration of the ndnSIM simulator, MATPOWER, and Real-Time Digital Simulator (RTDS). The results illustrate DRTP stands tall, significantly elevating the reliability of sensor data dissemination in power grids. The numerical results affirm a substantial advantage in reducing EEFR, coupled with a low end-to-end delivery time, particularly under severe link disruptions. A true game-changer when compared to existing standards such as IEEE C37.118.2, multi-path aggregation, and conventional hop-by-hop retransmission controls. The ripple effect extends its benevolence to State Estimation (SE) by bestowing superior grid observability and accuracy. This comprehensive exploration contributes valuable insights to the robustness and effectiveness of communication systems in power grids. Prepare to witness a paradigm shift in the landscape of power grid communication with the prowess of DRTP.

4.3.4 Core Mechanism of Resilient Sensor Data Dissemination (RSDD)

Navigating the intricacies of DRTP is akin to a high-stakes endeavor, particularly when treading the fine line of maintaining recursiveness during retransmission control, all the while adhering to the sacred covenant of correct arrival timeouts. The repercussions of any misstep in this delicate dance are profound. A breach in the timeouts for a retransmission request can send shockwaves through the cache realm, inducing a mismatch that becomes a catalyst for the dreaded request failure phenomenon.

Enter the scene of chaos: a hop triggering arrival timeouts, setting off retransmission instances in an unpredictable sequence. The control, caught in an unprotected parallel race condition, finds itself vulnerable to the very timeout violations it seeks to evade. A formidable challenge, indeed.

To address such a challenge, Resilient Sensor Data Dissemination (RSDD) emerges as the solution [Zhou et al., 2023a], a beacon of hope in implementing DRTP correctly and with unwavering verification. RSDD takes the reins, orchestrating concurrent retransmission instances with the precision of a maestro. The orchestration process stands guard, ensuring the exclusive execution of the same lost packet, each move meticulously choreographed based on timeouts.

Past trials have rendered RSDD victorious, validated as the epitome of a correct DRTP implementation. Its aura of exclusiveness and the stellar performance in QoS are not mere claims but tangible outcomes of rigorous experimentation. Our evaluation took place in a simulated network, for sensor data dissemination in power transmission grids. The results speak volumes, RSDD achieves an EEFR of 2.44% and an average End-to-End Payload Packet Delivery Time (EEDT) of 2.7 ms during full path disruption, featuring a 20% link loss rate in packets. In the intricate tapestry of DRTP, RSDD emerges as the linchpin, steering the course with correctness and performance as its guiding stars.

4.3.5 Less Fragile Routes with HeuMPSG

Embarking on the journey of RSDD to realize the grandeur of DRTP, the spotlight shifts to the crucial realm of MPSG generation algorithms. Here, precision is not just a preference but a necessity. The realm unfolds with two notable contenders: GenMPSG [Zhou et al., 2022a] and, the virtuoso, HeuMPSG [Zhou et al., 2023d].

GenMPSG, the pioneer, wields Dijkstra's algorithm to compute PP, unfurling RSPs within each PP. However, its narrative falls short of explicitly considering the fragility of MPSG or the constraints of maximum EEDT. The tale is told with time and space complexities donning the cloak of $O(\beta \cdot n^2 \cdot \log^2 n + \beta \cdot n \cdot \log^2 n + n \cdot \log n)$ and $O(n + \beta \cdot n \cdot \log n + (\beta + 1) \cdot \log n)$, respectively. An intricate dance, yet with space for optimization, particularly in the echoes of duplicated PP searching steps.

In the limelight enters HeuMPSG, a masterful algorithm design to compose MPSG with resilience against potential link failures. A strategic move to fortify the DRTP resilience. HeuMPSG achieves this by minimizing fragility, a ballet for the benefit of network resilience. The performance results manifest the mean EEFR and EEDT of DRTP, reducing by a maximum of 29.11% and 2.41 seconds, respectively.

Comparing the acts, HeuMPSG emerges as the efficient algorithm with a low time complexity of $O(n^2)$. Here, efficiency comes from HeuMPSG's adept use of heuristic and an elegant selection of candidate PPs, while searching for maximally disjoint RSPs, hop by hop. In the grand tapestry of DRTP, HeuMPSG stands as the brilliance, enhancing efficiency while composing a resilient dance against potential link failures.

4.4 Redundant Application Data Codes

4.4.1 Technical Challenges

To ensure that the transmission delay meets the requirements for the transmission of three-remote commands when the wireless channel is interfered with, the existing Long Term Evolution (LTE) channel reliability technology primarily improves the reliability of wireless channel transmission through Forward Error Correction (FEC) technology and Hybrid Automatic Repeat-reQuest (HARQ) technology. HARQ is used to compensate for FEC decoding failures, where FEC can recover lost bits from a Binary Erasure Channel (BEC) that eliminates multiple bits [3GPP TS 36.212, 2016]. FEC mainly adopts a fixed-length coding strategy that combines convolutional codes and Turbo codes to achieve optimal channel performance and certain anti-interference capabilities within a certain range. However, fixed-length coding, while reducing the coding rate, limits the channel's error correction ability.

Fountain coding is a type of variable-length coding where the sender can transmit an infinite length of code, and the receiver only needs to receive a fixed length of encoded data to obtain the original data. IETF RFC 6330 RaptorQ (RQ) implements linear time complexity for encoding and decoding rates, with throughput close to the theoretical capacity limit of the channel. It is currently one of the better coding and decoding technologies [Shokrollahi, 2006; IETF RFC 6330, 2011].

4.4.2 Redundancy Coding Methods

The latest fountain coding technologies mainly include Reed–Solomon (RS), Luby Transform, Raptor, and RQ [MacKay, 2005], with RQ currently being the most efficient encoding method. These coding techniques can tolerate a certain degree of channel loss or packet loss when facing interference [Shokrollahi, 2006]. Qualcomm has also proposed commercialized RQ encoding and decoding chips [Qualcomm, 2010]. However, RQ encoding strategy parameters are not tailored to the low data rate characteristics of power distribution network control operations. The LTE wireless channel itself cannot dynamically adjust RQ encoding strategy parameters based on the degree of channel interference to achieve low-latency reliable communication.

Some research has applied RQ coding technology to enhance communication reliability. Cui et al. introduced FMTCP technology, investigating a quality-of-service mechanism for multipath TCP based on Raptor coding in wireless networks. This involved dynamically adjusting TCP congestion control parameters on different paths to adjust data transmission amounts, with each path encoded using RQ [Cui et al., 2015]. However, FMTCP does not address the dynamic adjustment mechanism of RQ redundant encoding length under latency constraints, resulting in relatively low data transmission quality on each TCP path in the presence of noise interference. Subsequently, Kwon et al. proposed using RQ coding to correct channel errors in video streaming across multiple wireless networks, with RQ coding parameters dynamically adjusted based on RQ coding time complexity to maximize the throughput of video streams [Wu et al., 2016]. However, these works did not consider the problem of minimizing data packet loss rates, and the application scenarios differ from power distribution networks, which require high reliability for critical data like three-remote signals. Moreover, Kucera et al. analyzed how FEC in LTE environments can further reduce the redundant cost of packet reordering to achieve low-latency communication for high-bandwidth and low-latency applications like remote 3D graphics rendering [Kwon et al., 2016]. Yet, they did not address the problem of adapting RQ encoding redundant code length to changes in channel quality. Additionally,

Cao Zhen et al. proposed a two-tier FEC coding technology to address the poor recovery performance of RS error correction codes for continuous packet loss [CAO et al., 2011]. However, the two-tier FEC coding length is immutable and has lower recovery performance than RQ coding, without optimization for the low transmission rate characteristics of power distribution networks. Zhang Kuili et al. proposed an LTE hybrid automatic repeat request scheme [ZHANG et al., 2013], improving throughput and retransmission times, but did not propose an FEC technology optimization solution for wireless communication in power distribution networks.

4.4.3 Embarking on Adaptive RaptorQ Encoder and Decoder (ARED) Solution

The power distribution grids quest for reliable application data delivery in the LTE wireless network. Here, Adaptive RaptorQ Encoder and Decoder (ARED) [Zhou, 2019] emerges as the resolution, wielding reliability advantage. The essence of ARED lies in its artful utilization of RQ encoders and decoders, donning the cloak of adaptability by finely tuning RQ parameters in sync based on feedback from loss rates and throughput. ARED addresses the challenge in the non-linearity of the RQ failure probability model. This adaptability makes ARED orchestrate a harmonious interplay between RQ encoding and decoding, finely tuned to the rhythm of bandwidth shifts. Think of it as an ingenious tool that minimizes the length of RaptorQ redundant codes, adapting in real time. It ensures that the dance of data delivery remains resilient and reliable in the LTE wireless networks in grids. In the simulated world of NS-3 LTE, the results speak volumes. With LTE noise figures ranging from 18.6 to 20.5 dB and 25.6 to 28.7 dB, ARED showcases its prowess. The uplink and downlink loss rates in ARED witness a remarkable decrease – 64.35% and 89.92%, respectively – when compared to LTE alone.

4.5 Safeguarding Against LFAs

4.5.1 Technical Challenges and Solutions

For LFAs, the most defense solutions often fail to protect vulnerable PRs from speculation, especially during the renaissance attack stage.

Meanwhile, the emergence of the SDN paradigm has significant implications for network defense [Feamster et al., 2014]. SDN plays a crucial role in defending against LFAs by offering flexibility and advanced control over network traffic [Aydeger et al., 2016; Belabed et al., 2018; Ravi et al., 2020]. It enhances security not only within its own framework but also in conjunction with traditional

IP networks [Amin et al., 2018]. This capability allows for the programming of flow-level forwarding rules through the OpenFlow (OF) protocol [Open Networking Foundation, 2012], which includes the ability to measure packet reception rates at the host level [Yu et al., 2013].

Let us to discuss the challenges. In the realm of defending against LFAs in its initial attack stage, researchers have delved into the art of Moving Target Defense (MTD) and the craft of traffic throttling.

i) **MTD strategies**: A trio of methods takes center stage:
- **Host relocation**: Enter Venkatesan's proactive Proxy-based Moving Target Architecture (PROTAG) [Venkatesan et al., 2016]. This method orchestrates a proactive dance, seamlessly replacing the target with application-level proxies that are within reach. It then orchestrates a symphony of host shuffling, isolating bots and deftly mitigating the initial attack stage. However, a twist in the tale lies in the fact that some PRs connected to these proxies might remain unchanged, presenting a potential vulnerability for adversaries. This opens the door for adversaries to unleash effective flooding over these PRs connected to their designated proxies.
- **Route mutations**: Aydeger proposed a groundbreaking defense approach [Aydeger et al., 2016] that identifies traceroute patterns and calculates link-disjoint paths. These paths strategically avoid congested links. Aydeger later introduced a defense technique targeting illicit Internet Control Message Protocol (ICMP) probing, executing random route mutations for protection [Aydeger et al., 2021]. It's worth noting that this method doesn't address stealthier ICMP probing.

  Xu contributed to this field with an algorithm designed to mutate routes associated with a link or host under a DDoS attack [Xu et al., 2021a]. This mutation involves diverting routes to optional ones using a context-aware Q-learning algorithm. The algorithm adapts the mutation period and learning rate dynamically. This route mutation strategy has found application in defending against DDoS attacks in Vehicle Ad Hoc Network (VANET). For instance, Zhang proposed a hierarchical routing scheme based on route mutation by selecting a neighboring grid in the communication area within VANET [Zhang et al., 2022]. This mutation is executed using a multi-agent reinforcement learning approach based on the observation of nodes in the Markov process.

  Additionally, Zhang introduced an intelligent defense scheme against DDoS in the software-defined Internet of Vehicles in VANET [Zhang et al., 2023]. The scheme periodically mutates the network communication range and access capacity of roadside units based on deep reinforcement learning.

However, these solutions have certain limitations in defending against crossfire attacks:
- Defenses in VANET [Zhang et al., 2022, 2023] are designed based on ad hoc routes suitable for general networks.
- The solutions [Xu et al., 2021a; Zhang et al., 2022, 2023] do not detail how to identify links or hosts under attack.
- For all of them [Xu et al., 2021a; Zhang et al., 2022, 2023], finding the link-disjoint path or implementing route mutation is challenging for critical PRs in general networks.

Later, Steinberger developed a defense mechanism to alter interdomain paths and performed address hopping to set up a honeypot [Gupta et al., 2014]. However, these new paths may inadvertently include congested PRs.

- **Topology obfuscation**: Obfuscation techniques [Trassare et al., 2013; Achleitner et al., 2016] play a crucial role in providing deceptive paths to adversaries, preventing the exposure of Persistent Routes (PRs) during probing. Several notable approaches in this domain include:
  - **Meier's NetHide [Meier et al., 2018]**: This method modifies ICMP probing packets by introducing similar link delays to create a fake outward path.
  - **Ding's Linkbait [Ding et al., 2020]**: Linkbait reroutes the tracing flow, creating bait links to mislead potential adversaries.
  - **Hou's k-nearest neighbor analysis [Hou et al., 2020]**: This technique involves faking the topology of detected bots to confuse adversaries.
  - **Kim's EqualNet [Kim et al., 2022]**: Recognized for its effectiveness, EqualNet creates a virtual topology with nodes designed to equalize the tracing flow.

  However, some challenges persist. Approaches like [Meier et al., 2018; Kim et al., 2022] overlook the crucial detection of attacked links, a prerequisite for effective obfuscation. Additionally, techniques relying on ICMP probing flow detection [Hou et al., 2020; Ding et al., 2020] may face inaccuracies, especially with randomized flows. These approaches often neglect the need to change decoy addresses, leaving them vulnerable to persistent flooding.

ii) **Strategies for traffic throttling**: Traffic throttling aims to counter attacks through the following key methods:
- **Detecting flooding traffic**: Kang introduced the SPIFFY approach [Kang et al., 2016], which identifies bots by virtually increasing the bandwidth of Persistent Routes (PRs). However, detecting low-rate flooding consistently poses a significant challenge. Zheng later proposed RADAR, an SDN-based defense [Zheng et al., 2018], employing correlation analysis

on suspicious flows and changes in link utilization. Nevertheless, frequent adjustments in flooding rates by bots can undermine the accuracy of flow statistics sampling, rendering detection less effective. Both approaches face performance scalability bottlenecks in handling anomalies and lack divergence strategies for critical PRs.

- **Detecting ICMP traffic**: Sakuma suggested analyzing the concentration of traceroute packets near the target [Sakuma et al., 2017]. However, distinguishing sparse, long-term probing behaviors proves challenging. Wang later proposed LFADefender, a defender against link-flooding attacks [Wang et al., 2019]. It identifies anomalous traceroute packets using k-nearest neighboring estimation and then either blocks detected bots or diverges Persistent Routes (PRs). Yet, coping with randomized PR probing and critical PRs remains difficult.
- **Traffic diverging**: Belabed introduced an SDN-based defense mechanism that detects congested links by monitoring throughput increases. It then diverts the paths of decoys and the target under attack to new ones [Belabed et al., 2018]. However, detection struggles with identifying normal link loss events in high-throughput benign traffic. Ravi proposed a service-based hybrid SDN to locate congested links in crossfire attacks by computing the entropy of data flow throughput features [Ravi et al., 2020]. Unfortunately, this mitigation does not address bots with low-rate flooding flows, allowing the attack's flooding to persist.
- **Anomaly traffic detection based on programmable switches**: These approaches focus on harnessing switch abstraction to deliver a flexible defense. Zhang introduced Poseidon, providing switches with simple and modular tools to defend against Tbps-level attack traffic [Zhang et al., 2020]. Subsequently, Liu presented Jaqen, capable of detecting and mitigating volumetric DDoS attacks directly in switches at high line rates, handling TCP, UDP, and ICMP attacks [Liu et al., 2021]. Xing developed Ripple, offering a new abstraction to detect flow throughput through bloom filters entirely on the switch [Xing et al., 2021]. Their programmable APIs can be employed to support our proposed mechanism for achieving high performance.

Therefore, while these approaches offer robust switch-based defenses, they fall short in effectively countering the initial attack stage's stealthier PR probing. As a result, they may leave some PRs undefended and vulnerable to adversaries, as confirmed by our evaluation results in Chapter 11.3.6. It's essential to note that blocking the ICMP used by adversaries for discovering the maximum transmission unit for IPv6 [IETF, 2017] is challenging.

4.5.2 Uncharted Renaissance Defense

While significant strides have been made in defending against the initial attack stage, the renaissance attack stage has surprisingly received scant attention in research. Current defense solutions falter when it comes to anticipating and thwarting the adversary's speculation on vulnerable Persistent Routes (PRs) post-initial attack defense.

Firstly, the MTD approach, despite its merits, falls short as it often preserves unchanged routes involving PRs, failing to diverge critical PRs effectively [Venkatesan et al., 2016; Aydeger et al., 2016, 2021; Gupta et al., 2014]. Moreover, ingenious topology obfuscation methods [Meier et al., 2018; Ding et al., 2020; Kim et al., 2022; Hou et al., 2020] don't alter the underlying routes and host addresses, leaving the door open for potential reflooding of decoys.

Secondly, traffic throttling approaches struggle to detect randomized testing flooding at low rates [Kang et al., 2016; Zheng et al., 2018; Sakuma et al., 2017; Wang et al., 2019]. These shortcomings render vulnerable PRs susceptible to adversary speculation, especially when the PR is both reachable and poses a threat to the target. The adversary can gather information about PRs through long-term probing before congestion, making prevention a challenging endeavor. Surprisingly, defense strategies have largely overlooked this crucial aspect of the renaissance attack.

4.5.3 Unveiling Persistent Route Diversification Defense (PRDD) Solution

As Operational Technology (OT) networks draw nearer to Information Technology (IT) networks, the evolving landscape underscores the imperative to counteract the two-staged LFAs. This need is not merely a requirement; it stands as a linchpin for meeting the stringent QoS demands essential for the smooth orchestration of automation control within the intricate power grid.

In response to this imperative, this book unfurls the Persistent Route Diversification Defense (PRDD) [Zhou et al., 2022b] – a groundbreaking mechanism poised to render adversaries powerless in launching floods that could cascade into network congestion. The essence of PRDD lies in its innovative approach of diversifying victim routes and deploying decoys with the strategic creation of an ample array of network-layer proxies. This intricate move of diversification shields the routes and decoys, veiling them from the prying eyes of adversaries.

Extensive evaluations using NS-3 and Mininet unveil compelling results: (1) PRDD excels in mitigating a greater number of attacked PRs compared to existing solutions; and (2) its defense performance remains highly scalable in computation while maintaining an acceptable overhead.

PRDD emerges not just as a solution but as a scalable fortress, effectively mitigating a multitude of victim routes when compared with existing approaches. As we delve into the intricacies of PRDD, its role in fortifying network resilience becomes a beacon, guiding us toward a secure future in the realm of power grid automation control.

4.6 Summary

The five key research trends in creating highly available networks for smart power grids focus on critical challenges, including non-disjoint path aggregation, FRR, data transport protocols, redundant application data coding, and defenses against LFAs. These challenges are not just theoretical but have led to real-world incidents, significantly affecting grid operation and stability, with some even causing large-scale national or regional outages. For instance, disruptions due to LFAs and data transport failures have underlined the vulnerability of modern power grids to both technical and external threats.

Addressing these issues requires the ongoing development of robust solutions that ensure reliable data delivery, path stability, secure data transport, and complete application data receipt. Additionally, strong defenses against LFAs are essential to keeping resilience against relevant cyberattacks. Solutions that mitigate the risks posed by packet loss and excessive latency are crucial, particularly in the face of external conditions like network failures.

In Chapters 5–11, we will examine both established and advanced respective solutions, such as OPAM, TA-FRR, DRTP, and its related RSDD and HeuMPSG, ARED, as well as PRDD, that address these challenges and improve the resilience and reliability of networks within power grids.

Questions

4.1 In the context of power grid networks, what are the main challenges that the OPAM aims to address?

4.2 Provide an overview of the route model employed by TA-FRR to achieve both efficiency and robustness in network operation.

4.3 Explain the communication model utilized by DRTP and its role in fostering collaboration among power grid devices.

4.4 What challenges does the coding application data encounter, and how do these challenges impact the reliability and efficiency of data transmission in power grids?

4.5 Break down the rationale behind the two-stage approach observed in launching link-flooding attacks. What advantages or strategic benefits does this dual-stage process provide to attackers, especially in the context of power distribution networks?

4.6 Discuss the fundamental principles or strategies involved in constructing a resilient network infrastructure tailored to the unique demands of power grids. How can these principles be applied to mitigate challenges and enhance the reliability of communication networks in power grid scenarios?

Bibliography

3GPP TS 36.212. LTE; evolved universal terrestrial radio access (E-UTRA); multiplexing and channel coding (version 13.0.0 release 13), 2016.

Stefan Achleitner, Thomas La Porta, Patrick McDaniel, Shridatt Sugrim, Srikanth V. Krishnamurthy, and Ritu Chadha. Cyber deception: Virtual networks to defend insider reconnaissance. In *ACM CCS Workshop on Managing Insider Security Threats*, pages 57–68, 2016.

Rudolf Ahlswede. *Classical Work: Edge-Disjoint Branchings, Min-Max Theorems, and Shortest Connection Networks*, pages 413–433. Springer, 2019.

Rashid Amin, Martin Reisslein, and Nadir Shah. Hybrid SDN networks: A survey of existing approaches. *IEEE Communications Surveys & Tutorials*, 20(4):3259–3306, 2018.

A. Aydeger, N. Saputro, K. Akkaya, and M. Rahman. Mitigating crossfire attacks using SDN-based moving target defense. In *IEEE LCN*, pages 627–630, 2016.

Abdullah Aydeger, Mohammad Hossein Manshaei, Mohammad Ashiqur Rahman, and Kemal Akkaya. Strategic defense against stealthy link flooding attacks: A signaling game approach. *IEEE Transactions on Network Science and Engineering*, 8(1):751–764, 2021.

François Baccelli, Erol Gelenbe, and Brigitte Plateau. An end-to-end approach to the resequencing problem. *Journal of the ACM*, 31(3):474–485, 1984.

Gregor Bankhamer, Robert Elsässer, and Stefan Schmid. Local fast rerouting with low congestion: A randomized approach. *IEEE/ACM Transactions on Networking*, 30(6):2403–2418 2022.

Dallal Belabed, Mathieu Bouet, and Vania Conan. Centralized defense using smart routing against link-flooding attacks. In *Cyber Security in Networking Conference*, pages 1–8, 2018.

Aaron Bernstein. A nearly optimal algorithm for approximating replacement paths and k shortest simple paths in general graphs. In *ACM-SIAM SODA*, pages 742–755, 2010.

Zhen Cao, Degang Sun, Lingang Shi, Bihua Tang, and Gang Xie. Research on mechanism of reliable information transmission over unidirectional channel. *Computer Engineering and Applications*, 47(24):88–92, 2011.

Jiayin Chen, Si Yan, Qiang Ye, Wei Quan, Phu Thinh Do, Weihua Zhuang, Xuemin Sherman Shen, Xu Li, and Jaya Rao. An SDN-based transmission protocol with in-path packet caching and retransmission. In *IEEE ICC*, pages 1–6, 2019.

Jiayin Chen, Qiang Ye, Wei Quan, Si Yan, Phu Thinh Do, Weihua Zhuang, Xuemin Sherman Shen, Xu Li, and Jaya Rao. SDATP: An SDN-based adaptive transmission protocol for time-critical services. *IEEE Network*, 34(3):154–162, 2020.

Marco Chiesa, Andrei Gurtov, Aleksander Madry, Slobodan Mitrovic, Ilya Nikolaevskiy, Michael Shapira, and Scott Shenker. On the resiliency of randomized routing against multiple edge failures. In *ICALP*, 2016.

Marco Chiesa, Andrzej Kamisiński, Jacek Rak, Gábor Rétvári, and Stefan Schmid. A survey of fast-recovery mechanisms in packet-switched networks. *IEEE Communications Surveys & Tutorials*, 23(2):1253–1301, 2021.

Yong Cui, Lian Wang, Xin Wang, Hongyi Wang, and Yining Wang. FMTCP: A fountain code-based multipath transmission control protocol. *IEEE/ACM Transactions on Networking*, 23(2):465–478, 2015.

Xuyang Ding, Feng Xiao, Man Zhou, and Zhibo Wang. Active link obfuscation to thwart link-flooding attacks for Internet of Things. In *IEEE TrustCom*, pages 217–224, 2020.

Nils Dorsch, Fabian Kurtz, Hanno Georg, Christian Hägerling, and Christian Wietfeld. Software-defined networking for smart grid communications: Applications, challenges and advantages. In *IEEE SmartGridComm*, pages 422–427, 2014.

Theodore Elhourani, Abishek Gopalan, Srinivasan Ramasubramanian, Theodore Elhourani, Abishek Gopalan, and Srinivasan Ramasubramanian. IP fast rerouting for multi-link failures. *IEEE/ACM Transactions Networking*, 24(5):3014–3025, 2016.

Nick Feamster, Jennifer Rexford, and Ellen Zegura. The road to SDN: An intellectual history of programmable networks. *ACM SIGCOMM Computer Communications Review*, 44(2):87–98, 2014.

Klaus-Tycho Foerster, Yvonne-Anne Pignolet, Stefan Schmid, and Gilles Tredan. Local fast failover routing with low stretch. *ACM SIGCOMM Computer Communications Review*, 48(1):35–41, 2018.

Klaus-Tycho Foerster, Andrzej Kamisinski, Yvonne-Anne Pignolet, Stefan Schmid, and Gilles Trédan. Bonsai: Efficient fast failover routing using small arborescences. In *IEEE/IFIP DSN*, pages 276–288, 2019a.

Klaus-Tycho Foerster, Yvonne-Anne Pignolet, Stefan Schmid, and Gilles Tredan. CASA: Congestion and stretch aware static fast rerouting. In *IEEE INFOCOM*, pages 469–477, 2019b.

Klaus-Tycho Foerster, Juho Hirvonen, Yvonne-Anne Pignolet, Stefan Schmid, and Gilles Tredan. On the price of locality in static fast rerouting. *arXiv preprint arXiv:2204.03413*, 2022a.

Klaus-Tycho Foerster, Andrzej Kamisinski, Yvonne-Anne Pignolet, Stefan Schmid, and Gilles Tredan. Improved fast rerouting using postprocessing. *IEEE Transactions on Dependable and Secure Computing*, 19(1):537–550, 2022b.

Asaf Frieder and Liam Roditty. An experimental study on approximating k shortest simple paths. *Journal of Experimental Algorithmics*, 19(1):1–5, 2015.

Ivan Ganchev, Jacek Rak, Tibor Cinkler, and Máirtín O'Droma. *Taxonomy of Schemes for Resilient Routing*, pages 455–482. Springer, 2020.

J. J. Garcia-Luna-Aceves and Abdulazaz Ali Albalawi. Connection-free reliable and efficient transport services in the IP Internet. In *IEEE CNSM*, pages 1–7, 2020.

Erol Gelenbe. Steps toward self-aware networks. *Communications of the ACM*, 52(7):66–75, 2009.

Erol Gelenbe. *Self-Aware Networks: The Cognitive Packet Network and Its Performance*, pages 659–668. Springer International Publishing, 2017.

Arpit Gupta, Laurent Vanbever, Muhammad Shahbaz, Sean P. Donovan, Brandon Schlinker, Nick Feamster, Jennifer Rexford, Scott Shenker, Russ Clark, and Ethan Katz-Bassett. SDX: A software defined internet exchange. *ACM SIGCOMM Computer Communication Review*, 44(4):579–580, 2014.

John Hershberger and Subhash Suri. Vickrey prices and shortest paths: What is an edge worth? In *IEEE FOCS*, pages 252–259, 2001.

John Hershberger, Matthew Maxel, and Subhash Suri. Finding the k shortest simple paths: A new algorithm and its implementation. *ACM Transactions on Algorithms*, 3(4):45–es, 2007.

Tao Hou, Zhe Qu, Tao Wang, Zhuo Lu, and Yao Liu. ProTO: Proactive topology obfuscation against adversarial network topology inference. In *IEEE INFOCOM*, pages 1598–1607, 2020.

IEC. Communication networks and systems for power utility automation - Part 5: Communication requirements for functions and devices, 2013.

IEC. Industrial communication networks - high availability automation networks - Part 1: General concepts and calculation methods, 2016.

IEC 62439. High-availability automation networks, 2008.

IEEE. IEEE standard bus network dataset, 2025. URL http://www.ee.washington.edu/research/pstca/.

IETF. IETF RFC 8200: Internet protocol, version 6 (IPv6) specification. Standard, IETF, 2017.

IETF RFC 6330. RaptorQ forward error correction scheme for object delivery, 2011.

Farabi Iqbal and Fernando A. Kuipers. Disjoint paths in networks. In *Wiley Encyclopedia of Electrical and Electronics Engineering*, pages 1–11. John Wiley & Sons, Inc., 2015.

Babangida Isyaku, Kamalrulnizam bin Abu Bakar, Fuad A. Ghaleb, and Mohd Soperi Mohd Zahid. Path selection with critical switch-aware for software defined networking. In *IEEE ISWTA*, pages 22–26, 2021.

Chuan Jiang, Sanjay Rao, and Mohit Tawarmalani. PCF: Provably resilient flexible routing. In *ACM SIGCOMM*, pages 139–153, 2020.

Andrzej Kamisiński. Evolution of IP fast-reroute strategies. In *IEEE Workshop on RNDM*, pages 1–6, 2018.

Min Suk Kang, Virgil D. Gligor, and Vyas Sekar. SPIFFY: Inducing cost-detectability tradeoffs for persistent link-flooding attacks. In *USENIX NDSS*, 2016.

Jinwoo Kim, Eduard Marin, Mauro Conti, and Seungwon Shin. EqualNet: A secure and practical defense for long-term network topology obfuscation. In *USENIX NDSS*, 2022.

Oh Chan Kwon, Yunmin Go, and Hwangjun Song. An energy-efficient multimedia streaming transport protocol over heterogeneous wireless networks. *IEEE Transactions on Vehicular Technology*, 65(8):6518–6531, 2016.

Ziyong Li, Yuxiang Hu, Jiangxing Wu, and Jie Lu. P4Resilience: Scalable resilience for multi-failure recovery in SDN with programmable data plane. *Elsevier Computer Networks*, 208:108896, 2022.

Junda Liu, Aurojit Panda, Ankit Singla, Brighten Godfrey, Michael Schapira, and Scott Shenker. Ensuring connectivity via data plane mechanisms. In *USENIX NSDI*, pages 113–126, 2013.

Bin Liu, Huichen Dai, Wenquan Xu, Tong Yun, and Ji Miao. Scalable hardware content router: Architecture, modeling and performance. In *IEEE/ACM IWQOS*, pages 1–7, 2021.

David J. C. MacKay. Fountain codes. *IEEE Proceedings-Communications*, 152(6):1062–1068, 2005.

Spyridon Mastorakis, Alexander Afanasyev, and Lixia Zhang. On the evolution of ndnSIM: An open-source simulator for NDN experimentation. *ACM SIGCOMM Computer Communications Review*, 47(3):19–33, 2017.

Roland Meier, Petar Tsankov, Vincent Lenders, Laurent Vanbever, and Martin Vechev. NetHide: Secure and practical network topology obfuscation. In *USENIX Security*, pages 693–709, 2018.

Hideyoshi Miura, Kouji Hirata, and Takuji Tachibana. P4-based design of fast failure recovery for software-defined networks. *Elsevier Computer Networks*, 216:109274, 2022.

Open Networking Foundation. OpenFlow specification v1.3, 2012.

Miroslav Popovic, Maaz Mohiuddin, Dan Cristian Tomozei, and Jean Yves Le Boudec. iPRP—The parallel redundancy protocol for IP networks: Protocol design and operation. *IEEE Transactions on Industrial Informatics*, 12(5): 1842–1854, 2016.

Qualcomm. RaptorQ™technical overview. Technical report, Qualcomm, 2010.

Eman Ramadan, Hesham Mekky, Braulio Dumba, and Zhi-Li Zhang. Adaptive resilient routing via preorders in SDN. In *ACM Workshop on DCC*, 2016.

Nagarathna Ravi, S. Mercy Shalinie, and D. Danyson Jose Theres. BALANCE: Link flooding attack detection and mitigation via hybrid-SDN. *IEEE Transactions on Network and Service Management*, 17(3):1715–1729, 2020.

Gelli Ravikumar, Dan Ameme, Satyajayant Misra, Sukumar Brahma, and Reza Tourani. iCASM: An information-centric network architecture for wide area measurement systems. *IEEE Transactions on Smart Grid*, 11(4):3418–3427, 2020.

Kei Sakuma, Hiromu Asahina, Shuichiro Haruta, and Iwao Sasase. Traceroute-based target link flooding attack detection scheme by analyzing hop count to the destination. In *APCC*, pages 1–6, 2017.

Oliver Schweiger, Klaus-Tycho Foerster, and Stefan Schmid. Improving the resilience of fast failover routing: TREE (tree routing to extend edge disjoint paths). In *ACM ANCS*, pages 1–7, 2021.

Amin Shokrollahi. Raptor codes. *IEEE Transactions on Information Theory*, 52(6):2551–2567, 2006.

Apoorv Shukla and Klaus-Tycho Foerster. Shortcutting fast failover routes in the data plane. In *ACM ANCS*, pages 15–22, 2021.

Kai Su, Francesco Bronzino, K. K. Ramakrishnan, and Dipankar Raychaudhuri. MFTP: A clean-slate transport protocol for the information centric MobilityFirst network. In *ACM ICN*, pages 127–136, 2015.

Hai-Khun Tan and Tung-Wei Kuo. Optimistic fast rerouting. In *IEEE ICC*, pages 1692–1697, 2022.

Stelios Timotheou. The random neural network: A survey. *Computer Journal*, 53(3):251–267, 2010.

Samuel T. Trassare, Robert Beverly, and David Alderson. A technique for network topology deception. In *IEEE MILCOM*, pages 1795–1800, 2013.

Sridhar Venkatesan, Massimiliano Albanese, Kareem Amin, Sushil Jajodia, and Mason Wright. A moving target defense approach to mitigate DDoS attacks against proxy-based architectures. In *IEEE CNS*, pages 198–206, 2016.

Lan Wang and Erol Gelenbe. Real-time traffic over the cognitive packet network. In *Computer Networks*, pages 3–21, 2016.

Bin Wang and Jennifer C. Hou. Multicast routing and its QoS extension: Problems, algorithms, and protocols. *IEEE Network*, 14(1):22–36, 2000.

Zhaoxu Wang, Hongbin Luo, Huachun Zhou, and Jiawei Li. R2T: A rapid and reliable hop-by-hop transport mechanism for information-centric networking. *IEEE Access*, 6:15311–15325, 2018.

Juan Wang, Ru Wen, Jiangqi Li, Fei Yan, Bo Zhao, and Fajiang Yu. Detecting and mitigating target link-flooding attacks using SDN. *IEEE Transactions on Dependable and Secure Computing*, 16(6):944–956, 2019.

Jiyan Wu, Chau Yuen, Bo Cheng, Ming Wang, and Junliang Chen. Streaming high-quality mobile video with multipath TCP in heterogeneous wireless networks. *IEEE Transactions on Mobile Computing*, 15(9):2345–2361, 2016.

Wenfeng Xia, Yonggang Wen, Chuan Heng Foh, Dusit Niyato, and Haiyong Xie. A survey on software-defined networking. *IEEE Communications Surveys & Tutorials*, 17(1):27–51, 2014.

Jiarong Xing, Wenqing Wu, and Ang Chen. Ripple: A programmable, decentralized link-flooding defense against adaptive adversaries. In *USENIX Security Symposium*, 2021.

Changqiao Xu, Tao Zhang, Xiaohui Kuang, Zan Zhou, and Shui Yu. Context-aware adaptive route mutation scheme: A reinforcement learning approach. *IEEE Internet of Things Journal*, 8(17):13528–13541, 2021a.

Jiarui Xu, Sihao Xie, and Jin Zhao. P4Neighbor: Efficient link failure recovery with programmable switches. *IEEE Transactions on Network and Service Management*, 18(1):388–401, 2021b.

Jin Y. Yen. Finding the *k* shortest loopless paths in a network. *INFORMS Management Science*, 17(11):712–716, 1971.

Minlan Yu, Lavanya Jose, and Rui Miao. Software defined traffic measurement with OpenSketch. In *USENIX NSDI*, pages 29–42, 2013.

Kuili Zhang, Hansong Su, Gaohua Liu, and Peng Wang. New hybrid automatic retransmission scheme in LTE. *Computer Engineering and Applications*, 49(19):92–95, 2013.

Lixia Zhang, Alexander Afanasyev, Jeffrey Burke, Van Jacobson, Christos Papadopoulos, K. C. Claffy, Lan Wang, Patrick Crowley, and Beichuan Zhang. Named data networking. *ACM SIGCOMM Computer Communication Review*, 44(3):66–73, 2014.

Menghao Zhang, Guanyu Li, Shicheng Wang, Chang Liu, Ang Chen, Hongxin Hu, Guofei Gu, Qi Li, Mingwei Xu, and Jianping Wu. Poseidon: Mitigating volumetric DDoS attacks with programmable switches. In *USENIX NDSS*, 2020.

Tao Zhang, Changqiao Xu, Bingchi Zhang, Jiahao Shen, Xiaohui Kuang, and Luigi Alfredo Grieco. Toward attack-resistant route mutation for VANETs: An online and adaptive multiagent reinforcement learning approach. *IEEE Transactions on Intelligent Transportation Systems*, 23(12):23254–23267, 2022.

Tao Zhang, Changqiao Xu, Ping Zou, Haijiang Tian, Xiaohui Kuang, Shujie Yang, Lujie Zhong, and Dusit Niyato. How to mitigate DDoS intelligently in SD-IoV: A moving target defense approach. *IEEE Transactions on Industrial Informatics*, 19(1):1097–1106, 2023.

Jing Zheng, Qi Li, Guofei Gu, Jiahao Cao, David K. Y. Yau, and Jianping Wu. Realtime DDoS defense using COTS SDN switches via adaptive correlation analysis. *IEEE Transactions on Information Forensics and Security*, 13(7):1838–1853, 2018.

Jiaqi Zheng, Hong Xu, Xiaojun Zhu, Guihai Chen, and Yanhui Geng. Sentinel: Failure recovery in centralized traffic engineering. *IEEE/ACM Transactions on Networking*, 27(5):1859–1872, 2019.

Boyang Zhou. Adaptive RaptorQ coding to improve reliability of LTE communication in distribution power grid. *Computer Engineering and Applications*, 55(3):76–82, 2019.

Boyang Zhou, Qiang Yang, Yansong Wang, and Chunming Wu. Reliable communication in transmission grids based on nondisjoint path aggregation using software-defined networking. *IEEE Transactions on Industrial Informatics*, 15(2):842–855, 2019.

Boyang Zhou, Chunming Wu, Qiang Yang, and Xiang Chen. DRTP: A disruption resilient hop-by-hop transport protocol for synchrophasors measurement in electric transmission grids. *IEEE Access*, 10:133898–133914, 2022a.

Boyang Zhou, Chunming Wu, Qiang Yang, Xiang Chen, and Dong Zhang. A persistent route diversification mechanism for defending against stealthy crossfire attack. *Computer Engineering and Applications*, 2022(1):2566681, 1–20, 2022b.

Boyang Zhou, Chunming Wu, Yang Qiang, Qian Yaguan, and Nie Yinghui. Resilient sensor data dissemination to mitigate link faults in IoT networks with long-haul optical wires for power transmission grids. *IEEE Internet of Things Journal*, 11(9):15919–15939, 2023a.

Boyang Zhou, Chunming Wu, Qiang Yang, Zhen Zhu, Yinghui Nie, and Wenjie Yu. Traffic-aware fast reroute mechanism exploiting disjoint subpaths for named data networks. In *IEEE ICC*, pages 4013–4019, 2023b.

Boyang Zhou, Chunming Wu, Qiang Yang, Zhen Zhu, Yinghui Nie, and Wenjie Yu. Traffic-aware fast reroute mechanism exploiting disjoint subpaths for named data networks. In *IEEE ICC*, pages 4087–4093, 2023c.

Boyang Zhou, Tong Ye, Yinghui Nie, Wenjie Yu, and Chunming Wu. Generating routes less fragile to improve disruption-resilient transport protocol for phasor data in optical IoT networks of power transmission grids. In *IEEE ICPADS*, 2023d.

Boyang Zhou, Zhen Zhu, Yinghui Nie, Xiang Chen, Bingjing Yan, Qiang Yang, and Chunming Wu. Fast proactive re-route with detours on joint paths in named-data networks for communications of smart electric power transmission grids. In *COMSNETS*, pages 640–648, 2023e.

5

Empowering Network Robustness with Non-disjoint Path Aggregation

In electrical transmission grids, the redundant communication paths non-disjointly overlapping at links can be established between certain substations and the control center to guarantee reliable packet delivery under link failures. However, the generation of non-disjoint paths with multiplicative and concave constraints is with the NP-Complete complexity, and the failovers can lead to out-of-order packets.

In the pursuit of tackling the Non-Disjoint Path Aggregation (NDPA) problem, a groundbreaking mechanism takes center stage – the *OpenFlow-based non-disjoint path aggregation mechanism (OpenFlow-based Non-Disjoint Path Aggregation Mechanism (OPAM))*. This mechanism, rooted in the OpenFlow (OF)/Software-Defined Networking (SDN) paradigm [Farhady et al., 2015], introduces two integral components: the Multipath Subgraph Generation Algorithm (MPSGA) and the Packet Aggregation And Forwarding Module (PAFM), operating at the controller and switches, respectively.

When a data flow encounters a mismatch at an OF switch [Open Networking Foundation, 2012], the MPSGA springs into action. It dynamically generates a set of non-disjoint loop-free paths, minimizing the number of links while meeting Success Rate of Receiving a Packet (SRRP) and end-to-end delay constraints. The time complexity of this algorithm, in the worst case, is $O(\tilde{P} \cdot \log|R| \cdot |L|)$. These paths are then installed at the corresponding OF switches, each with a fixed expiration timeout [Open Networking Foundation, 2012]. Remarkably, the MPSGA only necessitates execution once during the entire flow's lifecycle, unless the paths in the switches expire. Concurrently, the PAFM, operating at the switches, receives the data flow from all corresponding ingress links, removes duplicate packets, performs packet reordering, and multicasts the organized packets to the egress links. The forwarding rules are stored in the switch's Static Random-Access Memory (SRAM), enabling efficient maintenance.

Communication Networks in Smart Power Grids, First Edition. Boyang Zhou.

Companion Website: www.wiley.com/go/SmartPowerGrids1e

Leveraging the OF/SDN paradigm, the OPAM demonstrates notable improvements in both Packet Out-of-Order Rate (POR) and SRRP. Consequently, it contributes to the reduction of Packet Inter-Arrival Time (PIAT) and Half Round-Trip Time (HRRT), while simultaneously decreasing memory usage for packet reordering at the switches.

To validate the efficacy of the proposed OPAM solution, comprehensive evaluations are conducted utilizing the Beacon controller [Erickson, 2013] and the OF switch simulator in NS-3 [Riley and Henderson, 2010]. The evaluation spans test networks of varying scales, including IEEE (14–300 buses), Texas (1725 buses), Pan European Grid Advanced Simulation and State Estimation (PEGASE) (1354–13,659 buses), and French (1888–6515 buses). The numerical results affirm the solution's effectiveness in enhancing the reliability performance of end-to-end Transmission Control Protocol (TCP) transmissions, particularly under link failures.

In the control plane, the MPSGA exhibits a substantial increase of 32.549%, 32.549%, and 45.97% in the Qualified Source-Destination Pairs (QSDP) compared to non-disjoint path algorithmic solutions proposed by Hershberger et al. [2007], Yen [1971], and Bernstein [2010], respectively, all within the IEEE 30-bus network. Furthermore, the execution time of MPSGA is drastically reduced by 81.51% compared to Hershberger in the Texas network. Relative to current disjoint algorithms like Maximally Disjoint Shortest & Widest Paths (MADSWIP) and Optimal Disjoint Paths Generation (ODPG), the MPSGA exhibits improvements of 30.59% and 23.91%, respectively. Meanwhile, the execution time results for the PEGASE and French networks align with the time complexity analysis of MPSGA for expected, worst, and best cases.

In the data plane, the packet reordering demonstrates substantial benefits, reducing PIAT by 58.63% and increasing TCP throughput by 121.62% compared to cases without packet reordering. The OPAM solution, as evidenced by these evaluations, presents a compelling stride toward advancing reliability in modern networking infrastructures.

An exemplary case with a strong link to SDN is the well-established Cognitive Packet Network (CPN), employing a distributed routing architecture [Gelenbe, 2017; Gelenbe, 2009; Wang and Gelenbe, 2016]. To further gauge OPAM's effectiveness, a comparative evaluation with CPN [Wang and Gelenbe, 2016] is conducted in a network scenario featuring a 2% link loss rate. The results underscore the significance of OPAM in mitigating routing blackholes caused by network failovers within CPN, leading to a 27.388 times higher HRRT and necessitating 3.267 times more memory usage. This stark comparison highlights the practicality of the OF/SDN architecture as a pivotal means for enhancing SRRP and HRRT in the face of network failures.

Expanding the horizon, OPAM seamlessly integrates with existing solutions, offering compatibility and synergy with various strategies:

- Employing path protection in a hot standby configuration on multiple disjoint paths, incorporating a primary set of forwarding paths and backup paths meeting bandwidth and reliability constraints [Chakrabarti and Manimaran, 2005]. This allows the non-disjoint loop-free paths to contribute to increased QSDP.
- Leveraging Raptor code, which decodes source packets with a success probability of $1/k^{o+\log_2 k}$ in scenarios with a packet loss rate below 50% [Shokrollahi, 2006]. This decoding capability is harnessed by Multi-Path TCP for packet delivery [Cui et al., 2012]. Here, non-disjoint paths play a crucial role in further enhancing end-to-end reliability.
- Integrating with iPRP [Popovic et al., 2016], a solution that transmits data packets along multiple Virtual Local Area Networks (VLANs). In this context, the single routing sub-path inside each VLAN can be replaced by the corresponding non-disjoint paths, contributing to an increase in SRRP.

These collaborative approaches underscore the versatility of OPAM, showcasing its ability to complement and enhance existing solutions across diverse networking scenarios. As networking landscapes continue to evolve, the adaptability of OPAM positions it as a key player in advancing reliability and performance in modern network infrastructures.

5.1 Problem Formulation

In this section, we delve into the intricacies of the OF/SDN communication model, breaking it down into two fundamental sub-problems of NDPA: the non-disjoint path generation and the packet reordering. Note that all mathematical symbols used throughout this section are listed in Table 5.1.

5.1.1 Reliability-Constrained Non-disjoint Path Generation

The non-disjoint path generation within the OF/SDN communication model is precisely formulated in the following manner. When switch r_i initiates the transmission of data to the controller c, the selection of G_P is orchestrated, taking into account the $\tilde{\gamma}$ and $\tilde{\tau}$ parameters requested by the current flow v.

Table 5.1 Mathematical symbols.

Symbol	Description
$G = (R, L)$	A directed graph for communication topology
R	A set of OF switches, $R = \{r_1, r_2, \dots, r_n, c\}$, where c is the control center
r_s	$r_s \in R$ exchanging messages with c
$\omega(r_i)$	A neighbored switch to r_i
L	A set of physical links, $L = \{l_1, l_2, \dots, l_m\}$
$l_{i,j}$	A physical link between r_i and r_j
P	A set of paths, $P = \{p_1, p_2, \dots, p_k\}$
$P(r_i, r_j)$	All sub-paths between r_i and r_j in P
G_P	A subgraph of $G = (R, L)$, $G_p = (R_p, L_p)$
$\gamma(r_s, G_P)$	Reliability ratio of G_P between r_s and c
$\eta(G, r_s, c)$	All possible paths from n_s to c in G
$\tilde{\gamma}$	User-defined min reliability ratio for r_s and c
$\tilde{\tau}$	User-defined max delay for r_s and c
$\tilde{P}$	User-defined max number of paths for P
$\phi(r_s, G_P)$	Enumerating all egress neighbors of r_s in G_P
$\psi(r_i, R_P)$	Enumerating all ingress neighbors of r_i in R_P
$\varepsilon(l_{i,j})$	Probability of failure events on $l_{i,j}$
$\tau(p_i)$	Cumulative delay of p_i
$\tau_p(l_{i,j})$	Packet transmission delay on $l_{i,j}$
$\tau_{\max}(G_P, r)$	Maximum delay of G_P between r and c
$B(l_{i,j})$	Link bandwidth of $l_{i,j}$
υ_{r_i}	A data flow sent from r_i, $\upsilon = (d_1, d_2, \dots, d_w)$, where d_j is the jth data packet
λ	Packet throughput in packets per second
H	Memory to record packet IDs
Q	Memory to reorder packets
$\tilde{H}$	Maximum memory limit for H
$\tilde{C}$	Maximum computation limit for C

Formally, the path selection, denoted as $p_i \in P$, relies on two key cost metrics:

- $\varepsilon(l_{i_u} \in p_i)$: This metric comes into play when data transmission encounters failures induced by link and node events, represented by $\varepsilon(l_{i_u})$. Here, i_u signifies the uth link in the ith path of L.

- $\tau(p_i) = \sum_{\forall l_{i_u} \in p_i} \tau_p(l_{i_u})$: This metric quantifies the cumulative transmission time along the selected path p_i, considering the individual transmission times $\tau_p(l_{i_u})$ for each link l_{i_u}.

Thus, the reliability metric $\gamma(r_s, G_P)$ is expressed in Eq. (5.1), achieved through the recursive calculation of one of its branch nodes within G_P. Here, $\tau_{\max}(G_P) = \max_{\forall p_i \in P} \tau(p_i)$ signifies the maximum cumulative transmission time across all paths in P. This meticulous formulation sets the groundwork for a comprehensive understanding of the intricacies associated with non-disjoint path generation in the OF/SDN communication model.

$$\gamma(r_s, G_P) = \begin{cases} 1 - \prod_{\forall \omega(r_s) \in \phi(r_s, G_P)} \varepsilon(l_{r_s, \omega(r_s)}) \cdot (1 - \gamma(\omega(r_s), P)) & |\phi(r_s, G_P)| \geq 2 \\ (1 - \varepsilon(l_{r_s, c})) \cdot \gamma(\phi(r_s, G_P), G_P) & |\phi(r_s, G_P)| = 1 \wedge r_s \neq c \\ 1 & r_s = c \\ 0 & r_s \notin R_P \end{cases} \tag{5.1}$$

$$\begin{aligned} &\operatorname{argmin}_{P \subseteq \eta(G, r_s, c)} |L_P|, \text{ subject to:} \\ &\gamma(r_s, G_P) \geq \tilde{\gamma} \wedge \tau_{\max}(G_P, r_s) \leq \tilde{\tau}, \\ &\text{given by } \varepsilon(l) \text{ and } \tau(p), \forall l, p \in G \end{aligned} \tag{5.2}$$

Consequently, the non-disjoint path generation unfolds as a well-defined optimization problem, succinctly articulated in Eq. (5.2). The objective is to identify a set of paths P that not only satisfies the specified $\tilde{\gamma}$ and $\tilde{\tau}$ but also minimizes the number of links involved, denoted as $|L_P|$.

This optimization problem can be characterized by two essential constraints:

- **Multiplicative constraint**: Ensures that $\gamma(r_s, G_P)$ adheres to the specified threshold $\tilde{\gamma}$.
- **Concave constraint**: Imposes a limit on $\tau_{\max} \leq \tilde{\tau}$. It's noteworthy that this constraint's concave nature contributes to the problem's NP-Complete complexity, as validated by previous research [Wang and Hou, 2000].

Navigating this optimization landscape demands a delicate balance between reliability, transmission time, and the efficient utilization of links. The inherent complexity of this problem underscores the intricacies involved in crafting a solution that optimally balances these considerations.

5.1.2 Packet Duplication, Loss, and Out of Order

A complementary sub-problem within NDPA involves packet aggregation for G_P, specifically at switches $r_i \in R_P$, satisfying the condition $|\psi(r_i, R_P)| \geq 1$.

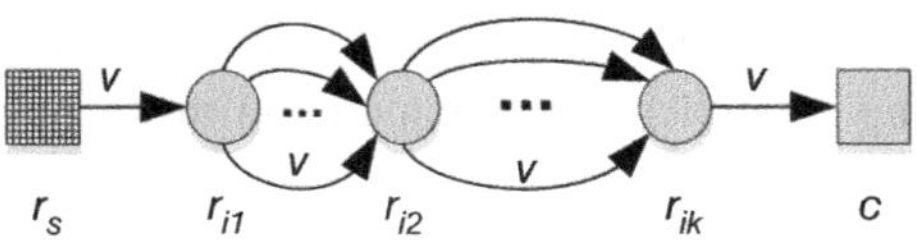

Figure 5.1 Model of packet aggregating points.

The intricate process unfolds, as depicted in Figure 5.1, with aggregating points denoted as $r_{i_1}, \ldots, r_{i_k}$ within G_P.

In this illustration, dashed squares, solid squares, and circles symbolize the switch r_s transmitting flow v, the controller c receiving flow v, and switches r_{i_j} aggregating and forwarding duplications of v, respectively. For simplicity, assumptions are made: the flow v is ordered, paths $P(r_s, r_{i_1})$ and $P(r_k, c)$ are free from losses, and packet forwarding on all ingress links in $\psi(r_{i_j}, G_P)$ operates independently.

Considering Figure 5.1, the reliability of the reverse path, where c transmits packets to r_s, mirrors that of the path where r_s transmits packets to c. Therefore, our analysis will focus solely on the latter path, unraveling the intricacies of packet aggregation within the context of NDPA.

To address the packet loss and duplications, an aggregation process is carried out at r_{i_j} as follows: (1) receiving all the duplications of v sent from $r'_{i_j} \in \psi(r_{i_j}, G_P)$ and temporarily storing them as $\tilde{v} = \cup v_{r'_{i_j}}$; (2) removing the duplicated $d_j \in \tilde{v}$ by filtering d_j out with its ID value forwarded once, to generate the output as $\rho(\tilde{v})$; and (3) multicasting $\rho(\tilde{v})$ to all the neighbors in $\phi(r_{i_j}, G_P)$. Hence, the error rate $\rho(\tilde{v})$ can be obtained as $\varepsilon(\rho(\tilde{v})) = \prod_{l_{i,j} \in \psi(r_{i_j}, G_P)} \varepsilon(l_{i,j})$.

In order to effectively tackle the challenges posed by packet loss and duplications within the context of NDPA, a meticulous aggregation process is enacted at r_{i_j}. The process unfolds in the following stages:

1) **Receiving duplicated flows**: All duplications of flow v sent from $r'_{i_j} \in \psi(r_{i_j}, G_P)$ are received and temporarily stored as $\tilde{v} = \cup v_{r'_{i_j}}$.
2) **Removing duplication**: Duplicated elements $d_j \in \tilde{v}$ are identified and filtered out by utilizing their Identification (ID) values, resulting in the output $\rho(\tilde{v})$.
3) **Multicasting**: The deduplicated output $\rho(\tilde{v})$ is multicast to all neighbors in $\phi(r_{i_j}, G_P)$.

Therefore, the error rate $\rho(\tilde{v})$ is calculated as $\varepsilon(\rho(\tilde{v})) = \prod_{l_{i,j} \in \psi(r_{i_j}, G_P)} \varepsilon(l_{i,j})$. This meticulous aggregation process stands as a strategic measure, effectively mitigating the adverse effects of errors stemming from packet loss and duplications within the overarching NDPA framework.

The intricacy of step 2 lies in the necessity for a memory H_t at time tick t to track recently sent packets in $\tilde{v}_{r_i}$. This stems from the computational complexity introduced by checking duplicated elements $d_j \in \tilde{v}$ against H_t, incurring additional demands on both time and space.

The challenge at hand is to identify an optimal memory policy that maximizes $\text{Prob}\{\tilde{v}(t) \in H_t\}$ while adhering to constraints such as $\varepsilon(\rho(\tilde{v}))$, the memory limit of $\tilde{H}$, and the computation limit of $\tilde{C}$ (refer to Eq. (5.3)). Here, Prob denotes the success probability for effectively filtering out duplications.

Navigating this optimization landscape requires a delicate balance between memory utilization, computational efficiency, and adherence to error rate constraints. Crafting an optimal memory policy is pivotal to the seamless execution of the aggregation process within the constraints posed by the NDPA framework.

$$\begin{aligned} &\text{argmax}_{H_t \subset \tilde{v}} \ \text{Prob}\{\tilde{v}(t) \in M_t\}, \text{subject to:} \\ &\varepsilon(\rho(\tilde{v})) \leq \tilde{\varepsilon} \wedge |H| \leq \tilde{H} \wedge |C| \leq \tilde{C} \end{aligned} \tag{5.3}$$

A potential challenge arises when v_{r_k} undergoes out-of-order sequencing before reaching the controller c. This can occur due to the loss of its associated duplication d_i sent from the ingress ports of $\omega(r_{i_k})$ and subsequent arrivals of duplicated d_i from other ports. Consequently, v_{r_k} takes the form $(d_{j_1}, d_{j_2}, \ldots, d_{j_k})$, where $j_1, \ldots, j_k$ represents the order of packets.

Notably, link failures consistently result in an out-of-order packet d_{j_p} at its current position u in v_{r_k}, characterized by $|j_p - \min{}_j(v_{r_k}) - u| > 2$. This out-of-order scenario leads to frequent TCP ACKs and retransmissions, as the TCP congestion window is significantly impacted. The number of packets to wait before sending ACKs is conventionally set to 2, contributing to a decrease in TCP congestion window and a consequential deterioration in TCP reliability and throughput [Leung et al., 2007].

An effective strategy to address this challenge involves reordering v_{r_k} before reaching c. This proactive reordering, as opposed to relying on TCP window enlargement at the receiver, not only conserves memory in terminal devices but also reduces the PIAT. This strategic approach contributes to the enhancement of overall system performance.

The reordering process applied to v_{r_k} necessitates an additional memory allocation denoted as Q. This memory must be not only sufficiently fast but also small enough to accommodate the potential size of v_{r_k}, considering the limited availability of SRAM resources in the switch.

Assuming the PIAT of all flows follows a Poisson distribution, where packets are independently sent with a fixed interval adhering to the on/off model in an IEC 61850 Manufacturing Message Specification (MMS) network. With a confident range ϵ for the subsequent reordering, and given that $\tau(p_i \in G_P)$ follows an Erlang distribution, the challenge lies in determining the optimal size of $|Q| = |v_{r_k}|$ to meet the conditions outlined in Eq. (5.4). This optimization task is contingent upon the measured values of $\tau_p(d_{j_l})$ and the arrival rate λ of the current flow at r_i. Striking the right balance in memory allocation is crucial to successfully address the out-of-order packet scenario while respecting the constraints imposed by the NDPA framework.

$$\epsilon = \text{Prob}_{\text{Erlang}[\text{hops}(p_i),\lambda]} \left\{ \max_{d_{j_l} \in v_{r_k}} \tau_p(d_{j_l}) \cdot \lambda \leq |Q| \right\} \tag{5.4}$$

5.2 Non-disjoint Path Aggregation

This section unfolds the design intricacies of the OPAM, elucidating its structure from two key perspectives: MPSGA alongside a comprehensive complexity analysis and PAFM.

5.2.1 Overview

In the context of the electric transmission grid, the proposed OPAM is crafted to enhance communication reliability by addressing both SRRP and PIAT concerns, especially under link failure scenarios. Illustrated in Figure 5.2, OPAM comprises two integral components: the MPSGA running at the controller, serving as an OF service for rule generation, and the PAFM embedded within the switches, responsible for executing packet aggregation.

The MPSGA exercises control over PAFM using the OF protocol 1.3 [Open Networking Foundation, 2012]. The coordination unfolds in the following steps:

1) **Packet arrival handling**: When a packet from a new ingress flow arrives at a switch, the switch checks the packet header against its flow table. If there's no match, the switch forwards the packet to the MPSGA via the PACKET_IN message in OF.
2) **Rule generation**: Upon receiving the packet, the MPSGA generates non-disjoint forwarding paths, creates table rules, and dispatches them to the corresponding PAFM within the switch via the PACKET_OUT message.
3) **Rule implementation**: Upon receiving the message, the switch establishes rules in both the flow and group tables and initiates the packet aggregation process.

It's important to note that the southbound reliability between the OF controller and its affiliated OF switches is beyond the chapter's scope, as it

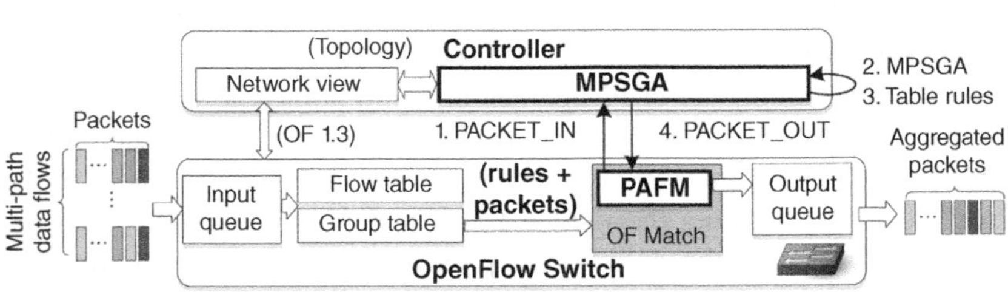

Figure 5.2 OPAM architecture.

can be addressed using controller replacement methods detailed in Tanha et al. [2017]. Detailed explanations of MPSGA and PAFM are presented in Sections 5.2.2–5.2.4.

5.2.2 Multi-path Subgraph Generation Algorithm

The non-disjoint path generation problem is NP-complete, making it computationally challenging to solve optimally. To tackle this, the MPSGA is designed to derive a near-optimal solution by combining a greedy approach with a heuristic path selection strategy, ensuring a balance between solution quality and computational efficiency.

The MPSGA operates through four core algorithms: *ReliaPath* (Algorithm 5.1), *mostReliaPathToExit* (Algorithm 5.2), *MPSGA* itself (Algorithm 5.3), and *ComputeSpurPath* (Algorithm 5.4). These algorithms work in coordination to generate a set of reliable, non-disjoint paths in the network.

The key notations used throughout the algorithms are defined as follows:

- G_P represents the current multi-path subgraph of the network.
- r_{spur} is the source switch in R, while *exits* refers to a subset of switches within R_P.
- Γ is a mapping that assigns reliability percentages to all nodes, indicating the likelihood of successful transmission.
- T_{spur} defines the set of delay constraints for each $r_i \in exits$.
- p_{k-1} and P_{spur} represent the current spur path and the set of all computed paths, respectively.

Algorithm 5.1 ReliaPath

function ReliaPath($G = (R, L)$, $G_P = (R_P, L_P)$, r_{spur}, $exits$, c, T_{spur})
    if $|exits| = 0$ **then return** $\emptyset$
    $r_{\max} = p_{\max} = null, \gamma_{\max} = \gamma(c), Q = \emptyset$
    $\tilde{G}_P = ((R \backslash R_P) \cup exits, L \backslash L_P), \Gamma = \{r_i \rightarrow \gamma_i | r_i \in R\}$
    $(\Gamma, prev) = \text{Dijkstra}(\tilde{G}_P, r_{spur}, exits)$
    for all $exit \in exits$ **do**
        $p_{spur}(r_{spur}, exit) = \text{buildPath}(prev, exit)$
        if $\tau[p_{spur}(r_{spur}, exit)] > T_{spur}[r_{exit}]$ **then**
            continue
        end if
        $\gamma_c = \text{MostReliaPathToExit}(G, exit, \Gamma)$
        if $\gamma_c > \gamma_{\max}$ **then**
            $\gamma_{\max} = \gamma_c, r_{\max} = exit, p_{\max} = p_{spur}(r_{spur}, exit)$
        end if
    end for
    return $p = \{p_{\max}, r_{\max}, \gamma_{\max}\}$
end function

Algorithm 5.2 MostReliaPathToExit

function MostReliaPathToExit($G = (R, L), exit, \Gamma$)
  $Q = \emptyset$, Q.push($exit$), $\gamma_{exit} = \Gamma(exit)$
  $reliaMap = \{r_i \rightarrow \gamma_i\}$, $parentMap = \{r_i \rightarrow r_j\}$
  repeat
    $r_{cur} = Q.\text{pop}(), \gamma_n = relia, \gamma_l = 1$
    if $r_p = parentMap[r_{cur}]$ **then**
      $\gamma_n = reliaMap[r_p]$
      $\gamma_l =$ reliability ratio of link(r_{cur}, r_p) in L
    end if
    $reliaMap[r_{cur}] = 1 - \frac{[1-reliaMap[r_{cur}]]*(1-\gamma_n*\gamma_l)}{1-reliaMap[r_p]*\gamma_l}$
    for $l \in$ egreeLinks(r_{cur}) **do**
      Q.push(egressSwitch(l))
      $parentMap$[egressSwitch(l)] $= r_{cur}$
    end for
  until $|Q| > 0$
  return $\gamma_c = reliaMap[c]$
end function

Algorithm 5.3 MPSGA

function MPSGA>($G = (R, L), r_s, c, \tilde{\gamma}, \tilde{\tau}, \tilde{P}$)
  $G_P(r_s, c) = (R_P = \emptyset, L_P = \emptyset)$
  $exits = \{c\}$, $P_{spur} = PriorityQueue(\tilde{P})$
  $p_0 =$ ReliaPath($G, G_P, r_s, exits, c, \{c \rightarrow \tilde{\tau}\}$)
  $G_P(r_s, c) = G_P(r_s, c) \cup p_0$
  if $\gamma(p, c) > \tilde{\gamma}$ **then return** G_P
  for $k = 2$ to $\tilde{P}$ **do**
    $P_{spur} =$ ComputeSpurPath($G, G_P, p_{k-1}, exits, P_{spur}$)
    if $|P_{spur}| = 0$ **then break**
    repeat $\tilde{p} = P_{spur}$.pop() **until** $hasLoop(G_P \cup \tilde{p})$
    $G_P = G_P \cup \tilde{p}$
    if $\gamma(\tilde{p}, c) > \tilde{\gamma}$ **then break**
  end for
  return $G_P(r_s, c)$
end function

Algorithm 5.4 ComputeSpurPath

function COMPUTESPURPATH($G, G_P, p_{k-1}, exits, P_{spur}$)
  for $r_{spur} \in p_{k-1}$ **do**
    $u = \text{IndexOf}(r_{spur}, p_{k-1})$
    $exits = \{r | r \in \text{descendents}(G_P, r_{spur})\}$
    $\tau_{pre} = \max\{\text{delay}(p_i[r_s \ldots r_{spur}]) | p_i \in G_P\}$
    $T_{reverse} = \{\tilde{e} \rightarrow \text{delay}(p[\tilde{e} \ldots c]) | \tilde{e} \in exits\}$
    $T_{spur} = \{\tilde{e} \rightarrow \tilde{\tau} - \tau_{pre} - T_{reverse}[\tilde{e}] | \tilde{e} \in exits\}$
    $p_{spur} = \text{ReliaPath}(G, G_P, r_{spur}, exits, c, T_{spur})$
    if $p_{spur} \neq \emptyset$ **then**
      $(p_{exit}, u) = \text{IndexOf}(p_{spur}[|p_{spur}|], P_{spur})$
      $p_{spur} = p_{exit}[1 \ldots u] \cup p_{spur} \cup p_{exit}[v \ldots |p_{exit}|]$
      $P_{spur} = P_{spur} \cup p_{spur}$
    end if
  end for
  return P_{spur}
end function

The workflows of these four algorithms are visualized in Figure 5.3. The specific roles of the algorithms are detailed below:

- *MPSGA* (Algorithm 5.3) is tasked with identifying a set of paths G_P from G, minimizing the length of $|L_P|$ (as defined in Eq. (5.2)), while satisfying reliability ($\tilde{\gamma}$, see Eq. (5.1)) and delay ($\tilde{\tau}$) constraints.
- *computeSpurPaths* (Algorithm 5.4) serves as a sub-function of *MPSGA*, computing candidate spur paths from each non-initial node on p_{k-1} to the *exits*, and subsequently adding these spur paths to P_{spur}.
- The *ReliaPath* function (Algorithm 5.1) computes the most reliable path from a source switch, r_{spur}, to all switches in a designated set *exits* $\subset R_P$ by maximizing node reliability values from Γ.
- *mostReliaPathToExit* (Algorithm 5.2) calculates the reliability ratio of each *exit* within the subgraph $G_P \cup p_{\text{spur}}$, updating their reliability values. The most reliable path from r_{spur} to an exit with the highest reliability value, $\gamma(exit)$, is then selected.

The execution of *MPSGA*, in coordination with the other algorithms (see Figure 5.3), results in the generation of G_P, the set of non-disjoint paths between the network nodes and the control center. These paths are selected to enhance network reliability while adhering to specified delay constraints.

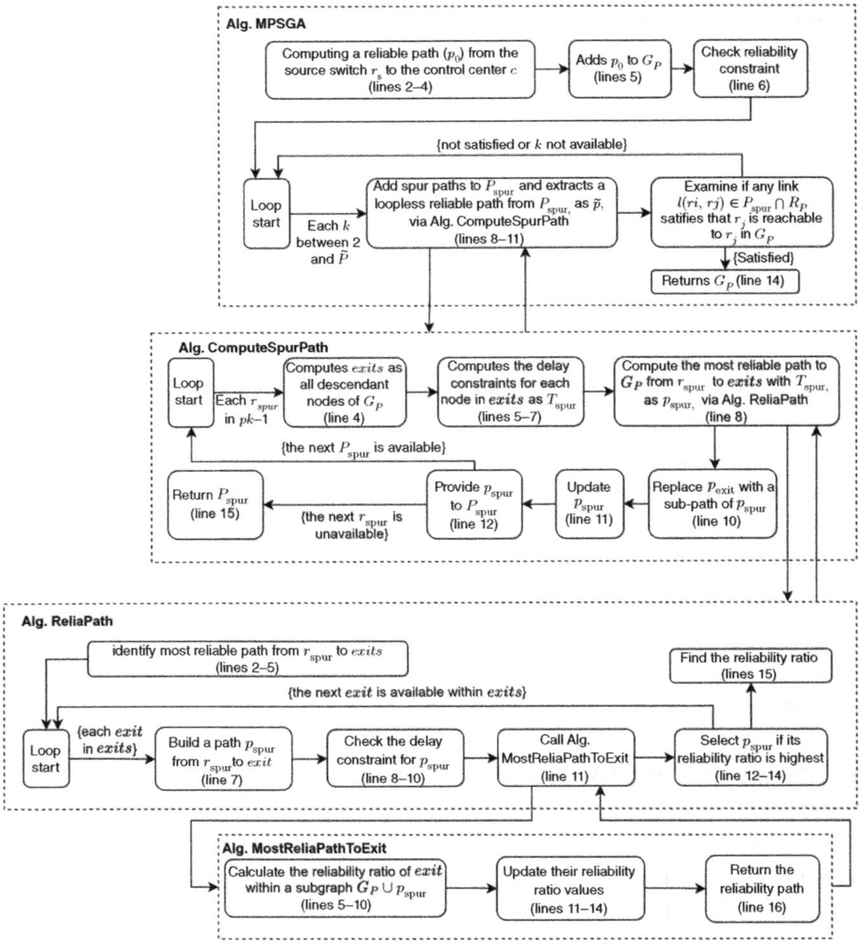

Figure 5.3 Workflows of OPAM's algorithms.

5.2.3 Algorithmic Analysis

We analyze the computational complexity of MPSGA, considering both time and space aspects. For Algorithm 5.1, we focus on a directed graph G that represents the communication network of a power transmission grid. This graph has an egress degree no less than 2, and $|L| = O(|R| \cdot \log |R|)$ [Soltan and Zussman, 2016]. Consequently, the time complexity of Dijkstra's algorithm using the Fibonacci heap is $|L| + |R| \cdot \log(|R|)$.

Moving on to Algorithm 5.2, within lines 2–16, the complexity is expressed as $\log(|R_P|) \cdot |\text{exits}|$. Given that $|\text{exits}| = O(|R_P|) = O(\xi \cdot \log(|R|))$, the overall complexity is effectively bounded by $\delta = O\{|L| + |R| \cdot \log(|R|) + \xi \cdot \log[\xi \cdot \log(|R|)] \cdot \log(|R|) + d\}$, where d is a constant. The space complexity is similarly bounded by $v = O\{|R| + \log[\log(\xi \cdot |R|)] \cdot \log(\xi \cdot |R|) + d\}$.

Now, considering Algorithm 5.3, its time complexity T_M is given by $O\{\xi \cdot \log(|R|) \cdot [\delta + |R_P| + \log(|R_P|)] + \delta + d\}$, while the space complexity is expressed as $O\{v + |R_P|\}$. Here, ξ represents the expectation of the generated paths in G_P for MPSGA, and $\xi \leq \tilde{P}$ due to line 12 of Algorithm 5.3.

In the worst-case scenario, where the maximum paths are iterated, the time complexity is $O[\tilde{P} \cdot \log(|R|) \cdot |L|]$, while in the best case, where no paths are iterated (see line 7 in Algorithm 5.3), the time complexity is $O[|L| + |R| \cdot \log(|R|)]$.

The complexity analysis of MPSGA is presented below through comparison with Yen, Hershberger, and Bernstein.

Yen. The time and space complexities of Yen's algorithm are expressed as $T_Y = O(\xi' \cdot g \cdot (|L| + |R| \cdot \log|R|))$ and $O(|L| + \xi' \cdot |R|)$, respectively [Yen, 1971]. Here, g is a function related to different cases indicating the length of the spur path, and ξ' is the expectation of generated paths in G_P for Yen, satisfying $\xi' \leq \tilde{P}$.

Comparatively, MPSGA showcases superior performance over Yen in three main aspects: (1) during candidate spur path selection (lines 7–15 in Algorithm 5.1), MPSGA opts for the path that contributes the most to overall reliability (lines 12–14 in Algorithm 5.1); (2) when selecting the spur source node (line 4 in Algorithm 5.4), MPSGA employs possible intermediate exit nodes rather than the destination node; and (3) MPSGA considers delay and reliability constraints during the selection of candidate paths.

The time complexity comparison between MPSGA and Yen unfolds in the following manner:

1) In the expected case, analyzing the time complexity of MPSGA, δ can be expressed as $O(|L|)$ based on Eq. (5.5), as $O(\xi \cdot \log(\xi \cdot \log|R|) \cdot \log|R|)$ is significantly less than $O(|R| \cdot \log|R|)$. Consequently, T_M equals $O(\xi \cdot \log|R| \cdot |L|)$ based on Eq. (5.6), given that $|R_P| \leq |R|$. Moreover, the optimal choice of the reliable path in Algorithm 5.2 results in a smaller minimum number of paths in G_P (i.e., ξ) required to meet $\tilde{\gamma}$ compared to Yen (i.e., ξ'). For instance, in evaluation, mean(ξ) and mean(ξ') in the IEEE 30-bus system are 2.646 and 3.647, respectively, using the simulation parameters in Section 5.3.1. For comparison, mean(T_Y) in Yen can be rewritten as $O(\xi' \cdot \log|R| \cdot |L|)$ based on Eq. (5.7) with $g = \log|R|$ according to Yen [1971]. Consequently, mean(T_M) is less than mean(T_Y), indicating that MPSGA is faster than Yen in this case.

$$\begin{aligned} \delta &= O(2 \cdot |L| + \xi \cdot \log(\xi \cdot \log|R|) \cdot \log|R| + d) \\ &= O(|L|) \end{aligned} \tag{5.5}$$

$$\begin{aligned} T_M &= O[\xi \cdot \log|R| \cdot (|L| + |R| + \log|R|) + |L| + d] \\ &= O[(\xi \cdot \log|R| + \xi + 1) \cdot |L| + \xi \cdot \log^2|R| + d] \\ &= O(\xi \cdot \log|R| \cdot |L|) \end{aligned} \tag{5.6}$$

$$\begin{aligned} \text{mean}(T_Y) &= O(\xi' \cdot \log|R| \cdot (|L| + |R| \cdot \log|R|)) \\ &= O(\xi' \cdot \log|R| \cdot |L|) \end{aligned} \tag{5.7}$$

2) In the worst case, examining the time complexity of MPSGA reveals that $\max(T_M) = O(\tilde{P} \cdot \log|R| \cdot |L|)$ (see Eq. (5.6)), where $\xi = \tilde{P}$ when the maximum number of paths are iterated in Algorithm 5.3. On the other hand, for the time complexity of Yen, $\max(T_Y) = O(\tilde{P} \cdot |R| \cdot |L|)$ can be derived by reducing Eq. (5.8) with $g = |R|$ and $\xi' = \tilde{P}$ [Yen, 1971]. Consequently, $\max(T_M)$ is significantly less than $\max(T_Y)$, signifying that MPSGA is faster than Yen in the worst-case scenario.

$$\begin{aligned} \max(T_Y) &= O[\tilde{P} \cdot |R| \cdot (|L| + |R| \cdot \log|R|)] \\ &= O(\tilde{P} \cdot |R| \cdot |L|) \end{aligned} \tag{5.8}$$

3) In the best-case scenario, both $\min(T_M)$ and $\min(T_Y)$ asymptotically converge to $O(\log|R| \cdot |L|)$ for MPSGA and Yen, where $\xi = 1, \xi' = 1$, and $g = \log|R|$. This implies that MPSGA and Yen exhibit similar performance in the best case.

In the aforementioned cases, the space complexity of MPSGA is denoted as $O(v + |R_P|)$, which is significantly less than $O(|L| + \tilde{P} \cdot |R|)$. Consequently, when compared with Yen, MPSGA can notably reduce both time and space complexities in the worst and expected cases while maintaining equivalent time complexity in the best case.

Hershberger. The time complexity of Hershberger's algorithm is given by $O[\xi' \cdot g' \cdot (|L| + |R| \log|R|)]$, where g' represents the number of invocation times for the path replacement routines. Notably, g' is constrained to be no more than g in Yen since the efficacy of the path replacement acceleration is contingent on the probability of encountering valid cut edges denoted as e' [Hershberger et al., 2007]. Specifically, the worst-case scenario for Hershberger aligns with that of Yen, both operating in $O(\xi' \cdot \log|R| \cdot |L|)$ where $g' = |R|$, a complexity that is asymptotically larger than MPSGA.

Additionally, the best-case scenario for Hershberger is in $O[|L| + |R| \log|R|]$, where $g' = 1$ and $\xi' = 1$. This complexity is lower than both MPSGA and Yen in the best case, albeit contingent on the graph having significantly higher edge density. However, the expected case complexity for Hershberger is $O[\xi' \cdot (|R| - e' \cdot |R| + e') \cdot (|L| + |R| \log|R|)]$, which surpasses that of MPSGA due to $\xi \leq \xi'$ resulting from optimal path selection.

In terms of space complexity, Hershberger utilizes $O(|R| \cdot |L|)$ when employing a directed graph for storing equivalent classes [Hershberger and Suri, 2001].

This space complexity is greater than that of both MPSGA and Yen across all cases.

Bernstein. The time complexity of the $(1 + e)$ Bernstein algorithm is expressed as $O(e^{-1} \cdot g'' \cdot (|L| + |R| \cdot \log|R|) \cdot \log^2|R|) = \tilde{O}(g'' \cdot |L|/e)$, where $e < 0$ is the acceleration parameter, and $g'' \leq \tilde{P}$. Notably, g'' is less than the corresponding parameter in MPSGA, making Bernstein more efficient. However, it is essential to acknowledge that Bernstein is an approximating algorithm that accelerates the path replacement of Hershberger at the expense of precision on the top K Shortest Paths (KSP), resulting in fewer QSDP than MPSGA.

The space complexity of Bernstein is denoted as $O(|R| \cdot \log|R|)$ [Frieder and Roditty, 2015], which is greater than that of MPSGA on average.

5.2.4 Packet Aggregation and Forwarding Module

The OF switch is enhanced with the PAFM, a logical block with a small memory, depicted in Figure 5.2. The PAFM records two types of flow states: (1) *States of Forwarding History (SFH)*, a small set $H = \{r | r = (ID_{\text{start}}, \text{len})\}$ storing the received IDs of packets (see IETF RFC 791) in the flow. Here, r indicates packets of the flow with ID starting from ID_{start} to $ID_{\text{start}} + \text{len}$ that have been forwarded; and (2) *States of Recent Out-of-order Packets (SROP)*, a priority queue $Q = \{p | p = (\text{ID}, \text{data})\}$ arranged in a sequence of ID. Both SFH and SROP need to be rapidly accessible for the PAFM, aligning with the forwarding throughput of the switch, and are typically stored in SRAM. Flow ID in the PAFM involves both header values in the IPv4 layer and port values in the transport layer. To handle flows restarting with $\text{ID} = 0$ due to sender failures and restarts, the PAFM treats such flows as new flows and expires the old ones.

When a new flow arrives at the switch, i.e., the flow is unmatched in the table, the PAFM operates as follows: (1) the switch forwards the first packet of the flow to its MPSGA; (2) the MPSGA generates flow rules with the group type of "select" [Open Networking Foundation, 2012] and dispatches the rules and the ports to the switch via a PACKET_OUT message; and (3) the switch updates all rules in the corresponding tables and assigns the SFH exclusively for the flow. For flows with their last hop to the destination node, the switch also designates the SROP exclusively for the flow. Subsequently, upon an incoming message m at a switch, the PAFM logic ensures that matched flows meet the following two requirements.

No duplicated packet (with the same field values in IP header) is forwarded in-flight for a flow. If a message m is matched by a switch, its PAFM operates as follows: (1) it searches for $h = (\text{ID}, \text{len}) \in H$ that is equal to either $m.\text{ID}$ or $m.\text{ID} - 1$; (2) if found, it updates h to $(\text{ID}, \text{len} + 1)$ and drops m when $m.\text{ID} \in h$; and (3) if not found, $h_{\text{new}} = (m.\text{ID}, 0)$ is added to H, the earliest element of H is cleared when memory H is full, and then m is forwarded.

$$1-\beta = e^{-\lambda\tau_{p_j}} \cdot \sum_{i=0}^{\tilde{p_j}-1} \frac{\lambda^i}{i!} \cdot \tau_{p_j}^i \tag{5.9}$$

Regarding the memory assumption, the PAFM systematically recycles unused $h \in H$ by deallocating $r \in H$ with ID_{start} smaller than $ID_{\text{threshold}} = m.\text{ID} - \max\{\tau_{p_j} | p_j \in P\} \cdot \lambda$, where m.ID is the ID of the latest packet, λ is the arrival rate of the current flow, β is the estimating probability of ID_{start}, and τ_{p_j} is the delay difference between p_j and the shortest path $p_s \in P$. The delay difference τ_{p_j} can be computed using Eq. (5.9), assuming the waiting time of a flow across multiple hops follows an Erlang distribution. The condition $|H| \geq 1$ holds only if at least one packet of the flow is lost. Additionally, $ID_{\text{threshold}}$ follows a Gumbel distribution due to its exponential tail of the Erlang distribution. Thus, $|H| \propto e^{-e^{-(t-\mu)/\theta}}$, where θ is a real constant representing the pattern of data packet arrivals.

It is also assumed that if a packet has an ID less than ID_{start}, such as an arbitrary delayed out-of-order packet, the forwarding to the next hop of the packet occurs with a negligible conditional probability of $1-\beta$. Furthermore, the time complexity using linear search is in $O(|H|)$, approximating to a constant level. This procedure ensures the proactively efficient and timely removal of duplicated m with the probability of β in memory.

No out-of-order packet is forwarded to destination node. The out-of-order packets v_d undergo reordering solely at the preceding switch before reaching the destination node, rather than at other intermediate switches along the forwarding path. This reordering is achieved through the use of a priority queue Q to ensure a processing time of $O(\log|Q|)$ for out-of-order packets and the subsequent release of cached packets $d_i \in Q$. The switch dynamically flushes Q under two conditions: either when the number of elements in Q reaches $\max_{\text{id}}\{Q\} - \min_{\text{id}}\{Q\} + 1$, or when the time since the most recent flush of Q exceeds $\tau_{\max}$, as defined in Eq. (5.10).

$$\tau_{\max} = \max\{\text{Prob}^{-1}_{\text{Erlang}[\text{hops}(p_i),\lambda]}(\epsilon/\gamma(r_d, G_P))\} \tag{5.10}$$

In this context, the allocation of Q can be implemented with a small memory in a few words, considering the following observations: (1) $\gamma(r_d, G_P)$ is generally closer to 1 at the destination node due to path aggregation, indicating that packet out-of-order occurrences have a low probability, not exceeding $1-\gamma(r_d, G_P)$; and (2) the Gumbel distribution exhibits a sharp drop to 0 when approaching $+\infty$. The size of $|Q|$ is a random variable (r.v.) following the Gumbel distribution, formally denoted as $|Q| \propto P_Q = e^{-e^{-(t-\mu)/\theta}}$, where $|Q|$ represents the extreme value of several Erlang-distributed r.v. $\tau(d_{j_i})$ (see Eq. (5.4)).

To estimate the parameters μ and θ in the Cumulative Probability Distribution Function (CDF) of the Gumbel distribution of P_Q, the following steps are performed at the switch: (1) allocating Q with a size of $\tau_{\max} \cdot \lambda$; (2) calculating the mean and variance of $|Q|$ as $\mu - \theta \ln(\ln 2)$ and $\pi^2\theta^2/6$; (3) computing θ and $|\tilde{Q}| = P_{\tau_{\max}}^{-1}(\mu, \theta)[\epsilon/\gamma(r_d, G_P)] \cdot \lambda$; and (4) releasing Q with the unused bytes of $|Q| - |\tilde{Q}|$.

As a result, the end-to-end data delivery performance can be enhanced during link failures with the utilization of $\gamma(r_d, G_P)$. Moreover, the setup delay of OF flow rules for the first packet is circumvented by proactively configuring all flow rules on the OF switch side when an r_s is initially connected to the c.

5.3 Performance Evaluation

This section assesses the performance of OPAM through comparisons with existing path generation algorithms and CPN across a variety of test networks.

5.3.1 Experimental Settings and Metrics

The performance evaluation of OPAM is conducted through simulations using NS-3 [Riley and Henderson, 2010] under conditions of frequent link failures. The simulation steps are illustrated in Figure 5.4, integrated with the delivery of OF messages. Specifically, the Beacon OF controller [Erickson, 2013] undergoes modification with two new logics: a new service implementation of MPSGA and the OF protocol stack for managing the "select" field in the group table. The controller operates externally to NS-3, establishing connections to OF switches through the OF stub via a TCP socket. The stub functions as a proxy running within NS-3. In simulations, NS-3 parses transmission grid topology files to derive corresponding communication topologies, generates the SDN data plane and the stub, and monitors simulation outcomes. The intra-substation network is simulated as an OF switch connected to a client node. The simulation parameters are detailed in Table 5.2.

Specifically, the test network topologies are selected from four datasets with varying network scales (utilizing only the greatest connected component in the

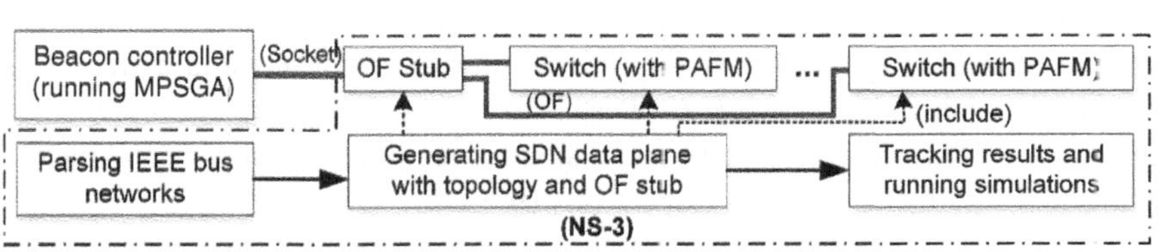

Figure 5.4 Simulation steps for OPAM in NS-3.

Table 5.2 Simulation parameters.

Simulation topology and OPAM					
Parameter	**Value**	**Parameter**	**Value**	**Parameter**	**Value**
$\tau_p(l_i)$	5 ms	$B(l_{i,j})$	10 Mbps	MTU	590 B
$\varepsilon(l_{i,j})$	2%	$\tilde{\tau}$	40 ms	$\tilde{P}$	5
$\tilde{\gamma}$	99%	e	1/16	ϵ	99.8%

TCP stack parameter	**Value**
Nagle algorithm	Disable
Size of sending and receiving buffers	131,072 B
Timeout for delayed ACK	200 ms
# packets to wait before sending ACK	2

bus topology graph for each bus network): (1) IEEE test networks with 14, 30, 57, 118, and 300 buses [IEEE, 2025], (2) the Texas bus network from the UIUC GRID DATA project with 1725 buses [ICSEG, 2016], (3) the PEGASE networks with 1354, 9241, and 13,659 buses [Josz et al., 2016], and (4) the French systems with 1888, 1951, 2383, 2736, 2746, 2848, 2869, 3012, 3374, 6468, and 6515 buses [Zimmerman et al., 2011]. These topologies are commonly employed in numerous studies to facilitate numerical result comparisons. Furthermore, the complexity analysis of OPAM demonstrates the scalability of the proposed solution for large-scale network scenarios.

Furthermore, the topology is configured with the link transmission delay $\tau_p(l_i)$, bandwidth $B(l_{i,j})$, and a Maximum Transmission Unit (MTU) of 5 ms, 10 Mbps, and 590 B, respectively. Consequently, the states of an extremely unstable network are emulated through frequent link failovers to conduct the performance evaluation of our solution in terms of SRRP and PIAT for end-to-end MMS packet delivery in TCP. The specific simulation parameters are as follows: (1) $\varepsilon(l_{i,j})$ is set to 2% to emulate link failovers; (2) $\tilde{\tau}$ is set to 40 ms, as the IEC 61850 standard mandates the complete transfer time of packets for a Type-2 MMS message to be less than 100 ms, considering the uncertainty in the number of packets and extra delays in the intra-substation network; (3) $\tilde{P}$ is set to 5, acknowledging that in the IEEE 30-bus test network, non-disjoint and disjoint algorithms can generate at most 3 paths and 5 paths for all pairs, respectively (see Table 5.3); and (4) $\tilde{\gamma}$ is set to 99%, indicating high communication reliability (IEC 61850 does not specify reliability requirements under link failures [IEC, 2013]).

Table 5.3 Count of pairs generated by MPSGA, MADSWIP, and ODPG.

Pairs with different # paths	1	2	3	4	5
MPSGA	45	269	1	35	85
MADSWIP	142	293	0	0	0
ODPG	142	278	13	2	0

In the control plane, SRRP is assessed by comparing MPSGA with non-disjoint and disjoint algorithms for all 435 node pairs in the IEEE test network with 30 buses and 41 links. In this context, one node is selected as the destination (the control center), while the remaining nodes act as sources (substations). The performance of MPSGA is further evaluated by comparing it with available non-disjoint algorithms for all pairs in both the IEEE and Texas test networks. For the sake of comparison, benchmark algorithms, including Hershberger, Yen, Bernstein, MADSWIP, ODPG, and Single Shortest Path Generation Algorithm Based on Dijkstra (SSPG), are configured with the same parameters as MPSGA, and also with a concave constraint of $\tilde{\tau}$. In the case of Bernstein, e is set to $1/15$ as specified in Bernstein [2010]. Additionally, the scalability of MPSGA is assessed by examining different execution times for all pairs in the PEGASE bus networks and the French systems.

In the data plane, PIAT is evaluated in a communication graph (G_P) generated by MPSGA, utilizing the IEEE 30-bus test network with $\gamma(r_s, G_P) = 99.877\%$. In detail, a source r_s (the left square node in Figure 5.5) sends 2MB packets (to stabilize the congestion window) using TCP to simulate the transmission of a series of IEC 61850 MMS data packets. To mimic the generic communication of an electric device, the TCP stacks of both the server and the client are configured with standard parameters given in Table 5.2. Node c (the right square) listens to the TCP port to accept the TCP connection from r_s. The out-of-order probability ϵ (see Eq. (5.4)) is set to 99.8%, indicating the likelihood of out-of-order packet arrivals at c.

Moreover, the CPN for real-time traffic [Wang and Gelenbe, 2016] is implemented in NS-3, extending the raw CPN [Gelenbe, 2017; Gelenbe, 2009] with a packet reordering buffer. The NS-3 CPN incorporates policies for individual CPN nodes, including: (1) initializing the excitation probability of each Recurrent Neural Network (RNN) neuron as 50%; (2) defining the RNN reward to consider both delay and loss equally, formally expressed as $R = 1/[norm(\tau_p) + norm(\varepsilon_p)]$,

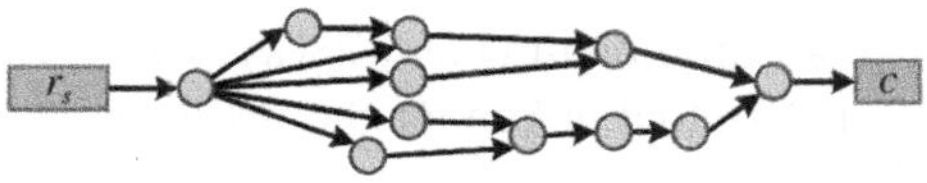

Figure 5.5 Non-disjoint paths of MPSGA in IEEE 30 bus network.

where $norm(x)$ normalizes x to the range $[0, 1]$, τ_p is the HRRT between the current node and the receiver, and ϵ_p is the Cognitive Packet (CP) loss rate; and (3) configuring the packet reordering buffer of CPN with different lengths during evaluation. The topology of CPN nodes and the sending rate mirror those of PAFM in Figure 5.5, where the source node sends IP packets encapsulated as CP, and vice versa. Two scenarios with different link loss rates (0% and 2%) are compared for both CPN and OPAM. Additionally, QoS measurements, in terms of SRRP $(1 - \epsilon_p)$ at the receiver side and HRRT (τ_p) for both CPN and OPAM, are evaluated to showcase improvements in packet loss rate and end-to-end delay compared to the results in Gelenbe [2009].

5.3.2 Comparison with Non-disjoint Path Generation

The performance of MPSGA is compared with the non-disjoint path generation algorithms of Hershberger and Yen in terms of SRRP and QSDP, as well as scalability. Firstly, Figure 5.6(a) presents the Probability Density Function (PDF) of SRRP for MPSGA, Hershberger et al. [2007], Yen [1971], and Bernstein [2010]. Additionally, their mean values are 99.23%, 98.38%, 98.38%, and 98.354%, respectively. In Figure 5.6(a) and 5.6(b), SRRP of most pairs is distributed closer to 1, as MPSGA selects paths that most contribute to $\gamma(r_s, G_P)$, while others choose paths with the minimum number of edges, resulting in higher SRRP and QSDP than others. Figure 5.8 shows that QSDP of MPSGA, Hershberger, Yen, and Bernstein is 77.7%, 58.62%, 58.62%, and 53.23%, respectively. Additionally, QSDP of SSPG is 0% due to its highest SRRP being less than $\tilde{\gamma}$. For pairs with SRRP lower than $\tilde{\gamma}$, MPSGA chooses a more nearly optimal sub-graph with $\gamma(r_s, G_P) > \tilde{\gamma}$ due to the optimization on path selection in each step to maximize Eq. (5.1).

Figure 5.7 shows the percentiles of the execution time of MPSGA, Hershberger, Yen, and Bernstein for the IEEE 14, 30, 57, 118, and 300 bus test networks, as well as the Texas 1725 bus network (the upper and lower whiskers represent the 75th and 25th percentiles). The observations confirm that: (1) the acceleration of Hershberger over Yen is less obvious in the small network due to overhead in path replacement execution [Hershberger and Suri, 2001], so the execution time of MPSGA is 59.86%, 158.18%, 77.4%, and 120% of that of Hershberger for the evaluated IEEE test networks. However, the execution time of MPSGA in the Texas bus network is reduced by 81.53% compared to that of Hershberger; (2) the acceleration of Bernstein is more obvious than Hershberger due to the multiplicative error of $1 - O(e)$ existing in each step for executing the path replacement [Bernstein, 2010]. However, it results in a 9.2% and 31.5% decrease in QSDP compared to Hershberger/Yen and MPSGA, respectively; and (3) the overall running time of MPSGA is 47.02%, 121.74%, 69.28%, 67.33%, 33.28%, and 13.11% of that of Yen for the IEEE test networks and the Texas network. This is due to the fact that the heuristic path selection of MPSGA in maximizing SRRP avoids searching for useless spur paths in Yen.

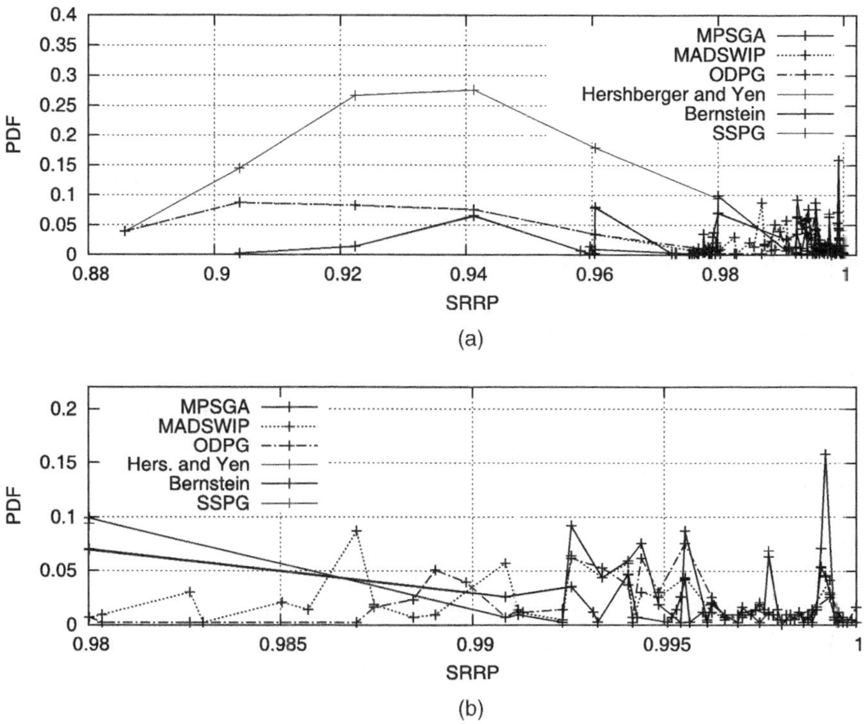

Figure 5.6 Performance evaluation of MPSGA. (a) Probability Density Function (PDF) for SRRP of MPSGA to current algorithms. (b) Zoomed Probability Density Function (PDF) for SRRP of MPSGA to current algorithms.

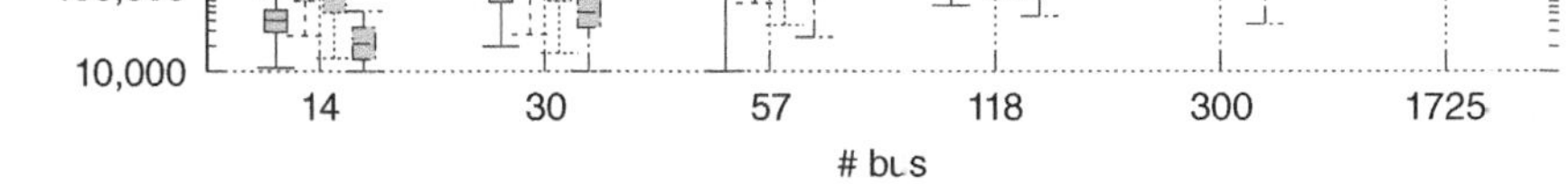

Figure 5.7 Execution time for non-disjoint path algorithms.

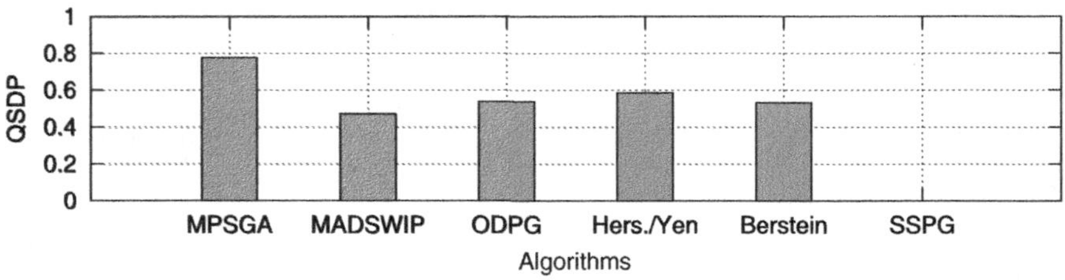

Figure 5.8 QSDP of MPSGA to disjoint and raw algorithms.

5.3.3 Comparison with Disjoint Path Generation

MPSGA is compared with SSPG [Cherkassky et al., 1996], MADSWIP [Taft-Plotkin et al., 1999], and ODPG [Pang and Goodwin, 2003] for each pair in the IEEE 30-bus test network. Figure 5.6(a) shows the PDF of SRRP for SSPG, MADSWIP, and ODPG, with mean SRRP of 93.57%, 99.23%, 96.93%, and 97.06%, respectively. Figure 5.8 shows that their QSDP are 77.7%, 0%, 47.13%, and 53.79%. The QSDP of SSPG is 0% due to its highest SRRP of 98%, which is less than $\tilde{\gamma}$. Table 5.3 shows the count of the tested pairs parameterized by the aforementioned algorithms of MPSGA, MADSWIP, and ODPG for different cases in the number of generated paths. The result indicates that MPSGA can exploit the links out of the disjoint paths generated by MADSWIP or ODPG, and hence improvements of SRRP and QSDP can be obtained.

5.3.4 Performance Scalability

The scalability of MPSGA is evaluated by examining its execution time in the PEGASE networks and French systems, with the number of buses ranging from 1354 to 13,659. Figure 5.9 presents the mean and maximum execution time of MPSGA against the number of network buses, along with their theoretical

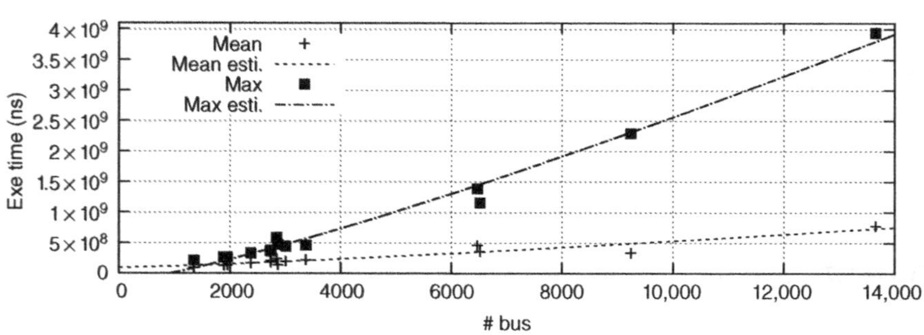

Figure 5.9 Mean and max execution time for MPSGA in PEGASE and French.

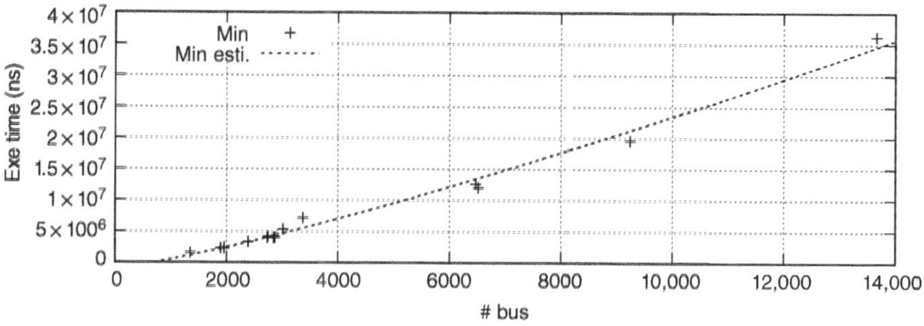

Figure 5.10 Min execution time for MPSGA in PEGASE and French.

estimation curves using the models mentioned in Section 5.2.3. Correspondingly, Figure 5.10 shows the minimum execution time of MPSGA. Firstly, $|L|$ is considered in $O(|R| \cdot \log|R|)$ according to Soltan and Zussman [2016]. Specifically, their execution times are close to $521.5 \cdot \log^2|R| \cdot |R| + 8.956 \times 10^7$, $3183 \cdot \log^2|R| \cdot |R| - 1.415 \times 10^8$, and $28.61 \cdot \log^2|R| \cdot |R| - 8.481 \times 10^5$, respectively. Therefore, it is confirmed that the time complexities of MPSGA are in $O(\xi \cdot \log|R| \cdot |L|)$, $O(\tilde{P} \cdot \log|R| \cdot |L|)$, and $O(\log|R| \cdot |L|)$ for the expected, worst, and best cases, respectively. The simulated communication network size is sufficiently large to support provincial power utilities, and the result confirms the scalability of MPSGA.

5.3.5 Enhancing TCP Performance

The performance of PAFM is evaluated through a comparison of two cases: one with a queue and one without a queue. In simulations, the number of dropped packets for the PAFM with a queue is 1149 in 13.27 seconds, while the scenario without a queue results in 1191 dropped packets in 31.64 seconds. At the destination, packet losses are 6 and 2 out of 3874 and 3875 packets, respectively, due to the path aggregation of PAFM, and SRRP approximates to $\gamma(r_s, G_P)$. Figure 5.11(a) shows that average throughputs for both PAFM scenarios are 285.36 and 128.76 pps, respectively, due to the absence of out-of-order packets at the receiver side, which are reordered at the edge switch by the PAFM with a queue. In contrast, the PAFM without a queue observes 13.94% out-of-order packets. Figure 5.11(c) presents the CDF of the congestion window (CWnd) for the PAFM. It illustrates that the average sizes of CWnds are 11,929.3 and 4771.6 for the two evaluated cases, respectively, influenced by the fact that 9.6% of packets have an out-of-order range greater than 2 packets (see Table 5.2).

Figure 5.11(d) presents the PDF of PIAT for the PAFM (with and without a queue). It shows that the average and Standard Deviation (STD) of PIAT are

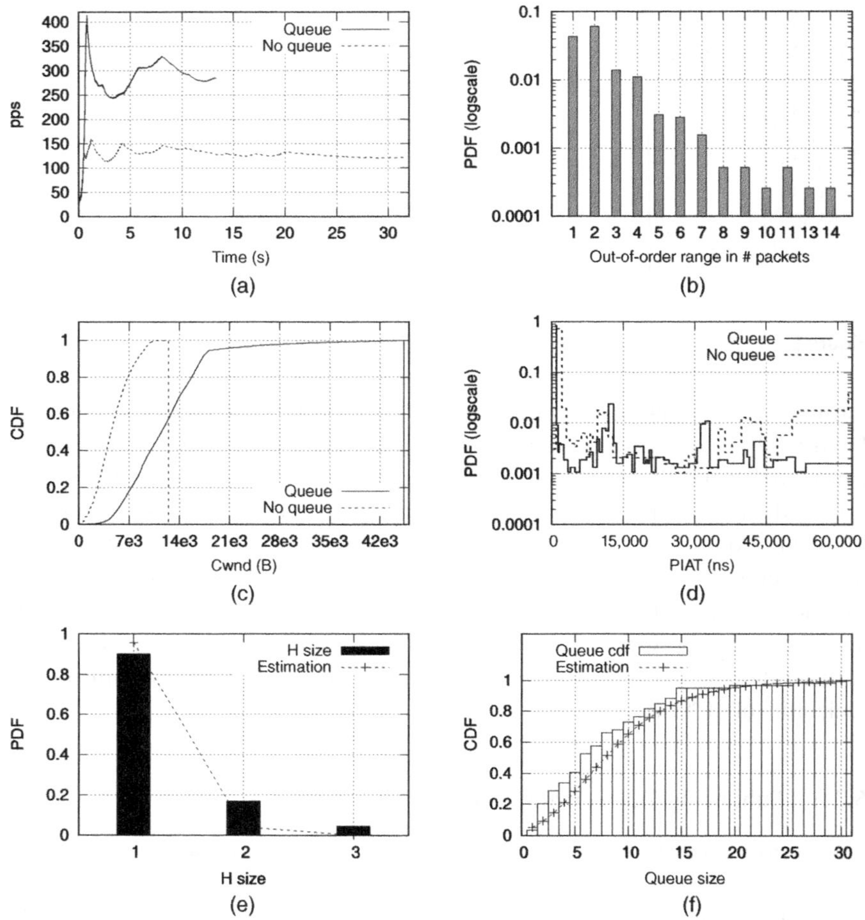

Figure 5.11 Performance evaluation of PAFM. (a) TCP throughput with and without queue. (b) PDF of out-of-order packet range. (c) CDF of TCP CWnd with and without queue. (d) PDF of PIAT with and without queue. (e) Size of forwarding history (H). (f) Changes of queue (Q) size.

(3528.1 μs, 9202.3 μs) and (8173 μs, 16, 083 μs), respectively. The latter average PIAT is significantly reduced by 56.83% for the PAFM with the queue due to the reduction of TCP re-transmissions and the enlargement of the TCP congestion window using the additive increase and multiplicative decrease policy [Leung et al., 2007]. Additionally, for the PAFM with a queue, the average, STD, and maximum of HRRT are 16, 605.2, 1403, and 27, 360 μs, respectively, all of which are less than $\tilde{\tau} = 40$ ms. The average is 3.707 and 1.032 times smaller than the PAFM without a queue and SSPG, respectively.

Finally, the results show that the size variations of Q and H are very close to their theoretical estimations (Section 5.2.4). Figure 5.11(f) illustrates the number of queue elements approximating Gumbel($\mu = 6.083, \theta = 4.629$) as the queue size (denoted as boxes) increases to the right. When $|Q| = 35$ with the minimum queue memory, $\epsilon = 99.8066\%$. Additionally, Figure 5.11(e) shows that the maximum and the expectation of $|H|$ are 3 and 1.113, respectively, which are closer to its estimation in Gumbel($\mu = 1.113, \theta = 0.281$). This occurs when $\tau_{\max} = \text{Gumbel}^{-1}_{(\mu=25.857,\theta=2.2)}(0.998) \cdot \lambda \approx 11.225$ and $\lambda \approx 284$ pps, confirming that the queue can remove the packets with an ID less than $m.ID - 11.225$ with a confidence range ϵ of 99.8% (see Section IV.D). Therefore, the memory usage is maintained at a low level.

5.3.6 Comparison with Cognitive Packet Networks

The CPN undergoes evaluation based on routing convergence time, packet reordering buffer, and QoS measurements. To observe the routing convergence process, we analyze the excitation process of RNN. Figure 5.12 displays the average potential changes per neuron (Δ_q) at node 0 of the CPN router topology (leftmost router adjacent to r_s in Figure 5.5). The RNN on node 0, with the highest degree in the graph, effectively reveals route changes. Formally, $\Delta_q = 1/m \cdot \sum_{i=1}^{m} q_i^r - q_i^{r-1}$, where m is the number of neurons of RNN at node 0, and q_i^t is the excitation probability of neuron i in the stable states of RNN of the rth round (when receiving the rth ACK at the current node) [Gelenbe, 2009]. In Figure 5.12, the convergence process of Δ_q approaches 0 after receiving a series of ACKs for both scenarios. Periodical spikes of Δ_q lead to changes in packet hops at the receiver side, indicating CPN's sensitivity to slight changes in traffic conditions. Overall, CPN routes take 770,729 and 430,786 µs to converge to the minimum path length of 4 for the two scenarios. In comparison,

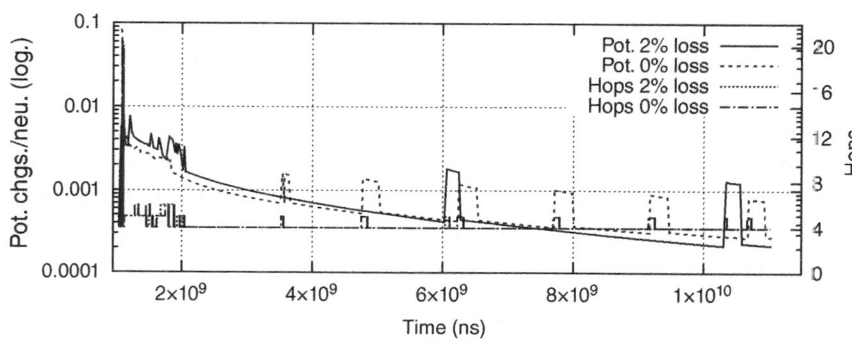

Figure 5.12 Average potential changes per neuron at node 0 in CPN.

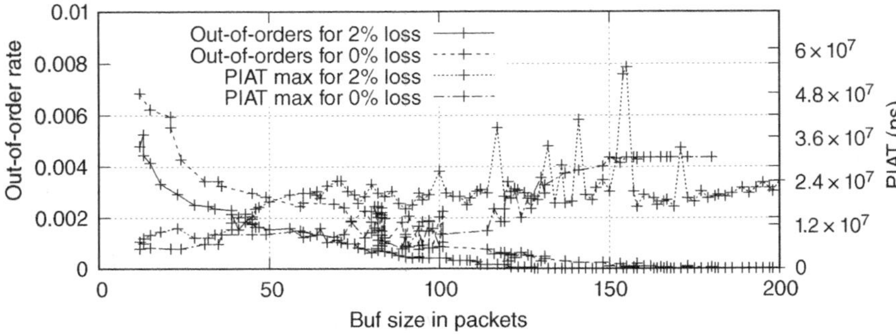

Figure 5.13 Performance of CPN with reordering buffer.

MPSGA computes non-disjoint multi-paths within a maximum time of 855 μs, significantly less than CPN's time.

Secondly, the performance of the packet reordering buffer is measured via examining the packet arrival statistics at the receiver side. Figure 5.13 shows that POR and PIAT against the size of the reordering buffer in packets for the two scenarios with 0% and 2% link loss rate. The results show that: (1) POR decreases along with the increases of buffer size in both the scenarios with the minimum buffer sizes of 168 and 128 to completely avoid the packet out-of-orders issue. In comparison, for the second scenario, the memory usage of PAFM is in the size of 30 packets (see Figure 5.11(f)) that is 76.563% less than that of CPN. It is because of the delay difference due to the randomness of forwarding path discovery and the slow convergence of routing states in the RNN. Hence, the less memory usage leads to better scalability in flows; and (2) PIAT is linearly increasing against the reordering buffer size, with average and max values of (470 μs, 30,481 μs) and (511 νs, 19,830 μs) in the minimum buffer size for two scenarios, respectively. In comparison, for the PAFM with queue version, the average and maximum values of PIAT are (472 μs, 472 μs) and (473 μs, 1888 μs) that are significantly less than the PIAT of CPN in sending IP packets. It is because that the neuron weight updates for ACK introduce amount of time in the routing blackholes. Compared with CPN, the packet reordering in OPAM uses significantly less memory with smaller PIAT and HRRT, due to the faster routing convergence and higher SRRP of MPSGA.

Examining the packet arrival statistics at the receiver side evaluates the performance of the packet reordering buffer. In Figure 5.13, POR and PIAT are presented against the reordering buffer size for scenarios with 0% and 2% link loss rate. The results reveal:

- POR decreases with increasing buffer size in both scenarios, reaching minimum sizes of 168 and 128 to entirely avoid packet out-of-order issues. In the second scenario, PAFM's memory usage (30 packets, Figure 5.11(f)) is 76.563% less than

CPN, attributed to the delay difference from the randomness of forwarding path discovery and slow convergence of routing states in RNN. This reduced memory usage enhances scalability in flows.

- PIAT exhibits a linear increase against the reordering buffer size, with average and max values of (470 μs, 30,481 μs) and (511 μs, 19,830 μs) in the minimum buffer size for both scenarios. In comparison, PAFM with the queue version demonstrates average and maximum PIAT values of (472 μs, 472 μs) and (473 μs, 1888 μs), significantly lower than CPN in sending IP packets. This is because neuron weight updates for ACK introduce time in routing blackholes.

Compared with CPN, OPAM's packet reordering uses significantly less memory with smaller PIAT and HRRT, benefiting from the faster routing convergence and higher SRRP of MPSGA.

QoS measurements, specifically SRRP and HRRT, are evaluated for CPN and OPAM. In the first scenario, both CPN and OPAM exhibit 100% SRRP. However, in the second scenario with a 2% link loss rate, CPN's SRRP is 91.921%, 7.97% lower than OPAM, indicating slower routing convergence in CPN. Figures 5.14(a) and 5.14(b) displays HRRT CDF for CPN and OPAM in both scenarios. Results reveal that average and maximum HRRT values for CPN are (26,448 μs, 105,034 μs) and (567,850 μs, 596,773 μs), respectively. In comparison, OPAM's values are (16,416 μs, 16,416 μs) and (20,003 μs, 32,936 μs), showing reductions of (37.931%, 84.371%) and (96.477%, 94.481%) compared to CPN. This is attributed to higher POR and longer routing churn time in CPN during link failures.

Figures 5.15(a) and 5.15(b) presents CDF of the number of forwarding hops for CPN and OPAM in the two scenarios. Numerical results indicate that average and maximum hop numbers in CPN are (5.166, 21) and (5.126, 17), respectively. In comparison, OPAM's numbers are (5, 5) and (5.039, 7), showing reductions of (3.2%, 76.19%) and (1.7%, 58.823%) compared to CPN. Although CPN and OPAM are similar in average forwarding hops, OPAM achieves higher SRRP and lower HRRT under link failures, attributed to its extra path learning process.

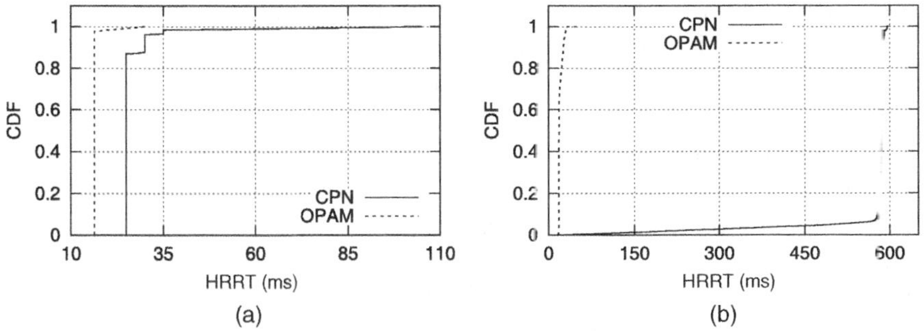

Figure 5.14 CDF of HRRT in CPN and OPAM. (a) No link loss and (b) 2% link loss.

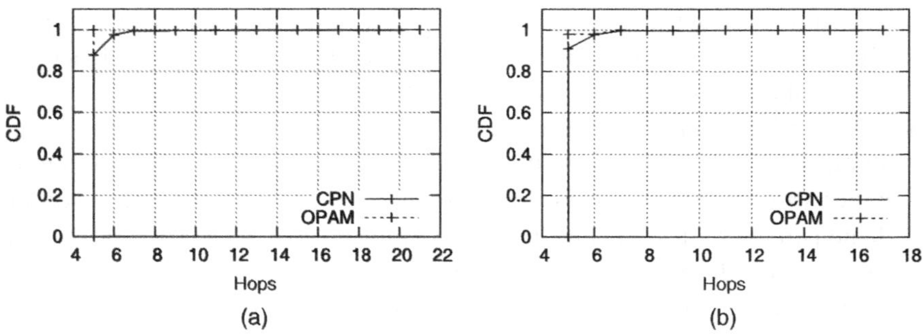

Figure 5.15 CDF of the number of hops in CPN and OPAM. (a) no link loss and (b) 2% link loss.

5.4 Summary

This chapter introduces a novel OF-based NDPA solution, comprising MPSGA and PAFM, aimed at enhancing IEC 61850 MMS packet delivery reliability in the presence of communication link failures. Simulation experiments on IEEE and Texas networks reveal that MPSGA significantly improves the success rate of receiving data packets compared to existing solutions. Additionally, experiments on Texas, PEGASE, and French networks demonstrate MPSGA's effectiveness in large-scale power networks. The queued PAFM reduces the inter-arrival time of TCP data packets during failures, meeting the stringent delivery time requirements of IEC 61850 MMS messages. OPAM further reduces memory usage in the packet reordering process and halves the Round-Trip Time (RTT), thanks to faster routing convergence and a higher success rate of packet receipt, compared to the CPN.

Questions

5.1 Elaborate on the trade-offs involved in the design choices made for the OPAM solution, particularly focusing on the balance between routing convergence time, memory usage, and packet delivery reliability.

5.2 Explain the significance of MPSGA's optimization on path selection in the context of power networks. How does it contribute to achieving higher SRRP and QSDP compared to other algorithms?

5.3 How does the excitation process of RNN in CPN contribute to observing routing convergence, and what does the average potential change per neuron signify?

5.4 Explain the significance of SRRP and HRRT in evaluating the performance of communication networks, and how do these metrics reflect the reliability and responsiveness of the proposed OPAM solution?

5.5 Compare the packet reordering buffer performance between CPN and OPAM, considering POR, PIAT, and memory usage. How do these factors contribute to the scalability and efficiency of each solution?

5.6 Analyze the results presented in Figure 5.11(a) regarding the performance of PAFM with and without a queue. How does the presence of a queue impact packet throughput, and what implications does this have for TCP performance?

5.7 Elaborate on the impact of link loss rates on the routing convergence time in CPN and OPAM. How do the observed differences in convergence time contribute to the overall reliability and responsiveness of the communication systems?

5.8 Compare the packet reordering behavior in CPN and OPAM under different scenarios. Discuss the reasons behind the differences in POR and PIAT, and how these factors influence the reliability of packet delivery.

Bibliography

Aaron Bernstein. A nearly optimal algorithm for approximating replacement paths and k shortest simple paths in general graphs. In *ACM-SIAM SODA*, pages 742–755, 2010.

Anirban Chakrabarti and G. Manimaran. Reliability constrained routing in QoS networks. *IEEE/ACM Transactions on Networking*, 13(3):662–675, 2005.

Boris V. Cherkassky, Andrew V. Goldberg, and Tomasz Radzik. Shortest paths algorithms: Theory and experimental evaluation. *Springer Mathematical Programming*, 73(2):129–174, 1996.

Yong Cui, Xin Wang, Hongyi Wang, Guangjin Pan, and Yining Wang. FMTCP: A fountain code-based multipath transmission control protocol. In *IEEE ICDCS*, pages 366–375, 2012.

David Erickson. The Beacon OpenFlow controller. In *ACM SIGCOMM workshop on HotSDN*, pages 13–18, 2013.

Hamid Farhady, HyunYong Lee, and Akihiro Nakao. Software-defined networking: A survey. *Elsevier Computer Networks*, 81(Supplement C):79–95, 2015.

Asaf Frieder and Liam Roditty. An experimental study on approximating k shortest simple paths. *Journal of Experimental Algorithmics*, 19(1):1–5, 2015.

Erol Gelenbe. Steps toward self-aware networks. *Communications of the ACM*, 52(7):66–75, 2009.

Erol Gelenbe. *Self-Aware Networks: The Cognitive Packet Network and Its Performance*, pages 659–668. Springer International Publishing, 2017.

John Hershberger and Subhash Suri. Vickrey prices and shortest paths: What is an edge worth? In *IEEE FOCS*, pages 252–259, 2001.

John Hershberger, Matthew Maxel, and Subhash Suri. Finding the k shortest simple paths: A new algorithm and its implementation. *ACM Transactions on Algorithms*, 3(4):45–es, 2007.

IEC. Communication networks and systems for power utility automation - Part 5: Communication requirements for functions and devices, 2013.

IEEE. IEEE standard bus network dataset, 2025. URL http://www.ee.washington.edu/research/pstca/.

Illinois Center for a Smarter Electric Grid (ICSEG). Texas bus network dataset, 2016. URL http://icseg.iti.illinois.edu/power-cases/.

Cédric Josz, Stéphane Fliscounakis, Jean Maeght, and Patrick Panciatici. AC power flow data in MATPOWER and QCQP format: iTesla, RTE snapshots, and PEGASE. *arXiv preprint arXiv:1603.01533*, 2016.

K. C. Leung, V. O. K. Li, and D. Yang. An overview of packet reordering in transmission control protocol (TCP): Problems, solutions, and challenges. *IEEE Transactions on Parallel and Distributed Systems*, 18(4):522–535, 2007.

Open Networking Foundation. OpenFlow specification v1.3, 2012.

Wanlin Pang and Scott D. Goodwin. *A Graph Based Backtracking Algorithm for Solving General CSPs*, pages 114–128. Springer-Verlag, Berlin, Heidelberg, 2003.

Miroslav Popovic, Maaz Mohiuddin, Dan Cristian Tomozei, and Jean Yves Le Boudec. iPRP—The parallel redundancy protocol for IP networks: Protocol design and operation. *IEEE Transactions on Industrial Informatics*, 12(5): 1842–1854, 2016.

George F. Riley and Thomas R. Henderson. The NS-3 network simulator. *Modeling & Tools for Network Simulation*, pages 15–34. Springer-Verlag, Berlin, Heidelberg, 2010.

Amin Shokrollahi. Raptor codes. *IEEE Transactions on Information Theory*, 52(6):2551–2567, 2006.

S. Soltan and G. Zussman. Generation of synthetic spatially embedded power grid networks. In *IEEE PESGM*, pages 1–5, 2016.

Nina Taft-Plotkin, Bhargav Bellur, and R. Ogier. Quality-of-service routing using maximally disjoint paths. In *IEEE IWQoS*, pages 119–128, 1999.

Maryam Tanha, Dawood Sajjadi, and Jianping Pan. Enduring node failures through resilient controller placement for software defined networks. In *IEEE GLOBECOM*, pages 1–7, 2017.

Lan Wang and Erol Gelenbe. Real-time traffic over the cognitive packet network. In *Computer Networks: 23rd International Conference, CN 2016, Brunów, Poland, June 14–17, 2016, Proceedings 23*, pages 3–21. Springer International Publishing, 2016.

Bin Wang and Jennifer C. Hou. Multicast routing and its QoS extension: Problems, algorithms, and protocols. *IEEE Network*, 14(1):22–36, 2000.

Jin Y. Yen. Finding the k shortest loopless paths in a network. *INFORMS Management Science*, 17(11):712–716, 1971.

Ray Daniel Zimmerman, Carlos Edmundo Murillo-Shez, and Robert John Thomas. MATPOWER: Steady-state operations, planning, and analysis tools for power systems research and education. *IEEE Transactions on Power Systems*, 26(1):12–19, 2011.

6

Navigating Network Resilience Using Traffic-Aware Fast Reroute

In the realm of smart grid communications, the widely adopted Named Data Network (NDN) has proven its prowess in data delivery efficiency. However, to enhance resilience in the face of link faults, NDN requires a robust Fast Reroute (FRR). The conventional approaches based on disjoint arborescences and paths face limitations in reroute options due to the scarcity of available disjoint end-to-end routes. Overcoming these constraints, we leverage Disjoint Subpaths (DSPs) between intermediate nodes in the Primary Forwarding Path (PP), extending rerouting possibilities beyond the conventional two-end linkage. Moreover, existing approaches designed for the data plane lack alignment with NDN. Notably, the swift reroute using detours, crucial for minimizing Link Recovery Time (LRT) and ensuring grid state estimation accuracy, has been insufficiently explored for NDN.

Motivated by the challenges in network resilience, this chapter introduces a novel Traffic-Aware Fast Re-Route (TA-FRR) mechanism tailored for NDN. This mechanism leverages a new route structure termed the Resilient Subgraph (RSG), which comprises a PP and a detour DSP between two routers in the PP for efficient detouring.

Designed as a control strategy embedded in each NDN router, TA-FRR operates with our extended NDN messages. For every communication pair, TA-FRR proactively constructs an RSG offline, optimizing the End-to-End Payload Packet Delivery Time (EEDT) and minimizing both the Reroute Failure Rate (RFR) and the total PP delay. This ensures detouring fault links with maximum resilience while maintaining a balanced PP length.

The RSG construction algorithm is designed for scalability, making it suitable for large networks. With the pre-installed RSG in routers, TA-FRR rapidly recovers fault links online within a short LRT. It identifies faults based on traffic arrival timeouts and selectively reroutes disconnected components in the PP.

The reroute process unfolds in stages: (1) Periodic fault detection and propagation of local link states across routers. (2) Detouring notified fault links on the

Communication Networks in Smart Power Grids, First Edition. Boyang Zhou.

Companion Website: www.wiley.com/go/SmartPowerGrids1e

PP using available Redundant Sub-Paths (RSPs) within the RSG. Detour choices are optimized to minimize steps, avoiding unnecessary rerouting actions within the connected component on the PP. The algorithm is scalable in both time and space complexity. Meanwhile, consumers efficiently re-sequence out-of-order Data packets using a priority queue.

In comparison to TI-LFA [Litkowski et al., 2022], TA-FRR stands out by not encapsulating paths into packets, minimizing overhead.

We conduct a comprehensive evaluation of the TA-FRR performance through simulations in ndnSIM [Mastorakis et al., 2017]. The evaluation employs an RSG selected with abundant joint paths, constructed from the IEEE 300 bus dataset network [IEEE, 2022], representing a real power transmission grid. Simulations introduce one, multiple, or complete link faults on the PP, varying in disruption degrees.

Numerical results demonstrate the efficiency of TA-FRR in swiftly recovering link faults, achieving a short LRT between 0.4 and 98.9 ms, with a maximum End-to-End Packet Loss Rate (EELR) of 2.03%. Additionally, the results showcase TA-FRR's capability to handle out-of-order Data packets during high-speed data delivery (200 to 8000 Hz) with a bounded end-to-end delay. These statistics underscore the effectiveness of rerouting in mitigating multiple link faults.

Furthermore, we test TA-FRR in ndnSIM [Mastorakis et al., 2017] using a synthesized Internet topology [Medina et al., 2001] under various link fault degrees. The outcomes reveal TA-FRR's ability to recover severe multiple link faults on the PP with a short LRT, constrained EEDT, and minimal message cost. The results align with the theoretical analysis of RSG construction time.

In a comparative analysis, we assess TA-FRR against existing approaches like arborescences and disjoint paths under random link faults in the network. TA-FRR demonstrates superior performance with significantly lower RFR and a maximum stretch cost comparable to the arborescences approach.

6.1 Route Model and Problem

Preliminaries on the TA-FRR are given as follows.

6.1.1 Multi-Path Sub-Graph-based Routes

In this communication network, let's explore the concept of an (RSG) from a producer to a consumer. Imagine it as a subgraph of our network, marked as $G_P \subset G$. This G_P can be defined as (R_P, L_P), breaking down into two vital parts:

a) The PP kicks off from the consumer and producer, taking the shortest route, denoted as $p_{\text{prim}} = (r_{d_1}, r_{d_2}, \dots, r_{d_b})$.

b) For each node on the PP, like r_{d_i}, a set of Redundant Sub-Paths (RSPs) link r_{d_i} with its upstream node r_{d_s} on the PP, written as $P_{\text{rsp}} = \{p_{\text{rsp}}(r_{d_i}, r_{d_s})\}$. Each RSP between r_{d_i} and r_{d_s} is distinct from the p_{prim} between r_{d_i} and r_{d_s}.

For instance, in Figure 6.1, you can see to an RSG manually crafted from the IEEE 300 bus dataset [IEEE, 2022], starting from the 165th node to the 100th node. The thick arrows represent the PP, while the thin lines represent the RSPs.

Creating such G_P is a breeze with existing algorithms like [Yen, 1971], Maximally Disjoint Shortest & Widest Paths (MADSWIP) [Taft-Plotkin et al., 1999], and the RSG Generation Algorithm (RSGA) [Zhou et al., 2019]. We can convert the RSG into a PP and its RSPs easily. First, compute the PP using the Dijkstra algorithm. Second, calculate all RSPs for each r_{d_k} of the PP. This process, demonstrated using IEEE bus topologies in Figure 6.2, showcases the ease of constructing such RSG. When generating 14 paths in the RSGA, the average RSPs for IEEE 30, 57, 118, and 300 bus topologies are 7.66, 10.63, 16.38, and 10.33, respectively. Abundant RSPs in these network grids affirm the simplicity of this approach.

6.1.2 Defining Problem

In NDN, imagine a consumer's journey to retrieve a data stream from a producer, as outlined by Zhang et al. [2014]. The process unfolds in a charming dance of communication. Initially, the consumer elegantly dispatches an Interest packet, adorned with a prefix name, toward the producer, following a designated path. In response to this intriguing request, the producer gracefully retorts with Data packets carrying the sought-after prefix, gracefully tracing the path in reverse.

Picture an NDN network as a captivating undirected graph, symbolized by $G = (R, L)$. Here, R embodies an ensemble of routers, each with a distinct personality: $\{r_1, r_2, \dots, r_{|R|}\}$. Meanwhile, L weaves connections between these routers, creating a mesmerizing tapestry: $\{l_1, l_2, \dots, l_{|L|}\}$. The anticipated fault rate of a link, a subtle nuance in this symphony, connecting r_x and r_w unfolds as $\epsilon(x, w)$.

Embark on a visual journey through the network in Figure 6.3, where an enchanting RSG unfolds between a producer (r_{d_1}) and a consumer (r_{d_b}). This magical RSG twirls with two mesmerizing components:

a) A PP, p_{prim}, emerges gracefully as $(d_1, d_2, \dots, d_b)$.
b) Each router on the PP, akin to a protagonist in this tale, like r_{d_k}, possesses a set of DSP, a loyal entourage denoted as $P_{\text{dsp}}[d_i]$. Every DSP carves a unique path between the current router and its upstream companion on the PP, away from the PP itself and other DSP voyages connected to the present router.

Guided by the magical RSG, each router pirouettes along the PP, elegantly rerouting any detected fault links toward its trusted DSP.

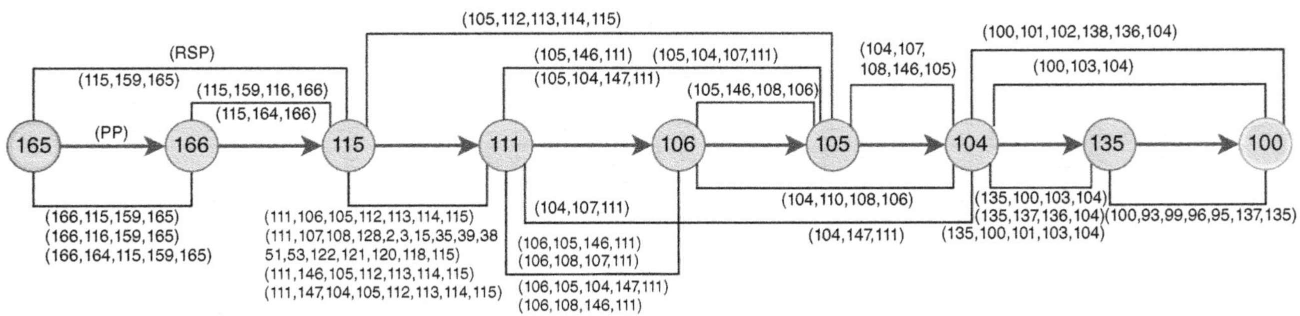

Figure 6.1 Example of to an RSG from 165th node to 100th node constructed from IEEE 300 bus dataset topology.

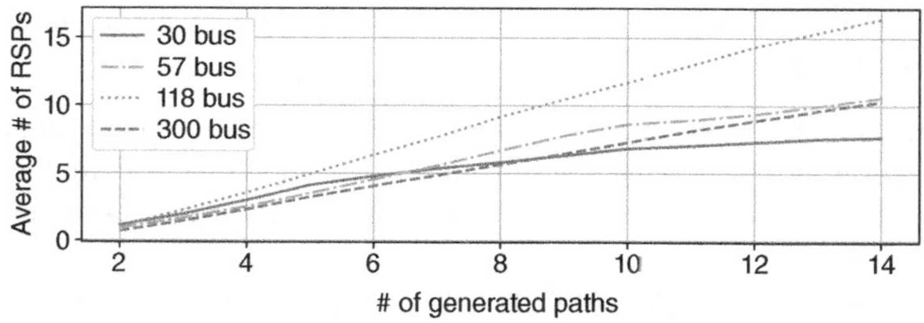

Figure 6.2 The average number of RSPs per node in the PP in a transmission grid constructed by RSGA according to the IEEE 300 bus dataset topology.

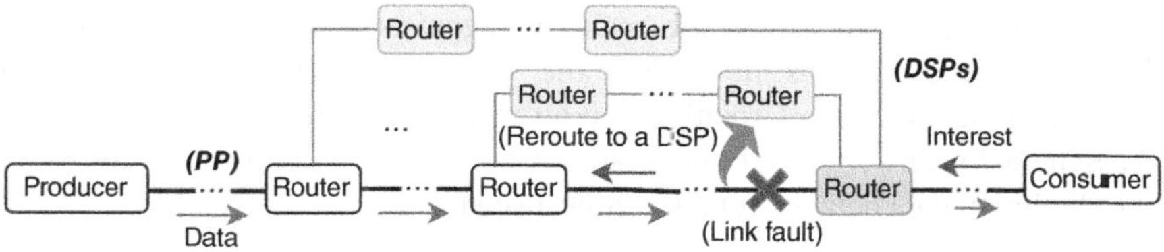

Figure 6.3 Reroute based on an RSG.

Concerning the network, the RFR at r_{d_i} gracefully unveils itself through the poetic expression in Eq. (6.1). This equation encapsulates the probability that all DSP stemming from r_{d_i}, along with the link (d_{i-1}, d_i) of the PP, gracefully dance together into the realm of faults.

Delving further into the symphony of the network, the graceful LRT takes center stage. It harmonizes the ballet of detection and detouring, a pivotal element in our narrative. The conundrum unfolds as an intricate dance – to minimize both the RFR and the fault rate of the PP, maintaining a delicate equilibrium within the bounds of a predefined EEDT of D.

Behold the crystallization of this challenge in Eq. (6.2). This equation artfully articulates the quest: to find the optimal balance, orchestrated by an adjustable parameter α, a maestro's touch to harmonize service needs. Amid this ballet, the rerouting unfolds with grace, guided by the constraint of a short LRT, a gesture to diminish the EELR.

$$l_r[d_i] = \prod_{p \in (d_{i-1}, d_i) \cup P_{\text{dsp}}[d_i]} \left(1 - \left(1 - \prod_{(r_{c-1}, r_c) \in p} [1 - \epsilon(d_{c-1}, d_c)]\right.\right.$$
$$\left.\left. \cdot \left(1 - l_r \left[p\left[|p|\right]\right]\right)\right) \tag{6.1}$$

$$\begin{aligned} \text{argmin}_{\gamma=(p_{\text{prim}}, P_{\text{dsp}})} \alpha \cdot \prod_{l \in p_{\text{prim}}} \epsilon(l) + (1-\alpha) \cdot l_r[d_b] \\ \text{subject to : EEDT}(\gamma) \leq D \end{aligned} \tag{6.2}$$

To gracefully waltz through the intricacies of LRT, a choreography unfolds, directed by two principal facets:

i) **Fault link sensitivity:** A dance of precision where fault link sensitivity takes the lead, ensuring a swift reduction in fault link detection time. Picture this as the heartbeat of the performance, orchestrating a rhythm that quickens the pace of recovery.
ii) **Avoidance of unnecessary rerouting steps:** Imagine a dance floor where unnecessary steps are skipped, a strategic move to minimize the convergence time for new routes. This dance, swift and purposeful, trims the fat, streamlining the journey to the grand finale.

These two acts emerge as the main protagonists in the narrative of LRT reduction. And lo, let's not forget the Consumer Router, the master of re-sequencing. In the aftermath of rerouting, a task at hand – to elegantly reassemble the out-of-order Data packets. A finale of sorts, a resolution to ensure the rhythm of data delivery remains untarnished.

6.2 Fast Proactive Reroute Mechanism

In the realm of TA-FRR, let's delve into the proposed mechanisms, adorned with discussions on the intricacies of reroute optimization and the graceful dance of computational complexity. It's a journey into the heart of efficiency and resilience, where algorithms take center stage and discussions unfold like a well-choreographed performance.

6.2.1 Design Overall Mechanism

Enter the realm of TA-FRR architecture, where innovation meets the well-trodden paths of NDN forwarding daemons. At its core, this design breathes life into the existing NDN infrastructure, enhancing it with a customized touch. Imagine the NDN daemon as a conductor orchestrating a symphony of Echo, Back, and Link State Advertisement (LSA) messages, each bearing a unique tune in the format of "/[content name]/[operation]/parameters" (see Table 6.1). These messages dance between routers, carrying the heartbeat of the network.

In this vibrant ecosystem, routers juggle two vital pieces of information: the intricate RSG mapping communication pairs and the ever-changing link states to neighboring routers, encapsulated in L_{states}. Figure 6.4 unveils the intricate ballet of TA-FRR, a dance of two synchronized processes. First, a meticulous exploration of link states, probing the local links entwined in the RSG and broadcasting their

Table 6.1 Messages extended by TA-FRR.

Operation	Parameters	Usage Description
Echo	Echo Identification (ID) and source node ID	A request for testing connectivity of a link
Back	Echo ID, source node ID, and destination ID	A feedback to indicate the connectivity of the link
LSA	Source node ID, update version, and link states	An advertisement for the link states of the source node

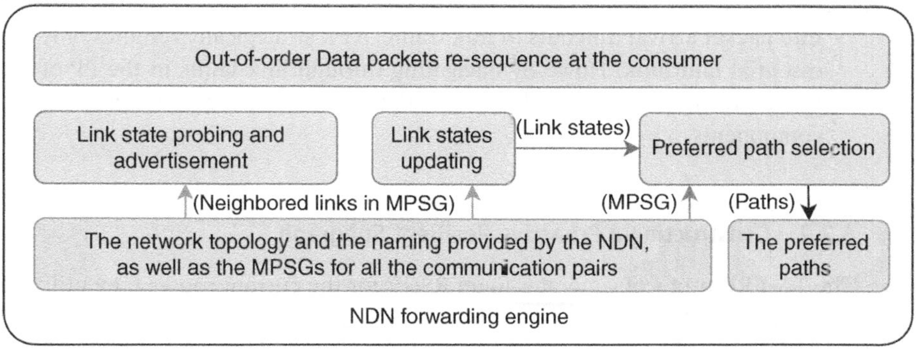

Figure 6.4 TA-FRR architecture.

states to neighboring routers. Then, a duet unfolds – link states update, synchronized to the rhythm of received LSA, and the selection of preferred paths, where routers navigate the network's dynamic tapestry. It's an architectural spectacle where collaboration and synchronization take center stage.

Figure 6.5 reveals the ingenious design of TA-FRR, drawing inspiration from the NDN forwarding daemon [NDN, 2021]. The magic happens with the network

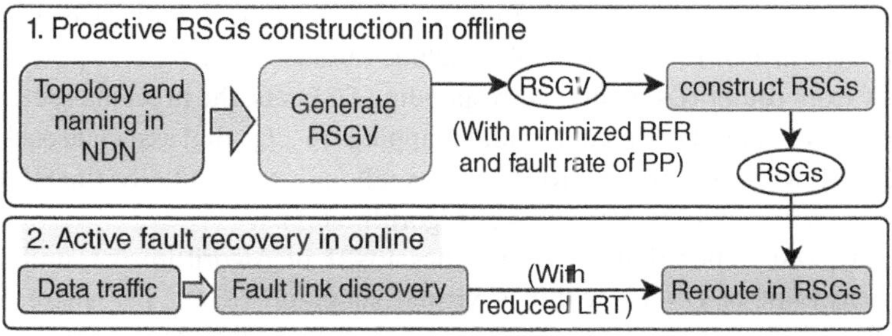

Figure 6.5 Overall design of TA-FRR.

topology discovery and naming, skillfully provided by NDN [Zhang et al., 2014; NDN, 2021].

TA-FRR takes on the challenge with a dynamic two-phase approach:

1) **Proactive RSG construction:** Picture this – in the network, RSG takes its roots from the present router and extends to others identified by their unique names. Each RSG is a blend of a PP and, at most, β DSP per node of the PP. The beauty lies in their once-computable nature, thanks to the overlapping optimization goals. We crunch all the RSG details into a vector (RSG vector (RSGV)) and then weave the intricate tapestry of each RSG accordingly.
2) **Active Fault Recovery (AFR) unleashed:** Enter AFR, the hero here to minimize LRT. Armed with the prowess to identify fault links by tapping into packet arrival timeouts in link traffic, AFR strategically reroutes only the essential fault link. How? By navigating through link faults in the PP of the related RSG, where a fault link holds the power to divide its PP into distinct components.

6.2.2 Constructing a Proactive Resilient Subgraph

The TA-FRR protocol generates to an RSGV for the current router r_i by utilizing the GenRSGV algorithm (see Algorithm 6.1) and its subroutine, the ComputeDSP algorithm (see Algorithm 6.2). This process ensures the network meets the EEDT constraint on delivery time D, using parameters α and β to optimize for scalability and efficiency. The goal is to minimize both the RFR and the fault rate along the PP.

The workflows of the GenRSGV and ComputeDSP algorithms are illustrated in Figure 6.6. GenRSGV computes each DSP under a time constraint τ_m, ensuring that it adheres to the delivery time D (see Algorithm 6.2). Each DSP is situated between r_v and its upstream router on the alternate path p', within a subgraph of G, denoted as $(R, L\backslash\hat{L})$, and avoids interference from links in the set $\hat{L}$.

As shown in Figure 6.6, GenRSGV requires the initialization of key variables in lines 2–3 of Algorithm 6.1. These variables include the size of the router set $R(n)$, a mapping X from router IDs to their corresponding DSP sets, the previous router ID (*prev*) on the PP, the EEDT record T, and mappings (S_p, L_r) that associate each router on the PP with its fault rate, RFR, and overall fault rate. Initially, the fault rates for router r_i are set to 0% (see line 4 of Algorithm 6.1).

Additionally, DSP generation occurs in lines 13–21 of Algorithm 6.2, under the delay constraint of $D - \tau'$, within a virtual subgraph $(R, L \subset \hat{L})$, where $\hat{L}$ includes all links of path p'. The newly generated DSP is added to the set P_{dsp},

Algorithm 6.1 Generate to an RSGV

1: **function** GenRSGV($G = (R, L), r_i, \alpha, \beta, D$)
2: $n = |R|, prev = [-1]_n, X = \{\}, T = [0]_n$
3: $S_p = [1]_n, L_r = [1]_n, L = [1]_n$
4: $S_p[i] = 0, L_r[i] = 0, L[i] = 0$
5: $q_1 = \text{FiboHeap}(L, \text{inc})$, **for all** $r_k \in R$ **do** $q_1 \leftarrow k$
6: **while** $|q_1| > 0$ **do**
7: $r_u = q_1.\text{poll}()$
8: **for all** $r_v \in \text{ExitNodes}(r_u) \land r_v \notin q_1$ **do**
9: $T' = T[u] + \tau_{\text{all}}^{\text{PP}}(r_u, r_v)$
10: **if** $T' > D$ **then continue**
11: $p' = \text{ConvertPath}(prev, r_i, r_u)$
12: $s'_p = S_p[u] \cdot \epsilon(u, v)$
13: $l'_r = 1 - (1 - L_r[u]) \cdot \epsilon(u, v)$
14: $P_{dsp} = \{\}, \hat{L} = \text{links}(p'), \tilde{L} = \{\}$
15: **for** $(i = 0; i < \beta; i = i + 1)$ **do**
16: $p_r = \text{ComDSP}(R, L, \hat{L}, \tilde{L}, v, p', \eta, D - T')$
17: **if** $p_r = \text{null} \lor p \in P_{dsp}$ **then break**
18: $P_{dsp} = P_{dsp} \cup p_r$
19: **if** $i < \beta - 1$ **then** $\tilde{L} = \tilde{L} \cup \text{links}(p_r.p)$
20: $l'_r = l'_r \cdot (1 - (1 - L_r[p_r[|p_r|]])) \cdot (1 - \epsilon(p_r))$
21: **end for**
22: $l' = \alpha \cdot (1 - s'_p) + (1 - \alpha) \cdot l'_r$
23: **if** $l' \geq L[v]$ **then continue**
24: $L[v] = l', q_1.\text{updateKey}(v)$
25: $T[v] = T', prev[v] = u$
26: $S_p[v] = s_p, L_r[v] = l_r$
27: **if** $|P_{dsp}| > 0$ **then** $X[v] = P_{dsp}$
28: **end for**
29: **end while**
30: **return** $RSGV = (prev, X)$
31: **end function**

ensuring maximum disjointness from links in $\tilde{L}$, which contains the last $\beta - 1$ DSPs from P_{dsp}. The RFR and overall fault rate are calculated between lines 13–20 and line 22 using Eqs. (6.1) and (6.2), respectively. The final overall fault rate, denoted as l', is determined in line 22.

Ultimately, GenRSGV returns the RSGV in line 30 of Algorithm 6.1, which provides the necessary data for constructing the RSG in the subsequent steps.

Algorithm 6.2 Compute DSP

1: **function** ComDSP($R, L, \hat{L}, v, p', \eta, \tau_m$)
2: $R = R \cup \{r_{\text{vir}}\}, L = L \cup \{l = (r_s, r_{\text{vir}}) | r_s \in p'\}$
3: $n' = |R| + 1, \kappa = [-\infty]_{n'}, \kappa' = [-\infty]_{n'}$
4: $\kappa[v] = 0, \kappa'[v] = 0, prev = [-1]_{1 \times n'}$
5: $q_2 = \text{FiboHeap}(\kappa, \text{dec})$, **for all** $r_h \in R$ **do** $q_2 \leftarrow h$
6: **while** $|q_2| > 0$ **do**
7: $w = q_2.\text{poll}()$
8: **for all** $r_x \in \text{ExitNodes}(r_w) \wedge r_x \notin q_2$ **do**
9: **if** $(r_w, r_x) \in \hat{L}$ **then continue**
10: **if** $T[w] + \tau(w, x) > \tau_m$ **continue**
11: $\kappa_1 = \log((1 - \epsilon(w, x))(1 - \epsilon(x, w)))$
12: $\hat{\kappa}_x = \kappa_1 + \kappa[w]$
13: **if** $\hat{\kappa}_x > \kappa[x]$ **then**
14: $prev[x] = w, \kappa[x] = \hat{\kappa}_x, \kappa'[x] = \kappa_1 + \kappa'[w]$
15: $q_2.\text{UpdateKey}(r_x)$
16: **end if**
17: **end for**
18: **end while**
19: $p = \text{ConvertPath}(prev, r_v, r_{\text{vir}}) \setminus \{r_{\text{vir}}\}$
20: $R = R \setminus \{r_{\text{vir}}\}, L \setminus = \{(r_s, r_{\text{vir}}) | r_s \in p'\}$
21: $\epsilon = 1 - e^{\kappa'[r_{\text{vir}}]}$
22: **return** (p, ϵ)
23: **end function**

Based on the generated RSGV for the current router r_i, an RSG is constructed between r_i and r_j under the EEDT constraint of D using the ConstructRSG algorithm (see Algorithm 6.3). The workflow of the ConstructRSG algorithm is shown in Figure 6.7.

6.2.2.1 Discussions

Let's delve into the intricacies of Algorithm 6.2's execution time complexity. Lines 2–5, 7, 8–18, 19, and 20–21 exhibit complexities of $O(|R|)$, $O(\log(|R|))$, $O(|L|/|R|)$, and $O(\log |R|)$, respectively. In the realm of Internet topology, $|L| = |R| \cdot \log |R|$ owing to the power-law degree distribution [Zhou and Mondragón, 2004]. Consequently, the overall execution time of Algorithm 6.2 can be encapsulated as $\nu = O(|R| \cdot \log |R| + |R| + \log |R|)$.

Next, we scrutinize the execution time of GenRSGV in Algorithm 6.1. Lines 1–5, 7, and 8–28 are bound by complexities of $O(\log |R|)$, $O(\log |R|)$, and $O(|L|/|R| \cdot (\log |R| + \beta \cdot \nu))$, respectively. Consequently, the overall execution

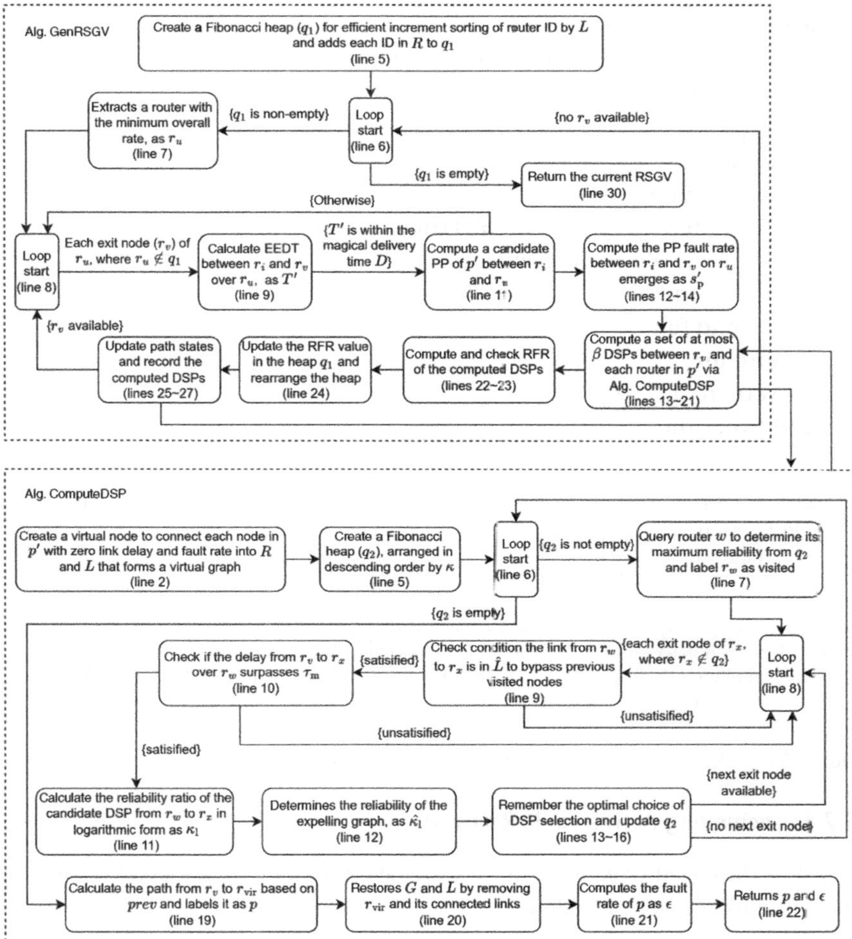

Figure 6.6 Workflows of TA-FRR's GenRSGV and ComputeDSP algorithms.

time can be expressed as $O((|R| + \log|R|) \cdot (\beta \cdot |R| + \log|R| + |R| + \beta + 1) + \beta \cdot \log^2|R| + \log|R|)$, which simplifies to $O(\beta \cdot |R|^2 \cdot \log^2|R|)$.

Meanwhile, the execution space of ComDSP unfolds as follows. Lines 2–18 and 19 pose complexities of $O(|R|)$ and $O(\log|R|)$, respectively. Thus, the overarching space complexity is $O(|R| + \log|R|)$. Moving on to the space analysis of GenRSGV, lines 2–4, 11, and 16 contribute complexities of $O(|R|)$, $\log|R|$, and $O(|R| + \log|R|)$, respectively. Consequently, the total space complexity materializes as $O(|R| + (\beta + 1) \cdot \log|R|)$.

Simultaneously, the execution space of ConstructRSG is intricately linked to the size of RSG in lines 2–3, i.e., $O(\beta \cdot \log|R| + |R|)$.

Algorithm 6.3 Construct RSG

1: **function** ConstructRSG(L, RRV, r_i, r_j, D)
2: $p_{\text{prim}} = \text{ConvertPath}(RRV.prev, r_i, r_j)$
3: $X_{\text{dsp}} = \{r_k \rightarrow X[r_k] | r_k \in p_{\text{prim}}\}$
4: **for all** $r_k \in p_{\text{prim}}$ **do**
5: **for all** $p_{\text{dsp}} \in X_{\text{dsp}}[r_k]$ **do**
6: **if** $p_{\text{dsp}} \cap p_{\text{prim}}(r_k, r_j) \neq \emptyset$ **then**
7: $X_{\text{dsp}}[r_k].\text{remove}(p_{\text{dsp}})$
8: **end if**
9: **end for**
10: **end for**
11: **return** $(p_{\text{prim}}, X_{\text{dsp}})$
12: **end function**

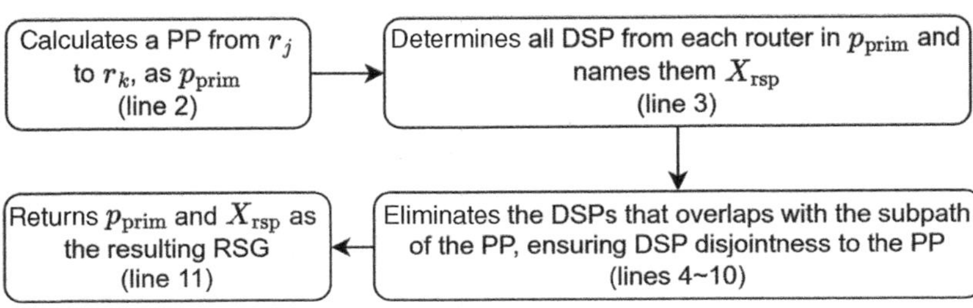

Figure 6.7 Workflow of TA-FRR's ConstructRSG algorithm.

6.2.3 Achieving Active Fault Recovery

Based on the constructed RSG, in AFR, we detail the aforementioned two steps of the fault link recovery as follows.

6.2.3.1 Identifying Faulty Links

In this phase, the router employs a two-fold approach to detect a fault link. It starts by broadcasting Echo messages to all its neighbors and patiently awaits corresponding Back messages. The waiting time is determined by the maximum Round-Trip Time (RTT) of the link, encompassing delays in transmission, queuing, and processing. Fault detection unfolds through two methods:

a) **Active detection:** This is triggered when the maximum expected Packet Inter-Arrival Time (PIAT) of incoming Data packets exceeds the timeout. The router determines this timeout by leveraging the Poisson distribution, modeling packet arrivals on the Internet [Karagiannis et al., 2004]. The router computes the mean throughput of each neighboring link l over the recent τ_s seconds as λ_l. Subsequently, it calculates the maximum PIAT with a Confidence

Ratio (CR) of h through $-\ln(1-h)/\lambda_l$. If no packet arrives within this maximum PIAT, the router flags the link as faulty.

b) **Periodic detection:** This is initiated at intervals configured by the operator, responding to conditions of link state flaps between normal and faulty. This active approach reduces the message count compared to the passive method, e.g., bidirectional forwarding detection [IETF RFC 5880, 2010]. Post-detection, the router broadcasts an advertisement containing the states of all neighboring links (LSA) to other routers in the RSG that share at least one neighboring link.

In the fault link discovery process, the time and message count for each router are $O(1)$ and $O(\log |R|)$, respectively.

Figure 6.8 illustrates the process through state transitions, involving the following steps. Each router probes the states of all links in the RSG in a period of $\tau_{\tau_{\text{check}}}$: (1) The router sends Echos with an incrementing Identification (ID) for each neighboring link, and (2) it waits for a corresponding Back message for each Echo ID. The waiting includes a timeout of τ_{echo}. These steps are repeated η times. If there is no Back message, the link is marked as failed; otherwise, it's deemed normal. The repetition prevents link state flapping caused by issues like link congestion. Following the probing, the router broadcasts an LSA to other routers, containing the states of the neighboring links with the current router ID at the current version.

6.2.3.2 Rerouting Strategies in Resilient Subgraph

When a router receives an LSA indicating a change in state, it dynamically reroutes through the feasible detoured paths within the RSG using the following four steps:

1) **Partitioning the PP:** The router divides the PP of each RSG into various connected components based on the fault link states of the PP.
2) **Virtual graph construction:** Create a virtual graph where each component becomes a vertex, and all non-fault DSPs and links from the PP become edges.
3) **Shortest detour computation:** Utilizing Dijkstra's algorithm, calculate the shortest detouring path from the two components containing the consumer and producer in the virtual graph.
4) **Rerouting:** Redirect each current path containing the fault links identified in the received LSA to its corresponding shortest detour path.

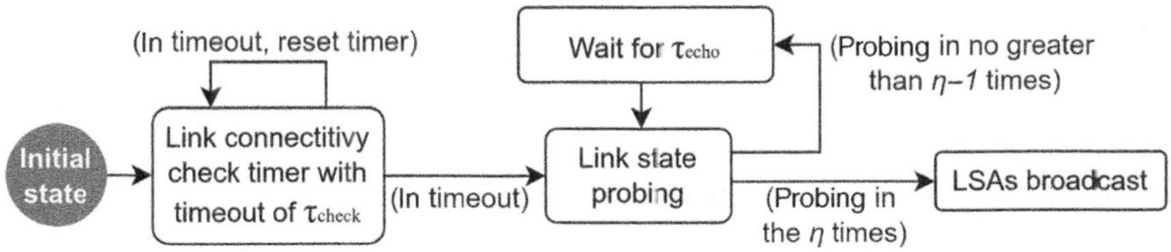

Figure 6.8 State transitions in the link state probing and advertisement.

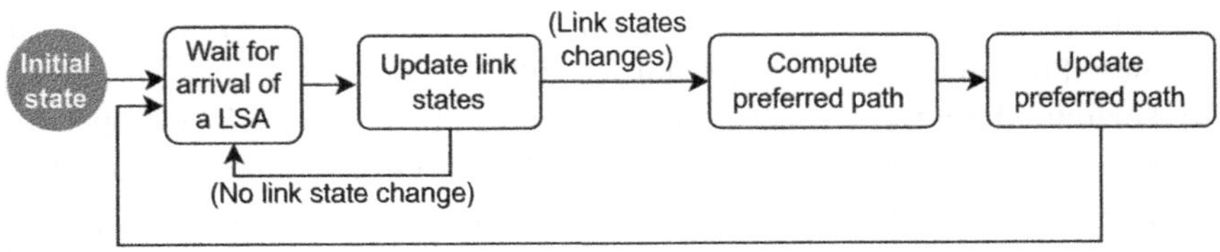

Figure 6.9 State transitions in the link state updating and path selection.

The execution time and space complexities for these steps are $O(|R_P| \cdot \log |R_P|)$ and $O(|R_P| + \log |R_P|)$, respectively. Here, $|R_P|$ represents the number of routers in the RSG, falling within $O(\beta \cdot \log^2 |R|)$. Consequently, the overall complexities are $O(\beta \cdot \log^2 |R| \cdot \log(\beta \cdot \log^2 |R|))$ and $O(\beta \cdot \log^2 |R| + \log(\beta \cdot \log^2 |R|))$, respectively.

Figure 6.9 visually illustrates the process through its state transitions. In practice, the process unfolds in the subsequent two steps.

6.2.3.3 Differential Link State Updates

After receiving the LSA, the current router r_k processes it using Algorithm 6.4 with the workflow presented in Figure 6.10. The workflow is specified with the following steps:

- **Check for changes**: In line 2, determine if any link states in the LSA have changed compared to the current ones. If not, discard the LSA to avoid unnecessary computation.
- **Path ID existence**: In line 3, if r_k has no path ID in the PP (only with RSPs), there's no need to update link states.
- **Current link states**: In line 4, obtain the current link states (L_{states}).
- **Update link states**: In lines 5 to 10, update the states of each link on the PP and each RSP of L_{states} to a failure state if the link or RSP contains at least one link in L'_{states}.

6.2.3.4 Implementing Distributed Preferred Path Selection

Based on L_{states} and G_P, the router r_k computes the preferred path through Algorithm 6.4 with minimized reroute actions. The workflow of this algorithm is visually presented in Figure 6.11.

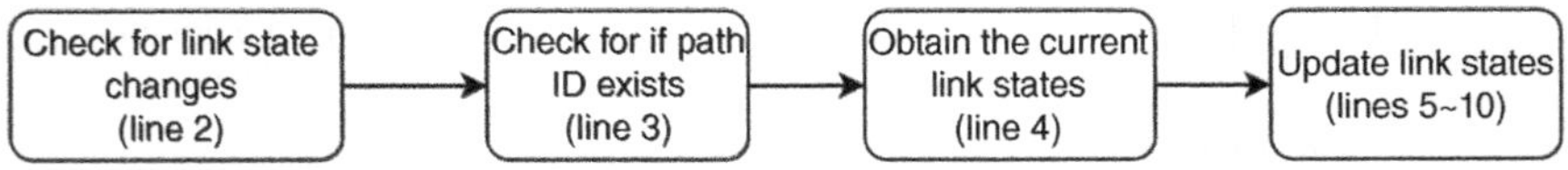

Figure 6.10 Workflow of link state updates.

Algorithm 6.4 Update Link States

procedure UPDATELINKSTATES(L'_{states}, r_k)
  if L'_{states} has no change **then return**
  if r_k is with only RSPs **then return**
  $L_{\text{states}} = \{l \rightarrow \text{t} | l \in p_{\text{prim}}\} \cup \{p \rightarrow \text{t} | p \in P_{\text{rsp}}\}$
  for $l_s \in L'_{\text{states}}$ **do**
    if l_s is in failure **then**
      $L_{\text{states}}[p] = \text{fail}, \forall p \in P_{\text{rsp}}[r_{d_i}] \wedge l_s \in p$
      if $l_s \in p_{\text{prim}}$ **then** $L_{\text{states}}[l_s] = \text{f}$
    end if
  end for
end procedure

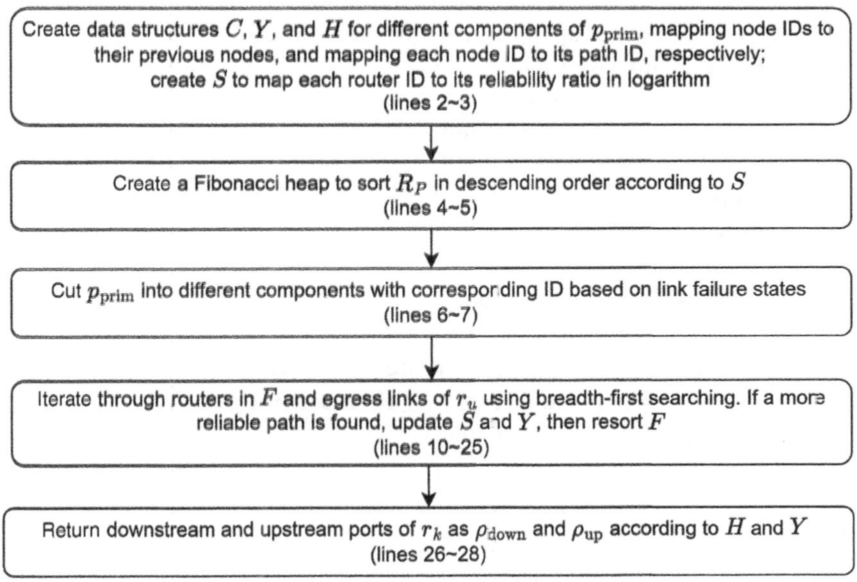

Figure 6.11 Workflow of distributed preferred path selection.

Discussions The algorithms presented above efficiently reduce collaboration steps during rerouting by avoiding unnecessary detours. Consider Figure 6.12, illustrating a reroute across the PP with routers from R1 to R6. In case of a fault in the link between R2 and R3, Algorithm 6.5 partitions the PP into two connected components based on fault links (lines 6 to 9). Subsequently, within a connected component like the second one, Algorithm 6.5 ensures no rerouting decision is

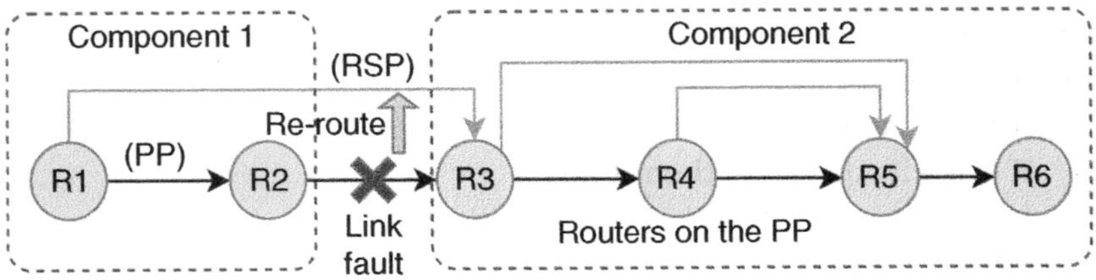

Figure 6.12 Example of rerouting over two components on the PP.

Algorithm 6.5 Computing Preferred Path

procedure PREFERREDPATH(G_P, r_k)
    $C = \{\}, c' = 0, Y = \{i \rightarrow \infty | r_i \in R_P\}, H = \{\}$
    $S = \{k \rightarrow 0\} \cup \{i \rightarrow -1 | r_i \in R_P \wedge i \neq k\}$
    F is a Fibonacci heap in descent order according to S
    for $r_i \in R_P$ **do** push(i, F)
    for $r_{d_i} \in (p_{\text{prim}} \setminus r_{d_1})$ **do**
        if $L_{\text{states}}(l(r_{d_{i-1}}, r_{d_i})) = \text{f}$ **then** $c' += 1$
        $C[d_i] = c'$
    end for
    while $|F| > 0$ **do**
        $u = \text{poll}(F)$
        for $l \in \text{EgressLinks}(r_u)$ **do**
            $r_v = \text{EgressNode}(l)$
            if $r_v \notin F$ or $L_{\text{states}}[l] = \text{fail}$ **then continue**
            if $v = C[u]$ and $(u, v) \notin p_{\text{prim}}$ **then continue**
            $s_{\text{alter}} = S[u] + \text{ReliaRatio}(l)$
            if $s_{\text{alter}} > S[v]$ **then**
                $S[v] = s_{\text{alter}}$
                $Y[v] = u$
                if $l \in p_{\text{prim}}$ **then** $H[v] = \text{LinkID}(l)$
                else $H[v] = \text{PathID}(l)$
                UpdateKey(F, v)
            end if
        end for
    end while
    $\rho_{\text{down}} = H[k]$
    if $Y[k] \neq \infty$ **then** $\rho_{\text{up}} = H[Y[k]]$ **else** $\rho_{\text{up}} = \infty$
    return $(\rho_{\text{down}}, \rho_{\text{up}})$
end procedure

---

made (line 15). This design results in rerouting exclusively between different components with fault links, minimizing the number of steps.

Additionally, let's analyze the computational complexity of Algorithm 6.5 in both time and space. The time costs of lines 1–4, 5, 6–9, 10–25, and 26–28 are $O(1)$, $O(|R_P|)$, $O(|P_{\text{prim}}|)$, $O(|R_P| \cdot (1 + |\log |R_P||))$, and $O(1)$, respectively. Thus, the time complexity is $O(|R_P| + \log |R_P| + |R_P| \cdot \log |R_P|)$. On the other hand, the space complexity, consumed in lines 2–4, is $O(|R_P| + \log |R_P|)$. This ensures scalability to large Resilient Subgraphs (RSGs) in extensive networks.

6.2.4 Efficient Resequencing of Data Packets

To handle out-of-order Data packets, the producer assigns each packet a sequential ID. The consumer then resequences these out-of-order packets based on their ID. The resequencing process operates through the following states:

- δ^{C}_{last}: the ID of the most recently arrived Data packet, initialized to -1;
- q: a priority queue used to temporarily store the out-of-order Data packets;
- E_{dr}: a set tracking the current events related to packet release from the queue;
- τ_{dr}: the resequencing delay, calculated as $\tau_{\text{dr}} = C_{\text{np}}/f$, where C_{np} is a constant that is always greater than or equal to 1, and f represents the packet frequency.

The resequencing process is managed by the Resequence Data Packets algorithm (see Algorithm 6.6) and its subroutine, the Release Queue algorithm (see Algorithm 6.7). The key function of the Resequence Data Packets algorithm is to sort the packets in q according to their ID, ensuring that the consumer receives them in the correct order. The Release Queue algorithm checks the delay τ_{dr} to release packets in the correct sequence while minimizing latency.

The workflows of these two algorithms are specified in Figure 6.13, showing how the packets are inserted into the queue, resequenced, and eventually released. The resequencing and release process ensures that data is delivered in the correct order despite potential packet reordering during transmission.

The computational complexity of these algorithms is analyzed as follows. For time, in Algorithm 6.6, lines 2–9, 10, and 11–15 take $O(1)$, $O(|q|)$, and $O(1)$, respectively. Hence, the total time complexity is in $O(|q|)$. In addition, the space complexity is in $O(|q|)$. Furthermore, in line 14, a synchronized scheduling for calling of ReleaseQueue is made. Hence, at runtime, the maximum queuing overhead time after receiving a Data packet is bounded by a constant of τ_{dr}. The above analysis demonstrates the efficiency in the Data packets resequence.

Algorithm 6.6 Re-sequence Data Packets

procedure RESEQUENCEDATAPACKETS(ξ, δ)
    if $\delta^{C}_{\text{last}} = -1$ **then**
        SendToApp(ξ), $\delta^{C}_{\text{last}} = \delta$
    else if $\delta^{C}_{\text{last}} = \delta + 1$ **then**
        SendToApp(ξ), $\delta^{C}_{\text{last}} = \delta^{C}_{\text{last}} + 1$
    end if
    if $|q| = \text{capacity}(q)$ **then**
        $\xi' = q$.pop(), SendToApp(ξ'), $\delta^{C}_{\text{last}} = \delta$
    end if
    ReleaseQueue(q, Now())
    if $|q| > 0$ **then**
        if $|E_{\text{dr}}|$ **then** $E_{\text{dr}} = \emptyset$
        $\tilde{\xi} = q$.top(), $t_1 = t_k(\tilde{\xi}) + \tau_{\text{dr}} - \text{Now}()$
        $E_{\text{dr}} \leftarrow$ Sched(t_1, ReleaseQueue, q, Now())
    end if
end procedure

Algorithm 6.7 Release Queue

procedure RELEASEQUEUE(q, t_{cur})
    while $|q| > 0$ **do**
        $\xi' = q$.top()
        if $\delta(\xi') = \delta^{C}_{\text{last}}$ **then**
            q.pop(), SendToApp(ξ'), $\delta^{C}_{\text{last}} += 1$
        else break
        end if
    end while
    if $|q| = 0$ **return**
    $\delta_{\text{old}} = \max(\{\delta(\xi') | t_{d_k}(\xi') < t_{\text{cur}} - \tau_{\text{dr}} \wedge \xi' \in q\})$
    while $|q| > 0$ **do**
        $\xi' = q$.top()
        if $\delta \leq \delta_{\text{old}}$ **then**
            q.pop(), SendToApp(ξ'), $\delta^{C}_{\text{last}} = \delta(\xi')$
        else break
        end if
    end while
end procedure

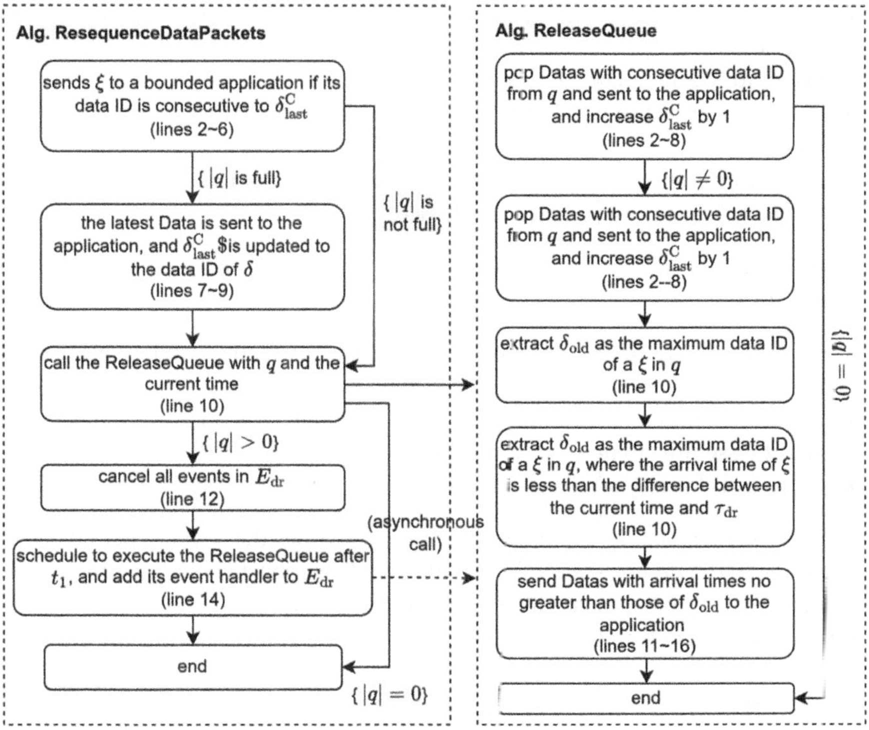

Figure 6.13 Workflow of data packet resequencing.

6.3 Performance Evaluation

We evaluated the performance of the proposed TA-FRR to demonstrate its effectiveness in rerouting the multiple link failures on the PP in the NDN for the grid communications.

6.3.1 Experimental Settings

6.3.1.1 Implementing TA-FRR

We've brought TA-FRR to life using the NDN forwarding engine [Mastorakis et al., 2017] embedded in the ndnSIM simulator [Mastorakis et al., 2017]. The source code for TA-FRR is available on GitHub [2022]. Let's dive into its configuration with the following parameters:

- The maximum time to send Echos is set at 3, preventing erroneous judgments for packet losses due to network congestion.

- The link check period is a swift 1 seconds, ensuring a rapid response to link state changes.
- α is set to 50% for a fair balance in Eq. (6.2).
- CR and τ_s are configured at 90% and 5 seconds, respectively, ensuring accurate AFR.
- D falls within the range [1, 10] ms to cater to general communication demands.
- β is set to 5, ensuring sufficient resilience.

Moreover, for efficient link fault detection and rerouting, TA-FRR is configured with:

- τ_{echo}, the maximum RTT of each link, set at 264 μs, encompassing transmission, queuing, and process delay.
- η is wisely chosen as 3 to prevent erroneous judgments on normal links with packet losses due to network congestion.
- τ_{check} spans from 7 to 100 ms, ensuring a rapid rerouting response.

For re-sequencing, τ_{dr} is thoughtfully set to 20 times the sending period corresponding to the producer's sending frequency. This robust choice ensures a sufficiently large τ_{dr} to handle path delay fluctuations caused by potential disruptions on the RSG and network congestion.

6.3.1.2 Network Topology Construction

In our ndnSIM experiments testing TA-FRR, we constructed the underlying network based on the IEEE dataset with 300 buses [IEEE, 2022] from a real electric power transmission grid. Each bus is represented as a substation equipped with an NDN router implementing TA-FRR. Optical lines connect these substations, providing a 10 Gbps bandwidth for generality [IEC, 2014]. Each link's port has a queue size of 100 packets and a maximum queueing delay of 30 μs. The link's transmission delay is 100 μs, considering a 30 km geographical distance between neighboring substations. Additionally, the port's processing delay is assumed to be 1 μs.

In this network, we use the RSG example discussed in Section 6.1 for testing. Each node on the PP has at least one RSP connecting to the upstream one on the PP. This setup simplifies the demonstration of fast rerouting validity. Simulating data traffic from the producer to the consumer, we focus on the synchrophasor measurement application [Ravikumar et al., 2020] with a high packet sending frequency of 50Hz [IEEE, 2011]. Additionally, we simulate extreme frequencies ranging from 200 to 8000 Hz to generate out-of-order Data packets. This frequency range aligns with the practical ceiling for wide-area measurement and control in the current power grid, considering high transmission delays in the network. Moreover, in our simulations, 1 to 7 links on the PP are randomly selected as faulty, and link states alternate between normal and complete fault states with packet loss rates of 0% and 100%, respectively.

For TA-FRR, we constructed NDN networks based on Internet topologies, synthesizing 100 to 1000 routers using BRITE [Medina et al., 2001]. Specifically, each link between two routers is modeled as a point-to-point optical line at 10 Gbps with a transmission delay of 100 μs [Bernstein et al., 2003]. Each link has a maximum queueing delay and processing delay of 30 and 1 μs, respectively. Thus, the maximum RTT of each link is 264 μs. The experiments were conducted on a platform equipped with an Intel Core i5-8250 CPU and 7.8 GB of memory.

6.3.1.3 Putting TA-FRR to the Test

Let's dive into the performance evaluation of TA-FRR, gauging its efficiency through the lenses of EELR and LRT metrics, showcasing its prowess in rerouting. We also dissect the messages – Echo, Back, and LSA – to unveil the underlying communication dynamics. To round it off, we introduce three additional metrics for assessing the quality of Data packet delivery at the consumer:

- The Out-of-Order Rate at the Receiver Side (OOR) scrutinizes the total received Data packets at the consumer.
- The EEDT measures the journey between the producer and the consumer.
- The Data PIAT at the consumer adds another layer of insight.

Each simulation spans 25 seconds. Between the 10th and 15th second mark, we spice things up by randomly throwing a wrench into the works – selecting some links of the PP and flipping them into fault states. This tumult is met with simulated data traffic using an on–off model, mimicking a random Poisson process for the Internet [Karagiannis et al., 2004] at an average throughput of 50 packets per second. To put AFR to the test, we configure a periodic fault recovery case, momentarily switching off the active detection with a set interval for checking.

Moreover, our trials extend to evaluating the False Positive Rate (FPR) in the fault link detection of AFR. We draw insights from a CAIDA trace captured from the vast realms of the Internet [CAIDA, 2016], encompassing a whopping 30.32 million packets over 60 seconds. These trace packets, spaced at intervals of 10, serve as our litmus test – sufficiently robust to counteract any packet time errors induced by capture buffering.

6.3.1.4 Comparative Studies

Let's pit TA-FRR against established FRR solutions designed for general networks in a synthesized 100-node Internet topology. The contenders are:

- CASA [Foerster et al., 2019b], a tactical player employing the k packing arborescences strategy seen in Foerster et al. [2019a], Shukla and Foerster [2021], and Tan and Kuo [2022].

- TREE [Schweiger et al., 2021], a nimble competitor leveraging a routing tree that extends multiple disjoint paths – a technique championed in Foerster et al. [2022], Shukla and Foerster [2021], Li et al. [2022], and Miura et al. [2022].

We throw curveballs at these contenders by randomly disrupting the network at varying degrees, gauged by the Fault Link Percentage (FLP) – the proportion of fault links to all links in the network. To evaluate their resilience, we turn to two widely accepted metrics:

- RFR, calculated as the ratio of disconnected routes to the total number of routes.
- The maximum stretch, representing the difference in the number of hops in a successful reroute under the worst-case scenario of the PP length.

6.3.2 Assessing Performance in OT Networks

Let us explore performance in two scenarios below, each with its own set of link states.

i) The links on the PP face disruptions of varying intensities, accompanied by a link check every 100 ms. The saga of results is unveiled in Figure 6.14. Witness the stable dance of End-to-End Failure Rate in Packet Delivery (EEFR) and LRT in Figures 6.14(a) and 6.14(b). Marvel at the EELR and LRT showcasing resilience at their peak, mean, and minimum – (2.63%, 2.09%, 1.4%) and (95.12, 45.53, 32.94 ms), respectively. The OOR stays put at 0%, as the path delay difference during rerouting graciously bows to the sending period. Rejoice as Figures 6.14(c) and 6.14(d) narrates the tale of EEDT and PIAT evolving with the number of disrupted links on the PP. Feast your eyes on the maximum and mean EEDT – 1.55 and 1.1 ms. The PIAT follows suit with a maximum of 520.72 ms and a mean of 20.42 ms. The pulse of Echos in Figure 6.14(e) holds steady at an average of 90.37. Meanwhile, the Backs perform a graceful descent, correlating with the disrupted links. Figure 6.14(f) unveils a rhythmic increase in the average LSA per PP router, a dance that remains consistent at 0.93. The overhead symphony maintains its harmony across all link faults on the PP.
ii) Now, brace yourself for the ultimate challenge – the complete disruption of each link on the PP, with a diverse selection of parameters for the link check period. Behold the unfolding drama in Figure 6.15. To decipher this performance masterpiece, delve into the specifics presented in Figures 6.15(a) and 6.15(b), showcasing the intricate dance of EELR and LRT over varying periods.

In the minutiae of the performance saga, witness EELR and LRT at their lowest, average, and peak, marked at (0.21%, 1.17%, 2.03%) and (0.4, 26.11, 98.9 ms), respectively. The plot thickens as Figures 6.15(e) and 6.15(b) unravels

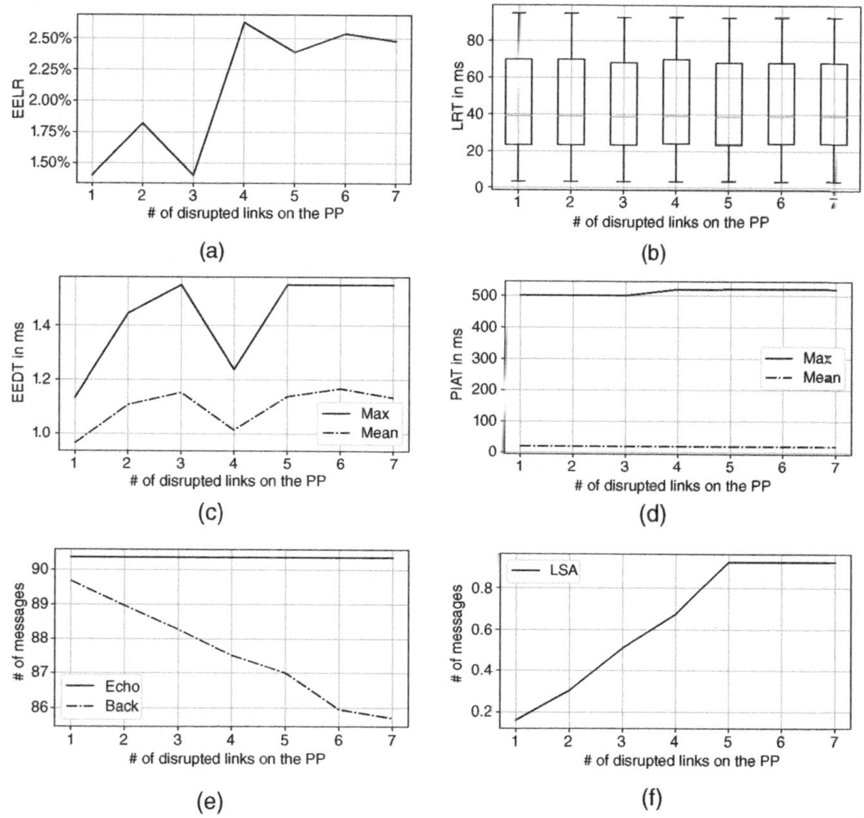

Figure 6.14 Performance of TA-FRR under different PP links disruption degrees. (a) EELR, (b) LRT, (c) EEDT, (d) PIAT, (e) the numbers of the Echo and Back messages, and (f) the number of the LSA messages.

the linear ascent of EELR and LRT with the ticking clock. This rise, a consequence of less sensitive link state detection, adds depth to both LRT and EELR. Remarkably, OOR remains resilient at 0% across all periods, a testament to the harmonious interplay of lower path delay differences.

Turn the pages to discover the metamorphosis in Figures 6.15(c) and 6.15(d), illustrating the changes in EEDT and PIAT with the ever-growing period. Delve into the details where the pinnacle and mean of all EEDT stand at 1.55 and 1.17 ms, respectively. Furthermore, the zenith and average of PIAT reach 520.72 and 20.23 ms, respectively. Witness the dwindling numbers of Echoes and Backs in Figure 6.15(e) as the period extends its dominion. Meanwhile, Figure 6.15(f) paints a consistent picture with an average LSA count of 0.93, unaffected by the passage of time.

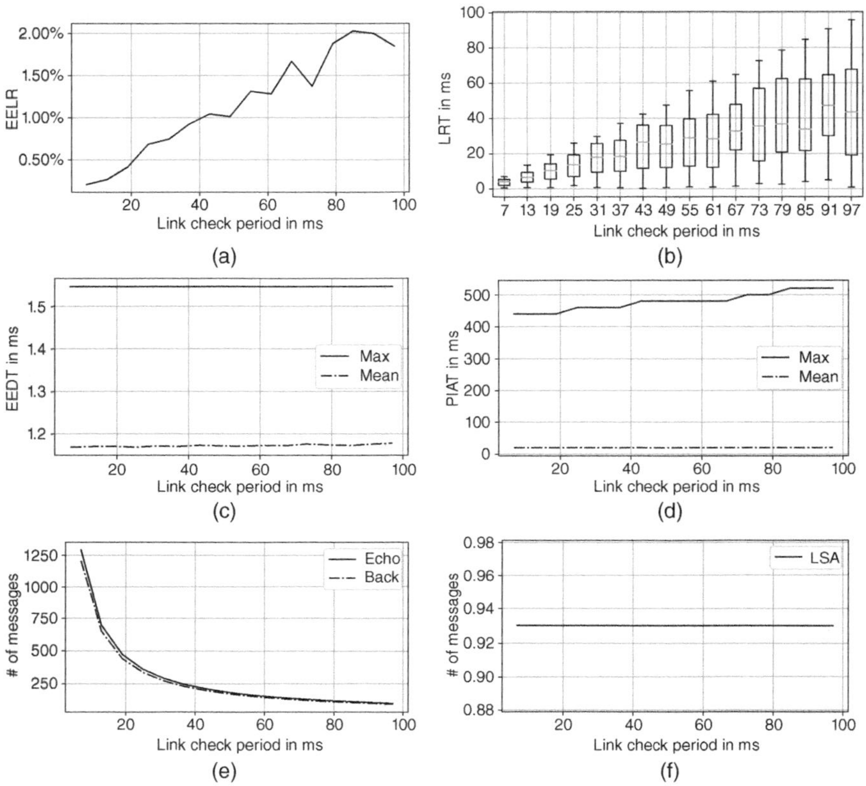

Figure 6.15 Performance of TA-FRR with different link check period parameter under the full PP disruption. (a) EELR, (b) LRT, (c) EEDT, (d) PIAT, (e) Echo and Back messages, and (f) LSA messages.

In the grand finale, these statistics unveil the orchestration of an acceptable message overhead through the proper adjustment of the period. The stage is set, and the period, a key player in this performance, takes a bow.

6.3.2.1 Reordering Performance

Delve into the rhythms of our high-frequency producer's performance, illustrated in Figure 6.16. Witness the dance of OOR in Figure 6.16(a), gracefully reaching a 0.015% upper limit as the sending frequency exceeds 2000 Hz. This subtle rise hints at the challenge posed by the upper limit on delay fluctuation across diverse paths within the RSG.

In the enthralling Figure 6.16(b), the saga of EEDT unfolds in two acts. Act I, before the re-sequence, paints a story with mean and maximum EEDT gracefully bounded by 1.17 and 1.55 ms, respectively. Act II, after the re-sequence, reveals a

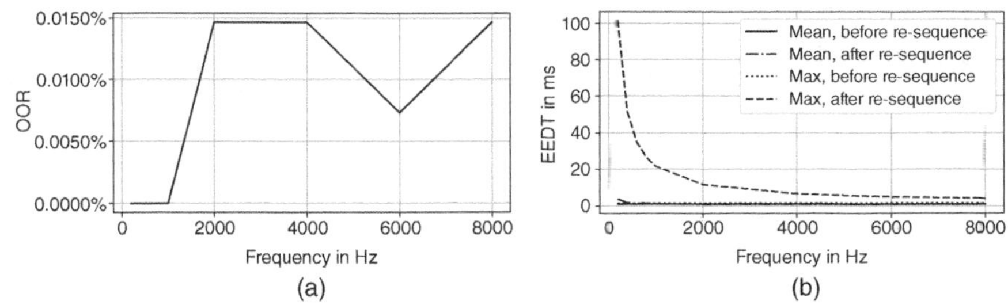

Figure 6.16 Performance of TA-FRR with different sending frequency at the producer under the full PP disruption. (a) OOR and (b) EEDT.

dynamic transformation with mean and maximum EEDT pushing their limits to 3.67 and 101.55 ms, respectively. These compelling results stand as a testament to the efficiency embedded in our reordering process.

In the grand finale, the symphony of these results harmonizes to echo the effectiveness of TA-FRR in navigating the labyrinth of multiple link failures within the NDN.

6.3.3 Assessing Performance in IT Networks

Behold the performance saga of TA-FRR as it unfolds in the mesmerizing 300-node topology of synthesized traffic, beautifully encapsulated in Figure 6.17. Let's embark on a visual adventure through the intricacies of EELR, LRT, EEDT, and PIAT captured in the captivating boxplots of Figures 6.17(a) and 6.17(b), Sections 6.17(c) and 6.17(d).

Marvel at the prowess of TA-FRR as it deftly navigates between 0 and 5 random fault PP links, with 75th percentile values of (0%, 0 ms, 0.51, 20.28) and (2.8%, 44.55 ms, 0.64, 20.32), respectively. Each LRT gracefully dances under the common standard of 50 ms, showcasing the swift rerouting prowess even in the face of serious PP faults.

Behold the symphony of messages in Figure 6.17(e), where the Cumulative Probability Distribution Function (CDF) of LSA messages unveils a tale of efficiency. For 95% of pairs, when fault PP links range from 1 to 5, the message numbers stand humbly at 0.7 and 2.5, a testament to the frugal elegance of TA-FRR.

In the realm of AFR, Figure 6.17(f) presents a nuanced ballet. As CR takes the spotlight from 90% to 99%, LRT of AFR gracefully waltzes in the range of [19.94, 65.44] ms. The echo of messages reverberates, with an average number of Echo and Back messages per node for each recovery decreased between 336.63 and 393.63.

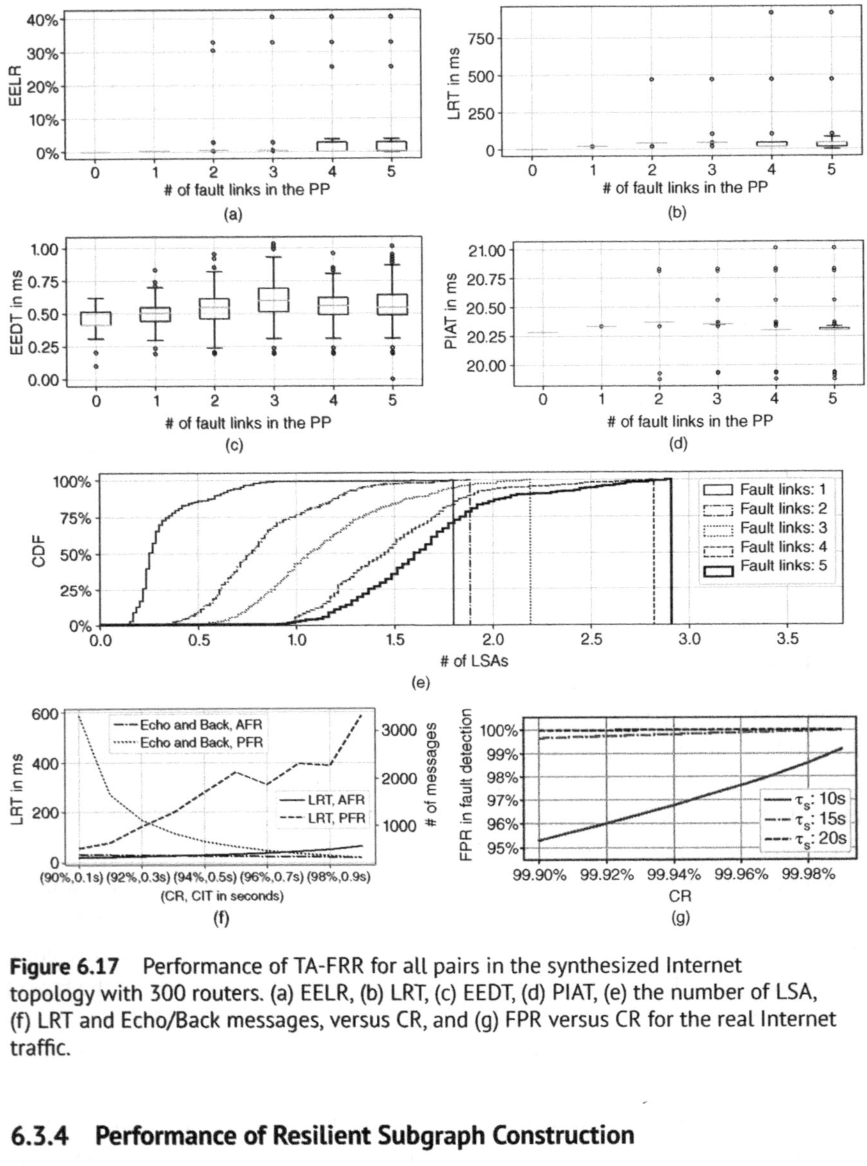

Figure 6.17 Performance of TA-FRR for all pairs in the synthesized Internet topology with 300 routers. (a) EELR, (b) LRT, (c) EEDT, (d) PIAT, (e) the number of LSA, (f) LRT and Echo/Back messages, versus CR, and (g) FPR versus CR for the real Internet traffic.

6.3.4 Performance of Resilient Subgraph Construction

Let's peek behind the curtain to explore the performance of proactive RSG construction. Join us in the enchanted land of algorithms and execution times.

Behold, in Figure 6.18, the dance of algorithms as they perform their intricate routines 50 times across various Internet topologies, ranging from 100 to 1000 routers.

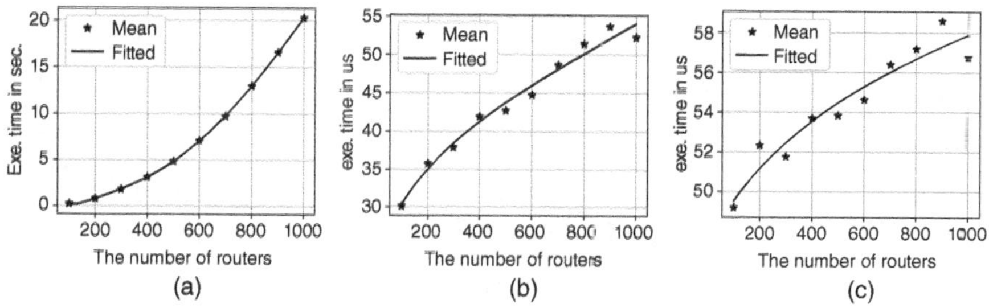

Figure 6.18 Algorithmic performance of TA-FRR for all pairs in the synthesized Internet topologies with 100 to 1000 routers. (a) the execution time of GenRSGV algorithm, (b) the execution time of ConstructRSG algorithm, and (c) the time to compute detour paths.

Marvel at the mystical GenRSGV, depicted in Figure 6.18(a), executing its spell with an execution time guided by the ethereal formula:

$$O(-0.01 \cdot \log^2|R| \cdot |R|^2 + 0.11 \cdot \log|R| \cdot |R|^2 - 33.41 \cdot \log|R| \cdot |R| \\ + 9826.96 \cdot \log^2|R| - 77374.46 \cdot \log|R| + 161094.25)$$

In Figure 6.18(b), witness the dance of ConstructRSG, whose execution time is elegantly estimated by the incantation:

$$O(4.83 \cdot \log|R| + 0.01 \cdot |R| - 6.64)$$

And now, in Figure 6.18(c), gaze upon the time taken to compute the detour path in AFR, guided by the mystical formula:

$$O(0.24 \cdot \log^2|R| \cdot \log(0.04 \cdot \log^2|R|) + 50.21)$$

These statistics unfold the scalability of TA-FRR.

6.3.5 Comparative Studies and Insights

In Figure 6.19, let's delve into the resilience comparison of CASA, TREE, and the dynamic TA-FRR across a 100-node network, exploring fault scenarios ranging from 0% to 25%. For FLP values of (10%, 15%, 25%), TA-FRR emerges victorious, showcasing improvements over CASA by (69.15%, 59.75%, 50.12%) and TREE by (52.3%, 42.6%, 33.84%) when D is set to 2 ms. The trend continues for D values of 4 and 10 ms, underscoring TA-FRR's robustness and adaptability in comparison to its counterparts. This highlights the power of TA-FRR in steering clear of network disruptions, thanks to its dynamic rerouting options via DSP.

Moving on to Figure 6.20, we unravel the quality of routes in CASA, TREE, and TA-FRR, adjusting D from 2 to 10 ms. Witness the maximum stretch and maximum number of PP links for CASA, TREE, TA-FRR with D set to 2 ms,

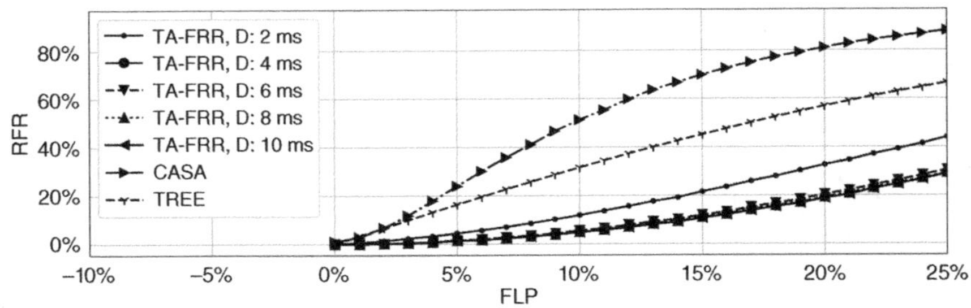

Figure 6.19 Comparison in RFR for a network with 100 routers.

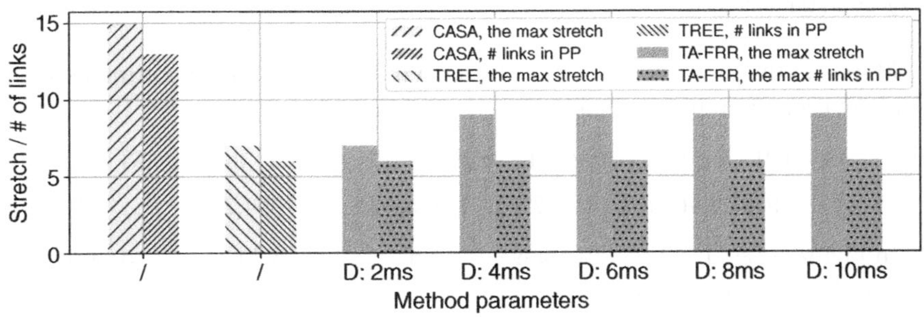

Figure 6.20 Comparison in the quality of routes for a network with 100 routers.

and TA-FRR with D ranging from 2 to 10 ms, showcasing (15, 13), (7, 6), (7, 6), and (9, 6), respectively. With a judiciously chosen lower D, TA-FRR effectively truncates overly long DSP, leading to a marked reduction in RFR. This distinct advantage positions TA-FRR as the go-to solution for fortifying networks against faults.

6.4 Summary

In conclusion, we introduce an innovative NDN-based FRR mechanism that leverages DSP to construct a resilient route structure and efficiently recovers from link faults using active cross-layer link fault identification. By incorporating DSP, our mechanism significantly enhances network resilience to link faults. Moreover, this fault recovery process not only reduces LRT but also incurs minimal message cost. Evaluation results highlight the mechanism's ability to maintain high resilience in synthesized Internet topologies, outperforming arborescence and disjoint path approaches with lower RFR and a comparable maximum stretch cost.

The mechanism proficiently handles multiple link faults through distributed link state probing, advertisement, and path selection based on joint paths. Optimized algorithms ensure path choices, avoiding unnecessary rerouting actions, while maintaining scalable computational complexity. Additionally, the mechanism excels in re-sequencing out-of-order Data packets at the consumer, effectively rerouting multiple link failures with short recovery times and minimal packet loss rates. It also efficiently manages out-of-order packets at high Data packet sending frequencies.

Questions

6.1 What are the core principles behind TA-FRR, and how does it leverage DSP for fault recovery?

6.2 How are EELR and LRT defined, and why are they crucial performance metrics in fault recovery mechanisms?

6.3 How does TA-FRR detect link states in the network, and why is this crucial for effective fault recovery?

6.4 Based on the presented results, discuss potential real-world scenarios where TA-FRR could be applied to enhance network reliability and resilience.

6.5 Explain the mechanisms employed by TA-FRR to control message overhead during fault scenarios. How does it achieve a balance between effective fault recovery and minimal message cost?

6.6 Analyze the impact of different link check periods on TA-FRR's performance, as depicted in Figure 6.15. What considerations should be taken into account when choosing an optimal check period?

6.7 In your opinion, what are the strengths and potential limitations of TA-FRR based on the presented results and insights? How might it perform in extremely large-scale networks?

6.8 Based on the findings, propose potential future directions for research and improvement in the field of fault recovery mechanisms, specifically in the context of TA-FRR.

Bibliography

Greg Bernstein, Bala Rajagopalan, and Debanjan Saha. *Optical Network Control: Architecture, Protocols, and Standards*. Addison-Wesley Longman, 2003.

CAIDA. The CAIDA UCSD anonymized Internet traces - equinix-Chicago.dira. 20160406-130000.UTC.anon, 2016. URL https://www.caida.org/catalog/datasets/passive_dataset.

Klaus-Tycho Foerster, Andrzej Kamisinski, Yvonne-Anne Pignolet, Stefan Schmid, and Gilles Trédan. Bonsai: Efficient fast failover routing using small arborescences. In *IEEE/IFIP DSN*, pages 276–288, 2019a.

Klaus-Tycho Foerster, Yvonne-Anne Pignolet, Stefan Schmid, and Gilles Tredan. CASA: Congestion and stretch aware static fast rerouting. In *IEEE INFOCOM*, pages 469–477, 2019b.

Klaus-Tycho Foerster, Andrzej Kamisinski, Yvonne-Anne Pignolet, Stefan Schmid, and Gilles Tredan. Improved fast rerouting using postprocessing. *IEEE Transactions on Dependable and Secure Computing*, 19(1):537–550, 2022.

GitHub, Inc. Traffic aware fast reroute mechanism (TA-FRR), 2022. URL https://github.com/zhouby-zjl/tafrr.

IEC 60794-4-10. Optical fibre cables - Part 4-10: Family specification - optical ground wires (OPGW) along electrical power lines, 2014.

IEEE. IEEE 30∼300 bus dataset, 2022. URL http://labs.ece.uw.edu/pstca/.

IEEE C37.118.2. IEEE standard for synchrophasor data transfer for power systems, 2011.

IETF RFC 5880. Bidirectional forwarding detection (BFD), 2010.

Thomas Karagiannis, Mart Molle, Michalis Faloutsos, and Andre Broido. A nonstationary Poisson view of Internet traffic. In *IEEE INFOCOM*, volume 3, pages 1558–1569, 2004.

Ziyong Li, Yuxiang Hu, Jiangxing Wu, and Jie Lu. P4Resilience: Scalable resilience for multi-failure recovery in SDN with programmable data plane. *Elsevier Computer Networks*, 208:108896, 2022.

Stephane Litkowski, Ahmed Bashandy, Clarence Filsfils, Pierre Francois, Bruno Decraene, and Daniel Voyer. Topology Independent Fast Reroute using Segment Routing. Internet-draft, IETF, 2022.

Spyridon Mastorakis, Alexander Afanasyev, and Lixia Zhang. On the evolution of ndnSIM: An open-source simulator for NDN experimentation. *ACM SIGCOMM Computer Communications Review*, 47(3):19–33, 2017.

A. Medina, A. Lakhina, I. Matta, and J. Byers. BRITE: An approach to universal topology generation. In *MASCOTS*, pages 346–353, 2001.

Hideyoshi Miura, Kouji Hirata, and Takuji Tachibana. P4-based design of fast failure recovery for software-defined networks. *Elsevier Computer Networks*, 216:109274, 2022.

Named Data Networking (NDN). NDN-0021, NFD Developer's Guide. Technical report, NDN, 2021. URL https://named-data.net/.

Gelli Ravikumar, Dan Ameme, Satyajayant Misra, Sukumar Brahma, and Reza Tourani. iCASM: An information-centric network architecture for wide area measurement systems. *IEEE Transactions on Smart Grid*, 11(4):3418–3427, 2020.

Oliver Schweiger, Klaus-Tycho Foerster, and Stefan Schmid. Improving the resilience of fast failover routing: TREE (tree routing to extend edge disjoint paths). In *ACM ANCS*, pages 1–7, 2021.

Apoorv Shukla and Klaus-Tycho Foerster. Shortcutting fast failover routes in the data plane. In *ACM ANCS*, pages 15–22, 2021.

Nina Taft-Plotkin, Bhargav Bellur, and R. Ogier. Quality-of-service routing using maximally disjoint paths. In *IEEE IWQoS*, pages 119–128, 1999.

Hai-Khun Tan and Tung-Wei Kuo. Optimistic fast rerouting. In *IEEE ICC*, pages 1692–1697, 2022.

Jin Y. Yen. Finding the k shortest loopless paths in a network. *INFORMS Management Science*, 17(11):712–716, 1971.

Lixia Zhang, Alexander Afanasyev, Jeffrey Burke, Van Jacobson, Christos Papadopoulos, K. C. Claffy, Lan Wang, Patrick Crowley, and Beichuan Zhang. Named data networking. *ACM SIGCOMM Computer Communication Review*, 44(3):66–73, 2014.

Shi Zhou and Raúl J. Mondragón. Accurately modeling the Internet topology. *APS Physical Review E*, 70(6):066108, 2004.

Boyang Zhou, Qiang Yang, Yansong Wang, and Chunming Wu. Reliable communication in transmission grids based on nondisjoint path aggregation using software-defined networking. *IEEE Transactions on Industrial Informatics*, 15(2):842–855, 2019.

7

Elevating Reliability: Disruption-Resilient Transport Protocol (DRTP) Revolution

In the intricate realm of modern power transmission grids, Internet-of-Things (IoT) networks rely on optical networks to disseminate sensor data over long distances to a central server. This process is crucial for accurate observation of grid states through a technique known as State Estimation (SE). The key to achieving this accuracy lies in minimizing End-to-End Failure Rate in Packet Delivery (EEFR).

To tackle the aforementioned challenge, this chapter delves into a novel approach – Disruption Resilient Transport Protocol (DRTP), facilitating hop-by-hop retransmission with the aid of Redundant Sub-Paths (RSPs), grounded in Named Data Network (NDN). Notably, our introduced routing model of the Multipath Subgraph (MPSG) seamlessly integrates RSPs, presenting an easily generatable solution tailored to the network topology of power grids (see Section 7.1.2).

In this exploration, we underscore a non-trivial distributed collaboration issue surfacing during retransmission under multiple link failures. This issue poses a risk of cache mismatching, potentially deteriorating the EEFR. Remarkably, to our knowledge, this intricate concern has been overlooked in existing hop-by-hop approaches. As we navigate this unchart territory, the significance of addressing distributed collaboration nuances comes to the forefront, emphasizing the need for meticulous considerations in enhancing the robustness of power grid communication systems.

Introducing a pioneering solution to grapple with the collaboration issue mentioned above, the crux of this solution lies in judiciously selecting retransmission opportunities, strategically avoiding cache mismatching through a best-effort approach, leveraging the fixed frequency. This innovative approach not only achieves significantly reduced EEFR but also ensures a bounded End-to-End Payload Packet Delivery Time (EEDT) under the umbrella of recursively controlled collaboration. As we unfold this solution, the promise of enhancing the reliability

Communication Networks in Smart Power Grids, First Edition. Boyang Zhou.

Companion Website: www.wiley.com/go/SmartPowerGrids1e

and efficiency of power grid communication systems becomes evident, charting a course toward more resilient and responsive networks.

We conducted a comprehensive evaluation of DRTP within the ndnSIM simulator [Mastorakis et al., 2017], exploring both the typical MPSG and the general MPSG constructed from authentic transmission grid topologies. Our numerical findings underscore the significant advantage of DRTP in reducing EEFR compared to existing approaches, accompanied by a low EEDT even under severe link disruptions.

The evaluation extended to the data dissemination process from Phasor Measurement Units (PMUs) [IEEE, 2011] and critical sensors within a simulated cyber–physical environment. This environment seamlessly integrated an optical Internet-of-things (IoT) network with a PMU-based SE process entailing real power transmission grids. Leveraging well-established tools such as ndnSIM [Mastorakis et al., 2017], MATPOWER [Zimmerman et al., 2011], and the Real-Time Digital Simulator (RTDS) [RTDS Technologies Inc., 2023], our experiments generated valuable insights, summarized as follows:

Further evaluations on the impact of transport protocol reliability using the standard IEEE 14-bus system [IEEE, 2022] highlighted DRTP's superior performance. Even under a full Primary Forwarding Path (PP) disruption with varying LLRs (10–100%), DRTP consistently ensured 100% Observable Grid Ratio (OGR) with impressively low-phase angle errors (0.16 milliradians) and voltage magnitude errors (1.52 milli power unit (p.u.)). Similarly, under full MPSG disruption at a 20% Link Loss Rate in Packets (LLR), DRTP achieved an OGR of 97.4%, with phase angle errors at 0.77 milliradians and voltage magnitude errors at 0.61 milli p.u. These results underscore DRTP's reliability advantage in supporting accurate power dispatches through robust SE.

7.1 Communication Model and Routes

Now, we detail and discuss the preliminaries on the DRTP involving models as well as related work as follows.

7.1.1 Communication Model

Let's step into the communication model within the transmission grid – think of it as a combination of routers and links, each with its unique role.

Treat this network as a directed graph, denoted as (R, L), where $r_i \in R$ is like the performer, representing the ith router stationed at critical substations, and $l_j \in L$ acts as the connecting threads between neighboring routers.

Now, let us analyze the maximum delay for these links – it's like assembling pieces. (1) There's the transmission delay, linearly related to physical distance. (2) The queueing delay, a bit like waiting in line, estimated using Little's law [Little, 2011] based on input and output buffer sizes. (3) And there's the maximum total processing time of connected routers. Formally, it's like figuring out the longest route and wait time along a path from r_a to r_b, denoted as $\tau^p_{\text{all}}(a, b)$ and $\tau^p_{\text{queue}}(a, b)$, respectively.

For simplicity, imagine PMUs and the Phasor Data Concentrator (PDC) as important parts – they connect directly to routers indoors with complete reliability and minimal delay. We keep things straightforward by assuming the flawless operation of PMUs and routers.

Now, let's explore the network's heartbeat. A PDC and a PMU engage in a rhythmic communication, like a dance. The PMU, a data maestro, generates information – phasor, frequency, and frequency change rates – from transformers or buses. The PDC, a conductor of sorts, harmonizes this data from several PMUs.

Figure 7.1 vividly depicts such a combination, following the IEEE C37.118.2 standard [IEEE, 2011]. Picture this: (1) PDC sends a Command to a PMU, asking for feedback in the form of a CFG frame; (2) PMU, upon receiving the Command, responds with a Configuration message, detailing its reporting capabilities; (3) PDC, now armed with Configuration, sends a Command to start data transmission; (4) PMU, upon receiving the Command, sends continuous Data frames with sampled statistics to PDC at a minimum frequency of f Hz; and (5) based on needs, PDC can send a Command to stop data transmission.

In the midst of this, step (3) introduces a sequence of packets, a digital dance of sorts, modeled as $(\xi_1, \xi_2, \ldots, \xi_\delta, \xi_{\delta+1}, \ldots)$, where δ takes on the role of an incremental data identifier (ID). For example, think of it as counting and tagging by the PMU's gateway. Step into this digital dance with us as we unravel the intricacies of communication in the transmission grid.

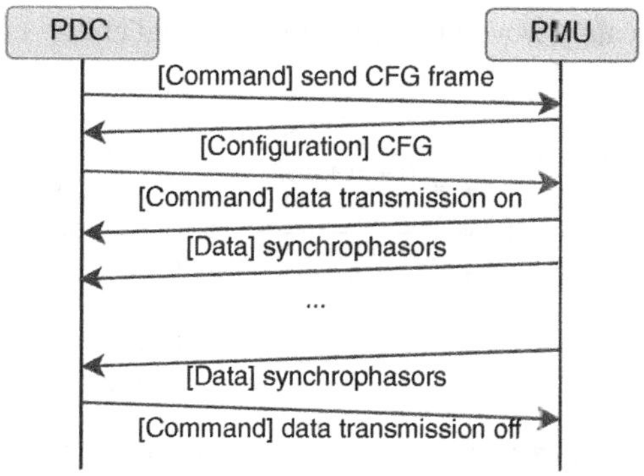

Figure 7.1 Communication pattern of IEEE C37.118.2.

In our network, we gauge the maximum disruption on a link by using the Link Loss Rate (LLR). This rate is determined by dividing the total lost packets on the link by the number of packets sent by the source router of that link. Simultaneously, we have the EEFR, which represents the average packet loss rate at the PDC per second.

Our goal in this chapter is clear: to minimize the EEFR while transporting PMU data packets, all under a specified LLR on all related links in the network. It's a quest for efficient and reliable data transport, ensuring the smooth flow of information in our interconnected world.

7.1.2 Routing on Multi-path Sub-graph

7.1.2.1 Routing Model

The MPSG unfolds as a route structure, with a PP leading from a PMU to the PDC, accompanied by its RSPs.

The PP embarks on a path, starting from r_{d_1} and reaching r_{d_b} through a sequence of router IDs: $\mathbf{p}_{\text{prim}} = (d_1, \ldots, d_s, \ldots, d_k, \ldots, d_b)$. Each r_{d_k} in the PP, where $k \geq 2$, plays host to a set of RSPs linking to its various upstream routers ($\{r_{d_s}\}$) of the PP. Each Redundant Sub-Path (RSP), in turn, is a sequence of router IDs: $\mathbf{p}_{\text{rsp}} = (d_k, \ldots, d_s)$. Importantly, each link in every RSP maintains link-disjointness from r_{d_1} to r_{d_k} in the PP, ensuring that losses on upstream PP links don't affect RSP links – an intentional design for robust retransmission control.

For a visual cue, Figure 7.2 illustrates an ordinary MPSG, manually constructed based on the IEEE 300-bus topology – a real-world model with 300 nodes and 409 links [IEEE, 2022]. In this depiction, the PP commences from node #138, reaching node #100, directly connecting a PMU and the PDC, respectively. Each node of the PP establishes connections to its upstream counterparts through diverse RSPs. For instance, node #104 links to nodes #136 and #138 through RSPs (104, 135, 137, 136) and (104, 103, 101, 102, 138), respectively. Meanwhile, the PP sequence is (138, 136, 104, 103, 100) – a network narrative where data traverses a path intricately woven for reliability and resilience.

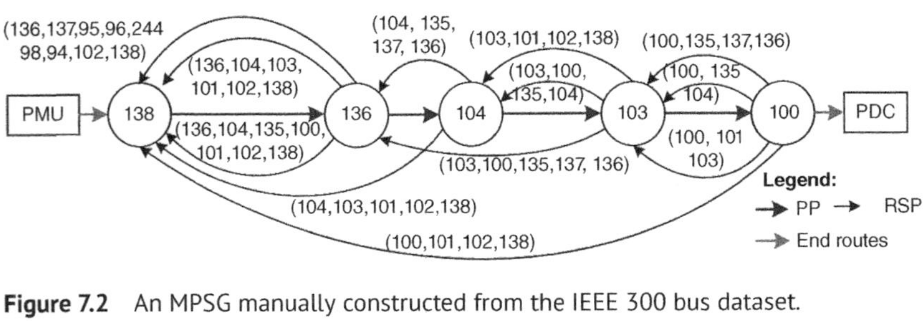

Figure 7.2 An MPSG manually constructed from the IEEE 300 bus dataset.

7.1.2.2 MPSG Generation

The MPSG can be easily generated from the above communication network. For example, Algorithm 7.1 presents an algorithm to generate the MPSG starting from the current router r_i to the rest of the network (GenMPSG). Furthermore, for each MPSG, each router in the PP of the MPSG contains at most β RSPs. GenMPSG is designed based on the classical Dijkstra's algorithm that uses a backtracking technique. GenMPSG first computes a PP, and then generates each RSP from an intermediate node to its upstream one, all in the PP, in the following steps.

In line 2, it computes a vector recording the previous router ID of the PP using the Dijkstra algorithm as $\mathbf{H}$. In line 3, it initializes an empty map from each destination node with respect to r_i to its corresponding MPSG as γ_{all}. In lines 5–23, it computes an MPSG with the source r_i for each destination r_j. In detail, first, in line 5, it computes the PP of the MPSG from r_i to r_j according to H. Second, in line 6, it initializes an empty map from each router in the PP to a set of RSPs of the router. Third, in line 7, it extracts all links in bi-directions of the PP as L_{PP}. Fourth, in lines 9–21, it computes at most β RSPs for each router in the PP, i.e., $\mathbf{p}_{\text{prim}}[g]$. Furthermore, it computes each RSP as follows. In line 9, it initializes a set to contain all upstream routers of $\mathbf{p}_{\text{prim}}[g]$ in the PP as P'. Meanwhile, it creates a set L_{RSP} to contain all links in all RSPs starting from $\mathbf{p}_{\text{prim}}[g]$. Afterward, in lines 11–20, it computes an RSP from $\mathbf{p}_{\text{prim}}[g]$ to each upstream router in the PP via the following steps:

1) In lines 11–13, it creates a virtual graph of $G' = (R, L')$. Here, R is added with a virtual router r_{vir}. In addition, the virtual router connects to each upstream router in the PP through a virtual link with a zero total delay in L_{vir}. Meanwhile, L' is L subtracted from each link contained in L_{PP} and L_{RSP}.
2) In line 14, it computes a vector containing the previous router ID of the shortest path from $\mathbf{p}_{\text{prim}}[g]$ through the Dijkstra as H'.
3) In line 15, it first obtains the shortest path from $\mathbf{p}_{\text{prim}}[g]$ to L_{vir} according to H'. Then, it computes an RSP as $\mathbf{p}_{\text{rsp}}$ by removing L_{vir} in the shortest path.
4) In line 16, it restores the virtual graph to the original G by removing v_{vir} and L_{vir} from R and L, respectively.
5) In line 17, it stops creating more RSP if $\mathbf{p}_{\text{rsp}}$ cannot be computed.
6) In lines 18–19, it adds all links in $\mathbf{p}_{\text{rsp}}$ to L_{RSP}.
7) In line 20, it adds $\mathbf{p}_{\text{rsp}}$ to a set found in X corresponding to the current router $\mathbf{p}_{\text{prim}}[g]$ in the PP.

Third, in line 23, it saves an MPSG for r_j consisting of $\mathbf{p}_{\text{prim}}$ and X into γ_{all}. Lastly, in line 25, it returns γ_{all} as all MPSG to each destination in the network.

The computational complexity of GenMPSGs can be explored in terms of time and space. Let's break it down for clarity.

Algorithm 7.1 Generate MPSG for all destination nodes

1: **function** GenMPSG($G = (R, L), r_i, \beta$)
2: $\quad$ $\mathbf{H} = \text{Dijkstra}(G, r_i)$
3: $\quad$ γ_{all} is a map from each destination node to its MPSG
4: $\quad$ **for all** $r_j \in R \setminus \{r_i\}$ **do**
5: $\quad\quad$ $\mathbf{p}_{\text{prim}} = \text{Convert}(\mathbf{H}, r_i, r_j)$
6: $\quad\quad$ X is a map from each PP node to its RSPs
7: $\quad\quad$ $L_{\text{PP}} = \{(r_{w-1}, w_w)_{\text{bi}} | r_w \in \mathbf{p}_{\text{prim}} \wedge w \geq 2\}$
8: $\quad\quad$ **for all** $g \in [2, 3, \ldots, |p_{\text{prim}}|]$ **do**
9: $\quad\quad\quad$ $P' = \{\mathbf{p}_{\text{prim}}[h] | 1 \leq h \leq g - 1\}, L_{\text{RSP}} = \emptyset$
10: $\quad\quad\quad$ **for all** $k \in [1, 2, \ldots, \beta]$ **do**
11: $\quad\quad\quad\quad$ $L_{\text{vir}} = \{(r_w, r_{\text{vir}})_{\text{two}} | r_w \in P'\}$
12: $\quad\quad\quad\quad$ $R = R \cup \{r_{\text{vir}}\}, L = L \cup L_{\text{vir}}$
13: $\quad\quad\quad\quad$ $L' = L \setminus (L_{\text{PP}} \cup L_{\text{RSP}})$
14: $\quad\quad\quad\quad$ $\mathbf{H}' = \text{Dijkstra}(G' = (R, L'), \mathbf{p}_{\text{prim}}[g])$
15: $\quad\quad\quad\quad$ $\mathbf{p}_{\text{rsp}} = \text{Convert}(\mathbf{H}', \mathbf{p}_{\text{prim}}[g], r_{\text{vir}}) \setminus r_{\text{vir}}$
16: $\quad\quad\quad\quad$ $R = R \setminus \{r_{\text{vir}}\}, L = L \setminus L_{\text{vir}}$
17: $\quad\quad\quad\quad$ **if** $p_{\text{rsp}} \neq \emptyset$ **then break**
18: $\quad\quad\quad\quad$ $L' = \{(r_{w-1}, r_w)_{\text{bi}} | r_w \in \mathbf{p}_{\text{rsp}} \wedge w \geq 2\}$
19: $\quad\quad\quad\quad$ $L_{\text{RSP}} = L_{\text{RSP}} \cup L'$
20: $\quad\quad\quad\quad$ $X[\mathbf{p}_{\text{prim}}[g]] = X[\mathbf{p}_{\text{prim}}[g]] \cup \{\mathbf{p}_{\text{rsp}}\}$
21: $\quad\quad\quad$ **end for**
22: $\quad\quad$ **end for**
23: $\quad\quad$ $\gamma_{\text{all}}[r_j] = (\mathbf{p}_{\text{prim}}, X)$
24: $\quad$ **end for**
25: $\quad$ **return** γ_{all}
26: **end function**

i) Execution time complexity is analyzed as follows:
   - **Lines 2 and 14**: $O(|L| + |R| \cdot \log |R|)$ for a Dijkstra algorithm using the Fibonacci heap.
   - **Lines 3, 5, 7, 11–13, and 15–19**: $O(\log |R|)$.
   - **Lines 6, 9, 20, 23, and 25**: $O(1)$.

   Hence, the total time complexity is $O(|R| \cdot \log |R| + \beta \cdot |R|^2 \cdot \log^2 |R| + \beta \cdot |R| \cdot \log^2 |R|)$, considering $|L| = |R| \cdot \log |R|$ in a power transmission grid.

ii) Space complexity is analyzed as follows:
   - **Lines 2 and 14**: $O(|R|)$.
   - **Memory for holding all MPSG required by lines 5, 6, 15, 20, and 23**: $O(|R| \cdot \log |R| \cdot \beta)$.

- **Line 7**: $O(\log|R|)$.
- **Memory for creating the virtual graph in lines 11–13 (de-allocated in line 16)**: $O(\log|R|)$.
- **Memory for L_{RSP} used in lines 9, 18, and 19**: $O(\beta \cdot \log|R|)$.

The overall space complexity is $O(|R| + \beta \cdot |R| \cdot \log|R| + (\beta + 1) \cdot \log|R|)$.

The feasibility of MPSG generation is demonstrated through Algorithm 7.1, employing various transmission grid topologies, including: (1) four IEEE datasets with 30, 57, 118, and 300 buses [IEEE, 2022], and (2) the South Carolina (SC) 500 bus dataset [Josz et al., 2016]. Figure 7.3 illustrates the features of the generated MPSG as follows.

In detail, Figures 7.3(a)–7.3(c) depicts Cumulative Probability Distribution Function (CDF) of the length of PP, the number of critical links in the PP, and the number of RSPs at each router in the PP of the MPSG. Furthermore, Figure 7.4 shows the maximum EEDT in the expectation computed using Eq. (7.9) mentioned later in Section 7.2.4.

Meanwhile, Figure 7.4 shows that the execution time of Algorithm 7.1 fits its theoretical estimation in $O(-8.22 \cdot 10^{-3} \cdot |R| \cdot \log|R| + 3.88 \cdot 10^{-7} \cdot |R|^2 \cdot \log^2|R| + 1.8 \cdot 10^{-3} \cdot \beta \cdot |R| \cdot \log^2|R| + 0.25)$. The above results indicate that the algorithm is feasible to construct MPSG in the network of the transmission grid.

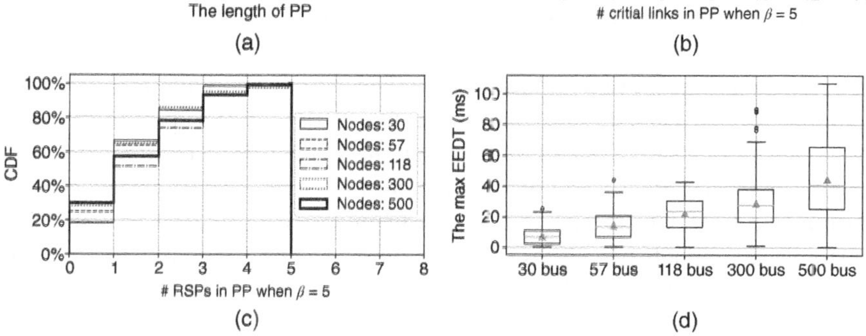

Figure 7.3 The features of the generated MPSG under different topologies. (a) The length of PP, (b) # critical links in PP when $\beta = 5$, (c) # RSPs at each router in PP when $\beta = 5$, and (d) the maximum EEDT in expectation

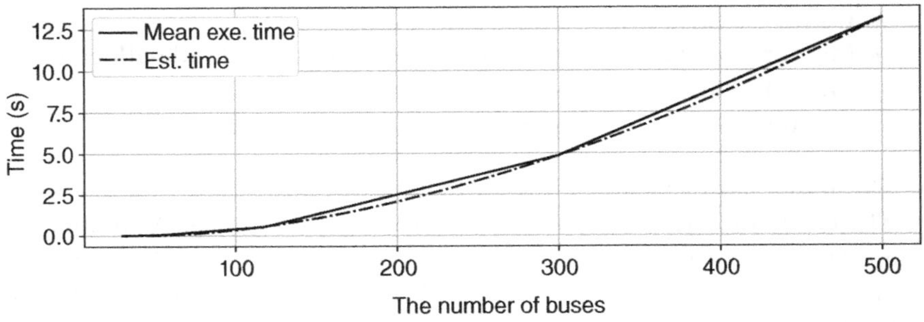

Figure 7.4 The mean execution time of GenMPSG.

7.2 Protocol Design and Its Issue

In this section, we will explain the general design and how DRTP collaborates to pinpoint issues during its retransmission.

7.2.1 Overall Design

The proposed DRTP serves as a forwarding strategy for the NDN router [Zhang et al., 2014], aligning with the previously discussed MPSG routing for transporting data from a PMU to the PDC. Let's delve into the design, focusing on three key aspects:

7.2.1.1 Message Design

DRTP messages are intricately crafted in NDN to carry packets during communication in the IEEE C37.118.2 framework (see Section 7.1.1). Figure 7.5(a) and 7.5(b) illustrates the formats of DRTP Interest and Data packets, respectively. At a low level, each message is a Type-Length-Value (TLV) collection encoded in a variable-size format. The TLV comprises three fields: (1) type (1 octet), (2) length (1 octet), and (3) value (length indicated by the length field). Some TLV may contain sub-TLV, and each sub-TLV may further nest [NDN, 2022]. The entire message originates from a TLV named LpPacket, serving as the link adaptation layer version 2 [Redmine, 2022]. The LpPacket includes a reserved tag (three octets), our customized tag (eight octets), and the fragment carrying Interest or Data. The customized tag holds two pieces of information: (1) an MPSG ID to differentiate communication pairs and (2) a path ID for identifying a PP or RSP to follow in forwarding on the MPSG.

On a higher level, the Interest comprises a variable-length name, CanBePrefix indicating the name as a prefix, nonce, and Interest life cycle time. Meanwhile, the Data includes a name, meta-information for the content type, a variable-length

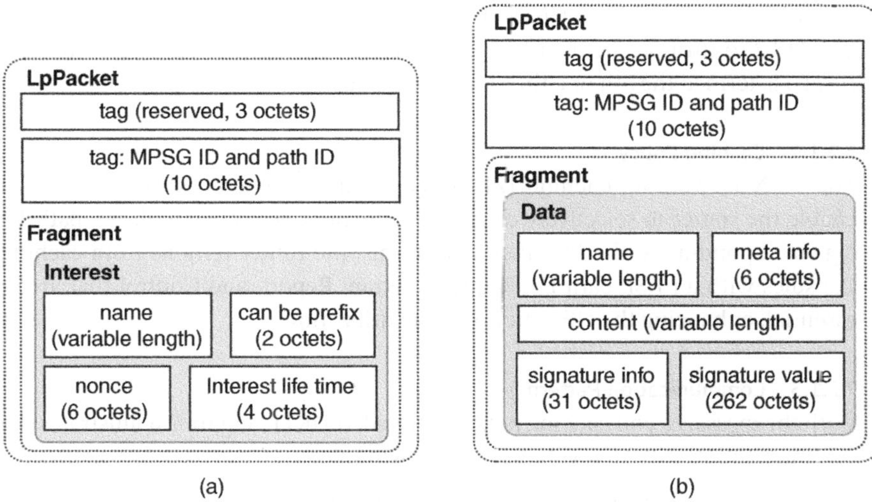

Figure 7.5 Packet format in DRTP. (a) Interest packet and (b) data packet.

content for holding a payload, and a Data signature. The signature is generated and verified by the producer and consumer, respectively. The maximum content length is constrained by the Maximum Transmission Unit (MTU) minus the length of the name and link layer, as well as the message overhead of 312bytes. In comparison, the length of the Data frame in IEEE C37.118.2 is determined by the total number of signals, including phasors, analog values, and digital status words from each monitored PMU. Encapsulating in NDN introduces a drawback of increased message overhead, potentially limiting the signals to be carried compared to IEEE C37.118.2. Furthermore, we expand the Data packet into four messages by leveraging their names as follows:

In the context of DRTP, several message types serve distinct purposes:

- **Capsule**: Used to carry the packet in its payload with a data ID designated by the producer.
- **Request**: Contains a retransmission request for the data ID of the detected lost Capsules.
- **Retran**: Carries a retransmitted payload marked with a data ID, where an empty payload indicates a mismatch during retransmission.
- **Report**: Indicates the data ID of the detected lost Capsules that have been requested for retransmission by an upstream router in the PP.
- **Control**: Mainly carries messages in IEEE C37.118.2 in its payload, including (1) the Command with the sending CFG frame and data transmission off, and (2) the Configuration with CFG.

Additionally, the Interest packet is utilized to carry the Command in IEEE C37.118.2 to turn on the data transmission. For simplicity, a message with data ID of δ is termed the δth one throughout the paper.

7.2.1.2 Caching Policy

The NDN forwarding daemon [Zhang et al., 2014] undergoes modification to enable the router to selectively cache Capsule messages solely with the path ID tag corresponding to the PP. In simpler terms, the router refrains from caching Capsules with an RSP path ID, Request, Retran, Report, and Control that aren't essential, enhancing the efficiency of cache utilization.

7.2.1.3 Communication Pattern

With the messages and caching policy outlined in DRTP, Figure 7.6 illustrates the communication pattern involving the following steps:

1) **Initiation**: The PDC sends Control to a PMU to request a CFG frame.
2) **Feedback**: The PMU responds with CFG via a Control message.
3) **Data transmission setup**: The PDC sends an Interest to the PMU to initiate data transmission.
4) **Data transmission**: Upon receiving the Interest, the PMU consistently sends synchrophasor results to the PDC at a minimum frequency of f.
5) **Data transmission shutdown**: The PDC sends Control to turn off data transmission.

In DRTP, the PDC streamlines its communication by sending only one Interest packet during the data transmission from the PMU to the PDC, effectively reducing Interest overhead compared to existing methods [Ravikumar et al., 2020]. The sequence from steps (3) to (4) collectively constitutes the packet delivery process. Throughout this operation, DRTP incorporates a hop-by-hop retransmission control process to ensure reliability in managing link disruptions. In Sections 7.2.2–7.2.4, we will delve into the intricacies of these two processes

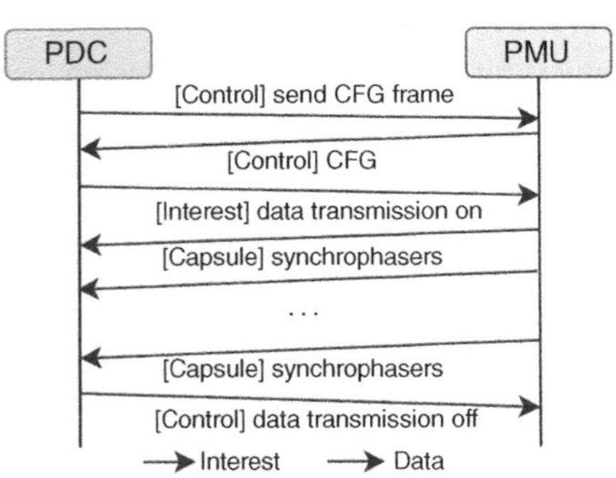

Figure 7.6 Communication pattern of DRTP.

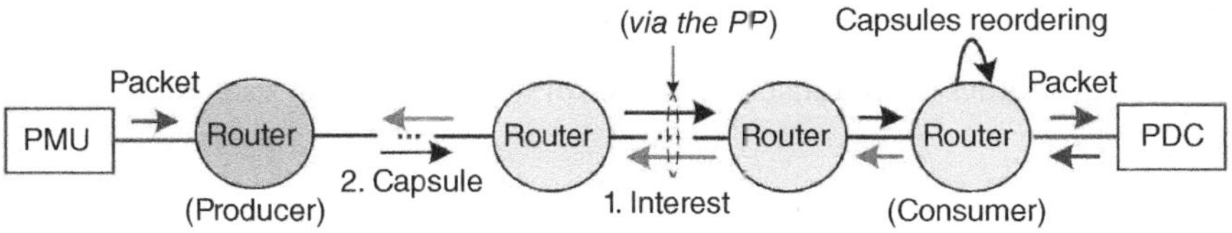

Figure 7.7 The normal packet delivery process of DRTP.

and shed light on a novel distributed collaboration issue that arises during the retransmission control process.

7.2.2 Packet Delivery Process

Packet delivery process: In a detailed breakdown illustrated in Figure 7.7, the packet delivery process unfolds through the following steps:

1) **Interest transmission**: The consumer initiates the process by sending an Interest to a producer along with the designated PP.
2) **Capsule sequence delivery**: Upon receiving the Interest, the producer responds by delivering a sequence of Capsules, each with an automatically incremented data ID, to the consumer along with the PP (depicted by black arrows). Each Capsule encapsulates a sampled packet transmitted from the PMU.
3) **Router handling**: As the Capsules traverse the PP, each router checks whether the Capsule has been forwarded previously. If affirmative, the router discards the Capsule; otherwise, it caches it into its local Content Store (CS) and forwards it to the next hop along with the PP.
4) **Efficient reordering at consumer**: The consumer efficiently organizes all received Capsules in sequence based on their data ID using a priority queue. The queue's size is determined by the maximum span of out-of-order Capsules received by the consumer.
5) **Capsule De-queuing**: The consumer de-queues each Capsule with the following criteria:
   - If the Capsule has a consecutive data ID to its previously received one, it is directly de-queued.
   - If nonconsecutive, the Capsule is initially waited for a maximum time for the arrivals of the missing Capsules and then de-queued.
6) **Consumer action**: Subsequently, the consumer decapsulates each de-queued Capsule into a packet and forwards the packet to the PDC.

This intricate process ensures the ordered and efficient delivery of packets from the PMU to the PDC within the DRTP framework.

7.2.3 Retransmission Control Process

Figure 7.8 outlines the retransmission control steps executed by each router within the PP:

i) The router identifies the data ID of lost Capsules based on the following criteria:
   - Non-consecutive to the previously received data Identifications (IDs).
   - Exceeding the *arrival timeout* of the router, expected but not arrived.

   If the data IDs are present in a previously received Report from upstream routers, no further action is taken, as they have already been retransmitted.
ii) The router sends the data IDs of lost Capsules to all downstream routers within the PP via a Report. These Reports instruct downstream routers to skip retransmission for the specified data IDs. Multiple Reports are sent ζ times to ensure retransmission state synchrony.
iii) The router sends Requests with the ID of lost Capsules to all upstream routers within the PP. These Requests are transmitted along both the upstream link of the PP and all RSPs of the router.
iv) Upon receiving a Request, each upstream router searches for a matching copy of each requested data ID in its CS. If a copy is found, the upstream router simultaneously sends the copy as a Retran to the router, along with its incoming path in reverse. In case of no match, an empty Retran is sent to indicate a mismatch during the current retransmission.
v) The router decapsulates the received non-empty Retran into a Capsule. If the Capsule has not been forwarded, it is forwarded to the next hop in the PP. If the Retran arrival times out, steps (iii) to (iv) are repeated at most η times until successful. The timeout is set as the maximum Round-Trip Time (RTT) of both the upstream link of the PP and all RSPs of the router.

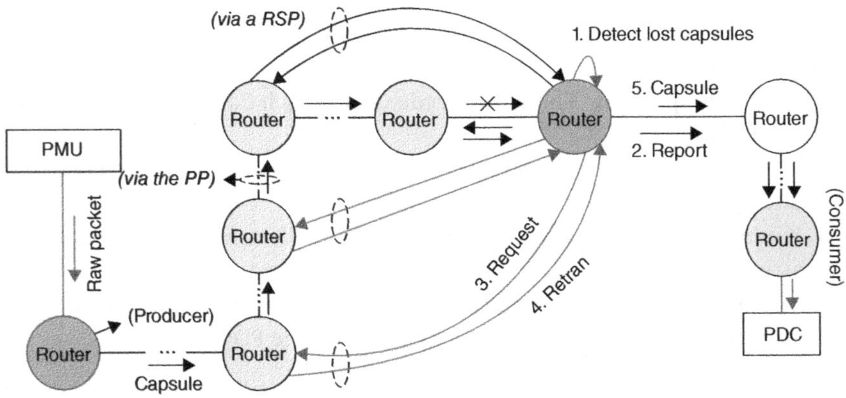

Figure 7.8 Retransmission control process of DRTP.

In the context of enhancing upstream link reliability during retransmission, steps (iii) and (iv) heavily leverage the RSPs. Furthermore, we estimate the complexity of the messages sent and received by each router within the PP during the delivery of a Capsule (Message Sent and Received by Each Router of the PP in Delivering a Capsule (MSR)) as follows: Steps (ii), (iii), (iv), and (v) involve $O(\zeta)$, $O(\eta)$, $O(\eta)$, and $O(1)$ messages, respectively. Therefore, the overall complexity is $O(2 \cdot \eta + \zeta + 1)$.

7.2.4 Distributed Collaboration Issue

In step (i), locating the Capsule in the arrival timeout state is non-trivial, involving computation to address the distributed collaboration issue in the presence of multiple failed links within the PP. This issue can contribute to an increased EEFR due to potential cache mismatching during retransmission.

To illustrate, consider the example in Figure 7.9, where r_{d_1} sends a Capsule to r_{d_b} along the PP from r_{d_1} to r_{d_b}. Initially, $r_{d_{k-1}}$ detects a lost Capsule (indicated by a cross) by waiting for its timeout for Capsule arrival. Subsequently, $r_{d_{k-1}}$ sends a corresponding Request for the Capsule in timeout to $r_{d_{k-2}}$ upstream. Upon receiving the Request, $r_{d_{k-2}}$ retransmits a matched Capsule to $r_{d_{k-1}}$.

However, the retransmitted Capsule is lost on the link $(r_{d_{k-1}}, r_{d_k})$ (cross). Consequently, r_{d_k} waits for a timeout, sends a Request to $r_{d_{k-1}}$, but the latter, having not received the corresponding Retran, finds no copy in its local CS – resulting in cache mismatching, and no copy is sent to r_{d_k}. Ultimately, r_{d_b} fails to receive the retransmitted copy, degrading the EEFR.

This example generalizes when a Capsule is lost on multiple links in the PP, achieved by combining any two adjacent links, such as $(r_{d_{k-2}}, r_{d_{k-1}})$ and $(r_{d_{k-1}}, r_{d_k})$.

To address this distributed collaboration issue, a solution is to determine a suitable arrival timeout at each r_{d_k} in the PP. With this timeout, the Request from r_{d_k} to $r_{d_{k-1}}$ can be sent immediately after the arrival of the requested payload, potentially from a Capsule or a nonempty Retran at $r_{d_{k-1}}$. In essence, each r_{d_k} waits for $r_{d_{k-1}}$ upstream to complete its retransmission in a recursive manner, termed "recursiveness." This ensures that the Request won't cause cache mismatch at $r_{d_{k-1}}$, reducing the EEFR.

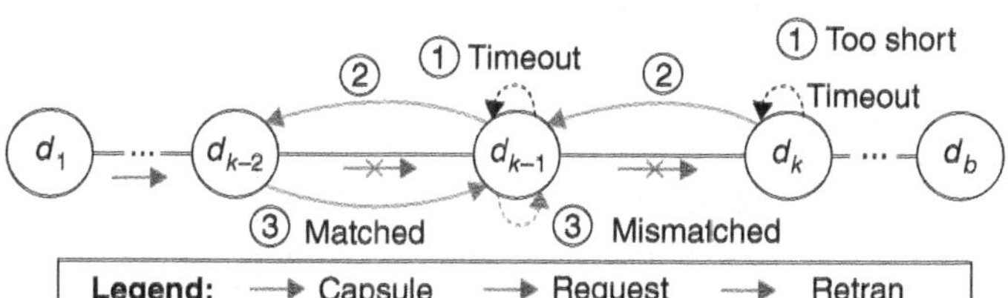

Figure 7.9 Distributed collaboration issue.

7.3 Computation of Arrival Timeouts

To tackle the distributed collaboration issue in DRTP outlined in Section 7.2.4, we introduce two critical assumptions related to the retransmission control process:

1) **Assumption 1**: Each router within the PP has successfully received at least one identical Capsule. If this condition is not met, the router discontinues successive retransmissions due to the absence of a reference time to trigger the arrival timeout.
2) **Assumption 2**: At least one router within the PP can successfully retransmit the lost Capsule in η attempts.

These assumptions serve as guidelines for identifying and calculating arrival timeouts under various retransmission scenarios.

7.3.1 Identifying Arrival Timeouts

In Step (i) of the retransmission control process, discussed in Section 7.2.3, we delve into the anticipated Capsule and its corresponding arrival timeout. The retransmission of the δth lost Capsule may result in either a success or failure scenario. In a successful situation, the router anticipates the arrival of the next δ_{next}th Capsule. This is the nearest untransmitted Capsule following the last successfully received one and is not included in previously received Reports, preventing duplicate retransmissions. In case of failure, the router expects the lost δth Capsule due to an unsuccessful retransmission.

In each of the two scenarios following the reception of the δth DRTP message, distinct arrival timeouts are defined:

- The maximum Capsule Inter-Arrival Time (CIAT) for the δ_{next}th Capsule.
- The maximum Capsule Delivery Time (CDT) between the new δth Capsule sent by its upstream router.

The DRTP message received can be identified by leveraging its receiving exclusiveness during retransmission. Specifically, each router in the PP can exclusively receive at most one of the Capsule, non-empty Retran, and empty Retran, each with their data ID of δ. Similarly, the router can exclusively receive at most one of the Report, the nonempty Retran, and the empty Retran containing their data ID of δ. This exclusiveness stems from the fact that when the δth Capsule or the δth Report is received in step (i), Requests will not be sent in step (iii). Therefore, no nonempty or empty Retran will be received. Table 7.1 summarizes the correspondence relationships of CIAT and CDT in different conditions for the received messages.

Table 7.1 Conditions of success and failure situations.

Condition (timeout type)	The δth DRTP messages recently received			
	Capsule	Non-empty Retran	Empty Retran	Report
#1 (CIAT)	Yes	No	No	No/yes
#2 (CIAT)	No	Yes	No	No
#3 (CDT)	No	No	No	Yes
#4 (CDT)	No	No	Yes	No

7.3.2 Capsule Inter-Arrival Time for Success Situation

Elaborating on the success condition in Table 7.1, the two conditions for computing the CIAT are as follows:

7.3.2.1 Condition #1

This scenario occurs when only the δth Capsule is received, both with and without the δth Report. Figure 7.10 illustrates this condition for routers r_{d_k}, $r_{d_{k+1}}$, and $r_{d_{k+2}}$ with the collaboration recursiveness maintained as follows. Assume the loss of the δ_{next}th Capsule on the link from $r_{d_{k-1}}$ to r_{d_k}. Consequently, r_{d_k} detects the loss after CIAT_1 and issues the Request for up to η times. Simultaneously, set CIAT_1

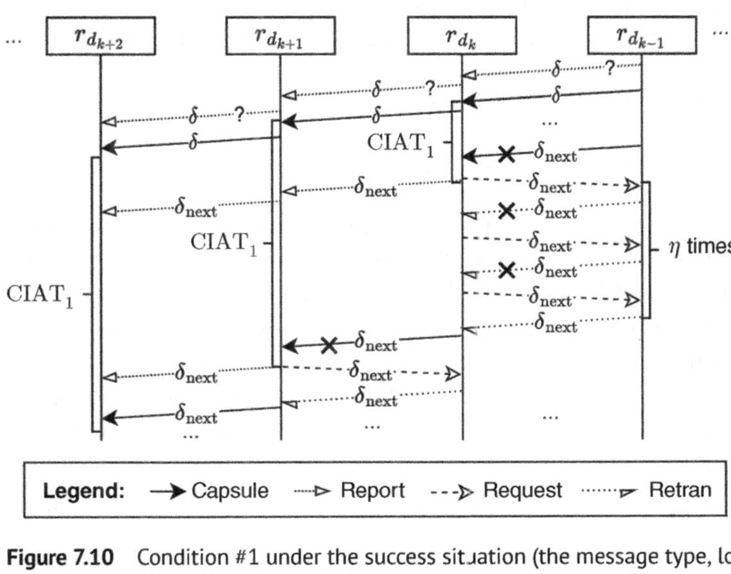

Figure 7.10 Condition #1 under the success situation (the message type, loss event, and data ID are indicated in different colors, cross, and arrow annotation, respectively).

for the downstream router of $r_{d_{k+1}}$ large enough to ensure that the Request arrives at r_{d_k} just before the feedback of the δ_{next}th Retran to r_{d_k}. Similarly, the CIAT_1 of $r_{d_{k+2}}$ should be sufficiently large for $r_{d_{k+1}}$. CIAT_1 for each router of r_{d_k} in the PP represents the time difference between the expected Capsule of ξ_{next} and the previously received Capsule of ξ, denoted as $t_k(\xi_{\text{next}}) - t_k(\xi)$. This difference satisfies the inequation given in (7.1) and is also proven in Theorem 7.1.

Theorem 7.1 *The time difference of $t_k(\xi_{\text{next}}) - t_k(\xi)$ satisfies the inequality given in Eq. (7.1). Wherein, $\theta[k]$ is given in Eq. (7.2), and $t_k(\xi)$ is the arrival time of ξ at r_{d_k}.*

$$t_k(\xi_{\text{next}}) - t_k(\xi) \leq (t_{k-1}(\xi_{\text{next}}) - t_{k-1}(\xi)) + \tau_{\text{queue}}^{\text{PP}}(d_{k-1}, d_k) + \theta[k] \tag{7.1}$$

$$\theta[k] = \eta \cdot \max_{p(d_k, d_s) \in P_{\text{rsp}}[k] \cup \{(r_{d_k}, r_{d_{k-1}})\}} (\tau_{\text{all}}^{p}(d_k, d_s) + \tau_{\text{all}}^{p}(d_s, d_k)) \tag{7.2}$$

Proof: Equation (7.3) illustrates that $t_k(\xi_{\text{next}})$ is no greater than the sum of the following times: (1) $t_{k-1}(\xi_{\text{next}})$, (2) the maximum time to forward ξ_{next} from $r_{d_{k-1}}$ to r_{d_k} over the PP (denoted as $\tau_{\text{all}}^{\text{PP}}(d_{k-1}, d_k)$), and (3) the request timeout of $\theta[k]$ computed through Eq. (7.2), i.e., η times the request timeout at r_{d_k}. The request timeout is the sum of all delays of the upstream link of the PP and its RSPs.

Simultaneously, Eq. (7.4) indicates that $t_k(\xi)$ is no less than the sum of the following times: (1) $t_{k-1}(\xi)$ and (2) the total delay from $r_{d_{k-1}}$ to r_{d_k} subtracted by its queueing delay over the PP. Subtracting Eq. (7.4) from Eq. (7.3) on both sides yields the CIAT between ξ and ξ_{next} at r_{d_k} in Eq. (7.1). This concludes the proof.

$$t_k(\xi_{\text{next}}) \leq t_{k-1}(\xi_{\text{next}}) + \tau_{\text{all}}^{\text{PP}}(d_{k-1}, d_k) + \theta[k] \tag{7.3}$$

$$t_k(\xi) \geq t_{k-1}(\xi) + (\tau_{\text{all}}^{\text{PP}}(d_{k-1}, d_k) - \tau_{\text{queue}}^{\text{PP}}(d_{k-1}, d_k)) \tag{7.4}$$

According to Theorem 7.1, CIAT_1 is computed as per Eq. (7.5). This computation is based on the fact that the CIAT at the consumer corresponds precisely to the period of the PMU traffic multiplied by the difference between δ_{next} and δ, i.e., $t_1(\delta_{\text{next}}) - t_1(\delta) = (\delta_{\text{next}} - \delta)/f$. Moreover, there is no retransmission performed between the successfully delivered δth and the unarrived δ_{next}th Capsule, leading to $\theta[m] = 0$. Additionally, by extending $t_{k-1}(\xi)$ to $t_1(\xi)$ iteratively on the right-hand side of Eq. (7.1), CIAT_1 is expressed in Eq. (7.5). Hence, the recursiveness is upheld.

$$\text{CIAT}_1 \leq (\delta_{\text{next}} - \delta)/f + \tau_{\text{queue}}^{\text{PP}}(d_1, d_k) \tag{7.5}$$

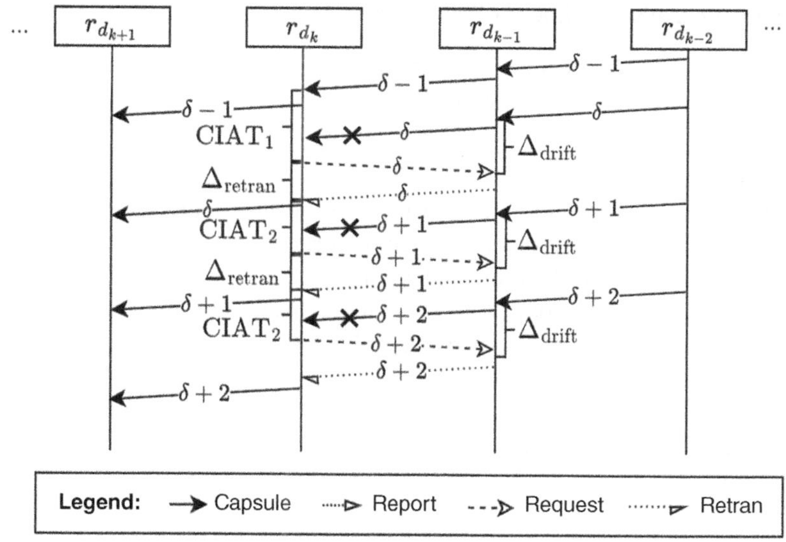

Figure 7.11 Condition #2 under the success situation.

7.3.2.2 Condition #2

This condition occurs when a router exclusively receives a nonempty Retran, indicating that retransmission is solely performed by the router. Illustrated in Figure 7.11, each router in the PP has received the $(\delta - 1)$th Capsule, and the $\delta \sim (\delta + 2)$th Capsules are all lost on the link of $(r_{d_k}, r_{d_{k-1}})$. Consequently, upon receiving the nonempty Retran, r_{d_k} detects the lost Capsule based on the CIAT computed through (7.6). Here, Δ_{drift} for the δth Capsule at $r_{d_{k-1}}$ represents the time difference between the Capsule arriving from $r_{d_{k-2}}$ and the δth Request arriving from r_{d_k}. Note that Δ_{drift} is piggybacked to r_{d_k} through the δth Retran.

Additionally, Δ_{retran} for the δth Capsule at r_{d_k} is the time difference between the sending of the δth Request and the receiving of the δth Retran. This is because the time between the δth Capsule and $(\delta + 1)$th Capsule is exactly CIAT_1 under the recursive collaboration process. Therefore, CIAT_2 is CIAT_1 subtracted by Δ_{drift} and Δ_{retran} and one round of the transmission delay between r_{d_k} and $r_{d_{k-1}}$ (Eq. (7.6)).

Moreover, this subtraction minimizes the unnecessary waiting time for the arrival of retransmission for each lost Capsule in cases of consecutive lost Capsules. Thus, this process adheres to the required recursiveness.

$$\text{CIAT}_2 = \text{CIAT}_1 - \Delta_{\text{drift}} - \Delta_{\text{retran}} + \tau_{\text{all}}^{\text{PP}}(d_k, d_{k-1}) + \tau_{\text{all}}^{\text{PP}}(d_{k-1}, d_k) \tag{7.6}$$

7.3.3 Capsule Delivery Time for Failure Situation

Meanwhile, in Table 7.1, the other two conditions for computing the CDT under the failure situation are detailed as follows.

7.3.3.1 Condition #3

This scenario occurs when the router only receives the δth Report, signifying that all corresponding retransmissions have been performed by its upstream routers. Depicted in Figure 7.12 for the routers of r_{d_k} and $r_{d_{k+1}}$, both routers have lost the δth Capsules retransmitted from $r_{d_{k-1}}$ and r_{d_k}, respectively, indicating retransmission failures. Consequently, the routers determine the lost δth Capsule based on the corresponding CDT relative to their δth Reports sent from $r_{d_{k-1}}$ and r_{d_k}, respectively. Thus, in accordance with Theorem 7.1, the CDT from an upstream router of r_{d_s} to r_{d_k} can be computed as per Lemma 7.1. Consequently, the CDT maintains recursiveness.

Lemma 7.1 *The CDT for retransmitting of the ξth lost Capsule from an upstream router of r_{d_s} at the current router of r_{d_k} of the PP can be computed through Eq. (7.7).*

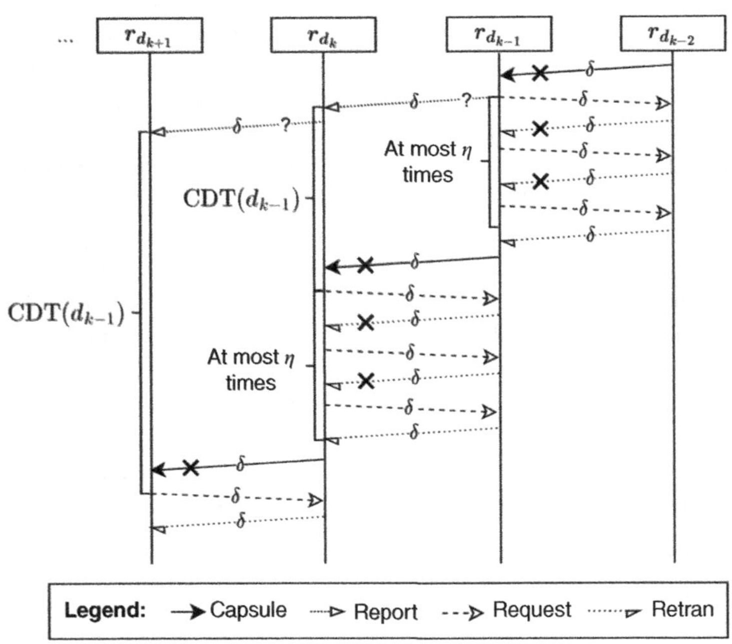

Figure 7.12 Condition #3 under the failure situation.

$$\mathrm{CDT}_k(d_s) \le \tau^{\mathrm{PP}}_{\mathrm{queue}}(d_s, d_k) + \sum_{s \le c \le k-1} \theta[c] \tag{7.7}$$

Proof: According to Eq. (7.1), we can derive Eq. (7.8). Specifically, we iteratively extend the right-hand side from t_{k-1} to t_s. Considering the collaboration recursiveness, the maximum time between the latest Capsule and the next one at r_{d_s} cannot exceed the maximum retransmission time at r_{d_s}, i.e., $t_s(\xi_{\mathrm{next}}) - t_s(\xi) \le \theta[s]$. Utilizing this inequality, we observe that the right hand is no greater than $\tau^{\mathrm{PP}}_{\mathrm{queue}}(d_s, d_k) + \sum_{s \le c \le k} \theta[c]$.

Furthermore, the CDT for ξ_{next} can be considered as $t_k(\xi_{\mathrm{next}}) - t_k(\xi) - \theta[k]$, where the subtraction for $\theta[k]$ is necessary to exclude the time required to perform the retransmission at r_{d_k}. Hence, the CDT is computed in Eq. (7.7) by setting $t_k(\xi) = 0$ and replacing ξ_{next} with ξ in Eq. (7.8).

$$\begin{aligned} t_k(\xi_{\mathrm{next}}) - t_k(\xi) &\le (t_{k-1}(\xi_{\mathrm{next}}) - t_{k-1}(\xi)) \\ &\quad + \tau^{\mathrm{PP}}_{\mathrm{queue}}(d_{k-1}, d_k) + \theta[k] \\ &\quad \dots \\ &\le (t_s(\xi_{\mathrm{next}}) - t_s(\xi)) \\ &\quad + \tau^{\mathrm{PP}}_{\mathrm{queue}}(d_s, d_k) + \sum_{s-1 \le c \le k} \theta[c] \\ &\le \tau^{\mathrm{PP}}_{\mathrm{queue}}(d_s, d_k) + \sum_{s \le c \le k} \theta[c] \end{aligned} \tag{7.8}$$

7.3.3.2 Condition #4

This scenario occurs when the router only has the empty δth Retran. This signifies that the δth Request is sent too early, causing a cache mismatch due to Report loss. Depicted in Figure 7.13 for the routers of r_{d_k}, $r_{d_{k+1}}$, and $r_{d_{k+2}}$, each router detects the lost δth Capsule according to its CDT. The CDT can be computed using $\mathrm{CDT}_k(d_1)$ according to Eq. (7.7). The reason is that the router is unaware of which upstream router has retransmitted the lost Capsule. Hence, the upstream router is assumed to be r_{d_1} to avoid the earlier request, maintaining recursiveness. Meanwhile, CDT also causes a longer retransmission time. Therefore, the condition should be avoided as much as possible by transmitting the duplicated Report for ζ times.

7.3.4 Discussions on Distributed Collaborations

With the assumptions in Section 7.3, the DRTP effectively addresses the distributed collaboration issue. This success is attributed to the CIAT and CDT

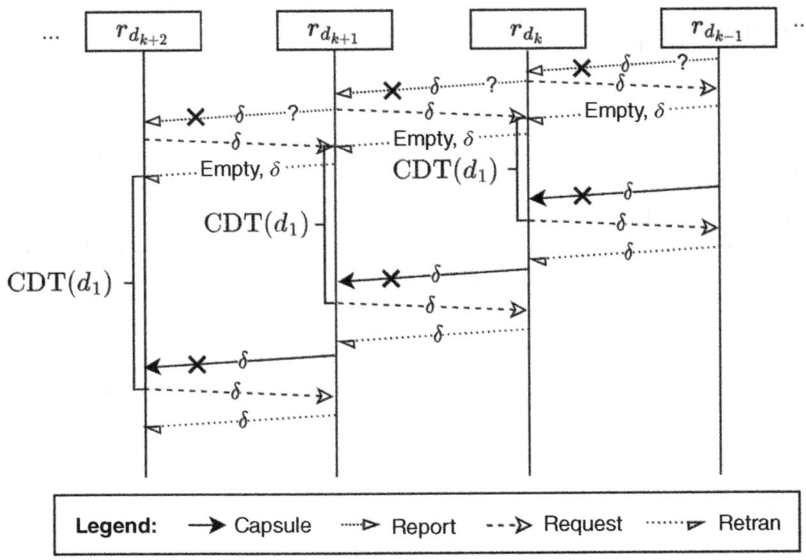

Figure 7.13 Condition #4 under the failure situation.

computations, which can prevent cache mismatching under all possible conditions listed in Table 7.1. The DRTP exhibits the following performance properties for these conditions:

i) **Much reduced EEFR**: The avoidance of cache mismatching leads to a significantly reduced EEFR. Moreover, the EEFR can be 0% when the last router in the PP encountering a lost Capsule can successfully retransmit the Capsule in η times. For instance, this occurs when there is at least a completely reliable RSP for each router of the PP and successful delivery of the Report message at each hop of the PP.
ii) **Bounded delivery time**: The maximum EEDT satisfies Eq. (7.9) in Lemma 7.2 when there is no cache mismatch. Therefore, the DRTP puts forth its best effort to retransmit the lost Capsules.

Lemma 7.2 *The EEDT for r_{d_b} satisfies Eq. (7.9) under the recursiveness of the DRTP when there is no cache mismatch.*

$$\text{EEDT}(d_b) \leq \tau_{\text{all}}^{\text{PP}}(d_1, d_b) + \tau_{\text{queue}}^{\text{PP}}(d_1, d_b) + \sum_{2 \leq c \leq b} \theta[c] \tag{7.9}$$

Proof: According to Theorem 7.1, we can derive Eq. (7.10) by replacing ξ_{next} with ξ and ξ with ξ'. Furthermore, by adding $t_k(\xi') - t_{k-1}(\xi)$ on both sides of Eq. (7.10), we obtain Eq. (7.11). In Eq. (7.11), we can iteratively write equations from t_{k-1} to t_2. Next, by independently adding the two sides of Eq. (7.11) and letting k be b, we get Eq. (7.12).

To prove Eq. (7.9), we consider the prerequisite when there is no cache mismatch under the collaboration recursiveness. The prerequisite indicates that either the success situation or Condition #3 must occur. Specifically, we need to verify that $t_b(\xi') - t_1(\xi')$, i.e., the EEDT of ξ' for r_{d_b} (termed as EEDT-P), on the right side of Eq. (7.12) is not greater than the total path delay of PP, i.e., $\tau_{\text{all}}^{\text{PP}}(d_1, d_b)$.

For the success situation, the EEDT of ξ' is analyzed based on whether ξ' has been retransmitted by any upstream router or not:

a) If ξ' is without retransmission, it implies that EEDT-P is not greater than the total path delay of PP.
b) Otherwise, if ξ' is a retransmitted Capsule, the nonempty Retran for ξ' must have arrived from an upstream router, assumed to be r_{d_s}. r_{d_s} is under condition #2 with CIAT_2. According to Eq. (7.6), CIAT_2 at an upstream router of r_{d_s}, denoted as $r_{d_{s'}}$, is computed from CIAT_1 by subtracting $\Delta_{\text{drift}} + \Delta_{\text{retran}} - (\tau_{\text{all}}^{\text{PP}}(d_{s'}, d_{s'-1}) + \tau_{\text{all}}^{\text{PP}}(d_{s'-1}, d_{s'}))$. The subtracted part represents the time difference between the original arrival of the lost ξ'th Capsule at r_{d_k} and the actual arrival of its retransmitted copy.

Therefore, EEDT-P is reduced by the total time of the subtracted parts of $r_{d_{s'}}$, which includes the time consumed in the retransmission carried out by all upstream routers of r_{d_s}. As a result, in the success situation, the EEDT is equal to Eq. (7.9).

Meanwhile, under condition #3 of the failure situation, the CDT from r_{d_1} to r_{d_s} and the Report sending time from r_{d_s} to r_{d_k} are no more than $\tau_{\text{all}}^{\text{PP}}(d_1, d_s)$ and $\tau_{\text{all}}^{\text{PP}}(d_s, d_k)$, respectively. Hence, the EEDT is the sum of these two times and the right side of Eq. (7.7), where EEDT is no more than Eq. (7.9). Therefore, the EEDT has an upper bound computed by Eq. (7.9) for the success situation and condition #3, completing the proof.

$$
\begin{aligned}
t_k(\xi) - t_k(\xi') \leq\ & (t_{k-1}(\xi) - t_{k-1}(\xi')) \\
& + \tau_{\text{queue}}^{\text{PP}}(d_{k-1}, d_k) + \theta[k]
\end{aligned}
\tag{7.10}
$$

$$
\begin{aligned}
t_k(\xi) - t_{k-1}(\xi) &\leq t_k(\xi') - t_{k-1}(\xi') \\
&\quad + \tau_{\text{queue}}^{\text{PP}}(d_{k-1}, d_k) + \theta[k] \\
t_{k-1}(\xi) - t_{k-2}(\xi) &\leq t_{k-1}(\xi') - t_{k-2}(\xi') \\
&\quad + \tau_{\text{queue}}^{\text{PP}}(d_{k-2}, d_{k-1}) + \theta[k-1] \\
&\cdots
\end{aligned}
\tag{7.11}
$$

$$t_2(\xi) - t_1(\xi) \leq t_2(\xi') - t_1(\xi') + \tau^{PP}_{queue}(d_1, d_2) + \theta[2]$$

$$t_b(\xi) - t_1(\xi) \leq (t_b(\xi') - t_1(\xi')) + \tau^{PP}_{queue}(d_1, d_b) + \sum_{2\leq c\leq b} \theta[c] \tag{7.12}$$

7.3.4.1 Feasibility Analysis

The performance assurance of DRTP, as discussed earlier, may face challenges under heavy link disruptions, primarily influenced by two factors:

i) **Violation of collaboration recursiveness**: This can occur due to the loss of Reports, leading to premature Request sending during retransmission. The loss rate of Reports tends to increase with a higher loss rate in the PP.
ii) **Retransmission failure rate**: The failure rate of retransmission at each hop of the PP is influenced by the presence of critical links without available RSPs at their exit points.

To enhance assurance, the structural characteristics of a typical MPSG under a certain LLR can be summarized as follows:

a) **Short PP length**: Keeping the PP length short ensures more reliable Report transmission in ζ times. Adjusting ζ to a higher value further increases reliability.
b) **Reduced critical links**: Minimizing the number of critical links is crucial. In contrast, a non-typical MPSG may be less assured, leading to a higher EEFR due to factors mentioned above.

7.4 Performance Evaluation

The proposed DRTP undergoes simulation testing in a cyber–physical setting to showcase its effectiveness during significant link disruptions. Furthermore, we assess DRTP by examining its implementation accuracy, comparing its performance to other transport protocols, and analyzing its impact on protocol reliability in the context of SE within grid systems.

7.4.1 Cyber–Physical Simulation Environment

Figure 7.14 depicts the cyber–physical simulation environment designed to replicate the data dissemination process of phasor measurements in an optical IoT network within a power transmission grid, as outlined in Chakrabarti et al.

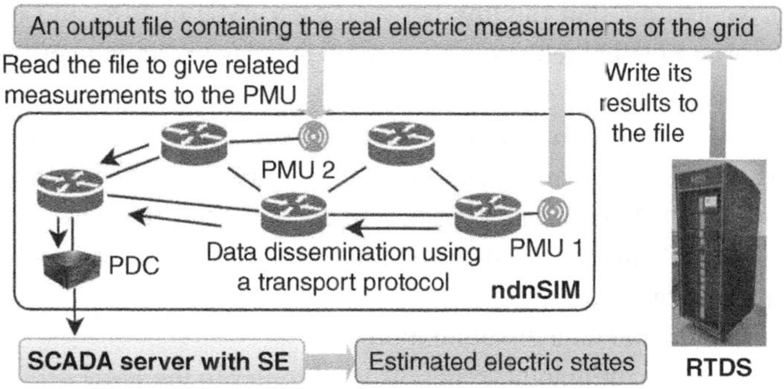

Figure 7.14 Cyber–physical simulation environment.

[2009; Dileep, 2020]. The simulation involves multiple data transport protocols. Let's delve into its key components.

i) **Network simulation**: We conducted a network simulation using ndnSIM to emulate an optical IoT network, incorporating two real transmission grid topologies with potential link disruptions. The network links substations using routers of a specific type, and these routers connect to adjacent PMUs or Phasor Data Concentrators (PDCs). The considered grid topologies are:
   - The South Carolina 500-bus system [Josz et al., 2016], suitable for current PMU deployment.
   - The IEEE 14-bus system [IEEE, 2022], whose physical part can be simulated by RTDS.

   Each substation's router is connected to its neighboring substation via optical lines, aligning with features described in the literature [Zhou et al., 2022]. Key network performance properties include:
   - Each link port operates at a maximum rate of 10Gbps, complying with the IEC 60794-4-10 standard [IEC, 2014].
   - The maximum queuing delay for each link is set at 2μs to prevent packet drops from queue overflow.
   - Message processing delay is assumed to be 1μs, corresponding to a rate of 1 Million Packets Per Second (MPPS). This delay is achievable due to the efficient processing of DRTP with $O(1)$ time complexity. The high-performance NDN data plane [Takemasa et al., 2017; Liu et al., 2021; Shi et al., 2020; Miguel et al., 2018; Li et al., 2014] and an efficient CS contribute to this efficiency. For example, hierarchical aligned transition arrays with 15 pipeline stages in an Field Programmable Gate Array (FPGA) can achieve 125MPPS with a delay of 0.12μs [Li et al., 2014]. The CS, implemented as NameTrie in software, can handle 3.56, 3.72, and 3.25 million name insertions, lookups, and removals per second, respectively [Ghasemi et al., 2018].

For simplicity, we assume that the link between each PMU or PDC and its adjacent router is completely reliable with negligible delay to facilitate effective performance evaluation.

ii) **Data dissemination using transport protocols**:The data dissemination process begins at the PMU and culminates at the PDC, both adhering to the IEEE C37.118.2 standard [IEEE, 2011]. This journey involves intermediate routers employing diverse transport protocols. Here's a step-by-step breakdown:

1) Command initiation and data dissemination:
   - The PDC triggers a command through its local router to the PMU.
   - Upon receiving the command, the PMU disseminates its data in payload packets, typically at a frequency of 50Hz, through the remote router to the PDC.
   - Each PMU and the PDC establish a direct connection with their respective adjacent routers, serving as proxies for PMUs and the PDC.
2) Transport protocol types:
   - The proposed DRTP is deployed on each NDN router, based on Multipath Subgraphs (MPSGs) created by the GenMPSG algorithm [Zhou et al., 2022]. The DRTP router is integrated into the NDN forwarding daemon [NDN, 2023], embedded in the ndnSIM simulator [Mastorakis et al., 2017]. The source code of DRTP and its relevant test environment is available online [GitHub, Inc., 2022].
   - Hop-by-hop transport protocols, utilizing the single shortest path computed via the Dijkstra algorithm, include Rapid and Reliable Transport Mechanism (R2T) [Wang et al., 2018], MFTP [Su et al., 2015], and Adaptive Transmission Protocol based on Software-Defined Networking (SDATP) [Chen et al., 2020].
   - Traditional IP forwarding utilizes the single shortest path, with two selective reliable transport protocols: (1) IEEE C37.118.2, using Transmission Control Protocol (TCP) for reliability [IEEE, 2011], and (2) non-disjoint path aggregation approaches, including TCP-OpenFlow-based Non-Disjoint Path Aggregation Mechanism (OPAM) and User Datagram Protocol (UDP)-OPAM [Zhou et al., 2019].
3) The configuration of DRTP includes parameters as follows:
   - The CS [Mastorakis et al., 2017] capacity set to 10,000 packets to prevent cache mismatches caused by cache replacement [Zhang et al., 2014].
   - η and ζ parameters configured with a value of 3 to address significant link disruptions.

4) Each TCP end typically configures the NewReno congestion control using default parameters:
   - Nagle algorithm disabled for sending timeliness.
   - Sending and receiving buffer sizes set to 13,1072B.
   - Timeout for delayed ACKs reached at 200 ms.
   - Number of packets to wait before sending ACK is 2.

iii) **Underlying MPSGs**: Considering the two degrading factors outlined in Section 7.3.4, we conduct experiments to investigate the performance of DRTP with both typical and general MPSGs:

The typical MPSG: The MPSG is structured according to Figure 7.2, derived from a real power transmission grid [IEEE, 2022]. Each node in the MPSG represents a DRTP-enabled router. The MPSG is characterized by zero critical links in the PP and a short PP length. We aim to verify that DRTP performs with 0% EEFR and a bounded EEDT not exceeding 15.696 ms under the assumptions in Section 7.3. Additionally, we evaluate the quality of the PMU packet delivery process using two metrics: (1) the out-of-order rate at the consumer (Out-of-Order Rate at the Receiver Side (OOR)), and (2) the Packet Reordering Time at the Consumer (PRT). The average MSR is also assessed to demonstrate the message overhead of DRTP. In this case, the absence of excessively long PP and critical PP links facilitates DRTP's collaboration recursiveness under PP disruption, validating the performance assurance mentioned in Section 7.3.4.

The general MPSG generated by GenMPSG: We test DRTP with MPSGs generated using the GenMPSG Algorithm 7.1 for two transmission grids modeled from real-world datasets: the IEEE 300 bus dataset [IEEE, 2022] and the SC 500-bus dataset [Josz et al., 2016]. For each grid, 150 communication pairs are randomly selected, and the corresponding MPSG is generated for each pair. These MPSGs vary in the length of the PP and the number of critical links in the PP (refer to Figure 7.3). This diversity is crucial to testing DRTP under conditions of recursiveness violation and retransmission failures, as discussed in Section 7.3.4. The performance of EEFR and EEDT, and the satisfaction ratio of EEDT (Satisfaction Ratio of EEDT (SRE)) are assessed. SRE represents the ratio of pairs with EEDT not exceeding the bound computed using Eq. (7.9) to the total number of pairs.

iv) **Simulating link disruptions for the protocols**: To evaluate the protocols, disruption events were simulated with varying Link Loss Rate in Packetss (LLRs) in two distinct scopes: PP and MPSG. The LLR of each link is modeled as a percentage of random packet drops, distributed uniformly. The two scopes were customized for tests in the corresponding protocols:

1) Full PP disruptions:
   - **Protocols evaluated**: DRTP, IEEE C37.118.2 in TCP, MFTP, R2T, and SDATP.
   - **Evaluation scenario**: Full PP disruptions were examined to assess the resilience of the mentioned protocols under severe link disruptions.
2) Full MPSG disruptions:
   - **Protocols evaluated**: DRTP, TCP-OPAM, and UDP-OPAM.
   - **Evaluation scenario**: Full MPSG disruptions were analyzed to gauge the protocols' performance when faced with disruptions across the entire MPSG.

   These targeted simulations provide insights into how each protocol responds to disruptions in specific contexts, aiding in the comprehensive assessment of their robustness and reliability.

v) **The simulation model of a power transmission grid**: To rigorously assess the accuracy of SE, we meticulously crafted a grid model utilizing the esteemed RTDS [RTDS Technologies Inc., 2023]. Renowned for its exceptional performance and precision in power system modeling, RTDS ensures reliability in our evaluation.

   Model Specifications:
   - Our model adheres to the standardized IEEE 14-bus system [IEEE, 2022], featuring a base voltage of 18 kV.
   - Dynamic loads, emulating periodic variations in electrical conditions experienced by buses and branches, were incorporated into the model. Each load dynamically adjusts its active and reactive powers within a uniform distribution spanning 50–200% of their predefined mean values.
   - To introduce further realism, we scripted bus #3's power load to toggle between on and off states at a frequency of 1 Hz.

   Simulation Duration: The grid simulation spanned 60 seconds, a chosen duration ensuring an accurate reflection of grid dynamics with minimal test bias.

   Data collection: Raw measurements from each bus and branch were collected at a typical frequency of 50 Hz, aligning with PMU measurements. The collected measurements encompassed phase, voltage magnitude, active power, and reactive power.

   Output and sensing data: The collected data was outputted into a file, serving as the sensed data for the corresponding PMUs in subsequent simulations.

   This comprehensive simulation model, rooted in the robust capabilities of RTDS, ensures a reliable foundation for assessing the precision and efficacy of SE in the context of a power transmission grid.

vi) **An integrated Supervisory Control and Data Acquisition (SCADA) server with PMU-Based SE**:
Our attention centers on SCADA due to its pivotal role in SE, with SE functionality provided by MATPOWER [Zimmerman et al., 2011]. This focused approach enables an exploration of how the completeness of received data influences SE, impacting its observability and accuracy. Key Components:
SE mechanism: Utilizing PDC to manage the most recent data measurements, SE employs classical methods such as the weighted least-squared method and the Gauss–Newton algorithm to estimate electric states for all grid buses and branches [Huang et al., 2012].
Assessment parameters: To evaluate correction capabilities, we extended the time window constraint for received data to infinity, shifting the focus from EEDT and OOR to assessing EEFR.
Simulation setup: We assumed a lossless link between the PDC and SCADA, mimicking practical scenarios where they are often co-located in the same computer facility. This simplification allows a concentrated focus on SE tests without delving into inner data transmission intricacies.
Experimental design: The experiment settings facilitate the exploration of various cases, enabling the evaluation of corresponding metrics. This comprehensive approach ensures a nuanced understanding of SE performance under different conditions.

7.4.2 Experiment Cases and Metrics

The evaluation of DRTP encompasses four distinct cases:

i) **Performance assessment of DRTP**: The comprehensive evaluation of DRTP is conducted for both typical and general MPSG scenarios, as previously discussed. Each simulation, spanning 33 seconds, is carefully crafted to capture the intricacies of retransmission dynamics. Commencing at 3 seconds, all link fault events are initiated. This timing ensures that each router in the PP receives at least one Capsule without retransmission, adhering to the first assumption.
Different levels of link fault events, measured by LLR – representing the possibility of randomly discarding arriving packets on the link are simulated. Faults with identical LLR for each link are introduced for a specified duration in two distinct ranges:

- **Full disruption on all Links of the PP**: This scenario serves to test the success situation, ensuring the existence of at least one reliable RSP for each PP router, thereby achieving 0% EEFR as outlined in Section 7.3.4.
- **Complete disruption of all links of the MPSG**: Here, we explore both success and failure situations for relatively smaller and higher LLR, respectively. The higher LLR is essential for triggering condition #3 or #4. However, caution is exercised, as excessively high LLR can lead to assumption violations, consequently impacting their corresponding EEFR.

ii) **Performance comparative studies of DRTP**. In this phase, we conduct a thorough comparison of DRTP against existing hop-by-hop control approaches, namely MFTP [Su et al., 2015], R^2T [Wang et al., 2018], and SDATP [Chen et al., 2020]. The comparative analysis involves configuring them with a maximum of three retransmission retrying times, aligning with the settings of DRTP, under PP disruption. The tests encompass both typical and general MPSG scenarios and are structured as follows:

- **For the Typical MPSG**: DRTP undergoes testing in EEFR, OOR, and EEDT to unveil its effectiveness and efficiency during retransmission. Additionally, we gauge the load capacity per router by measuring the mean number of messages in DRTP arrived at the router per second (Messages in DRTP Arrived at the Router Per Second (MAR)). This metric is crucial as each message must be processed within a limited time frame.
- **For the General MPSG**: DRTP is subjected to testing in EEFR, EEDT, and OOR to assess its feasibility and advantages in a generated MPSG with varying path-length features.

iii) **Assessing implementation correctness of DRTP**: This critical assessment involves verifying the correctness of our implementation under different LLR on each link within the PP or MPSG. The evaluation, focused on the South Carolina 500-bus system with 400 randomly selected pairs, hones in on specific metrics:

- **Recursiveness evaluation (The Number of Messages Sent Per Node of PP (NMN))**: The count of NMN should not exceed $2 \times \eta$ under heavy faults and η under light faults. Messages are categorized into two types. One is RequestS and RetranS (non-empty), where "S" indicates $|\Delta| = 1$ in σ_1, σ_3, and σ_4. The other is RequestM and RetranM (non-empty), where "M" means $|\Delta| \geq 1$ in σ_2.
- **Quality-of-Service (QoS) performance evaluation**: For EEFR and EEDT, consistency with previous experiments conducted in Zhou et al. 2022 is tested. To confirm EEDT in Eq. (7.9) under recursive assumptions in Section 7.3, we assess the satisfactory ratio between the number of pairs with their End-to-End Payload Packet Delivery Times (EEDTs) no greater than Eq. (7.9) and the total number of pairs in the network.

- **OOR calculation**: Calculated as the division of the number of distinct payload packets currently received with IDs less than the previously received one by the total number of distinct payload packets sent.

iv) **Comparing performance of DRTP**: This phase involves a detailed performance comparison between DRTP and the aforementioned transport protocols under corresponding disruption types. The evaluation is conducted on a randomly selected 150 pairs in the network designed for the South Carolina 500-bus system. Our attention is directed toward the following key metrics:
   - **EEFR**: Enables tests in transport protocol reliability.
   - **Equivalent EEDT**: Calculated with the assumption that EEDT for a lost payload packet is an impossible 15 seconds, while EEDT for others is real.
   - **OOR**: Assesses the orderliness of received payload packets.

7.4.3 Performance of DRTP Under Disruptions

We comprehensively evaluated the performance of DRTP based on different routes including both the typical MPSG and the general MPSG mentioned above.

7.4.3.1 DRTP Performance in Typical MPSG

Let's dive into the DRTP performance, especially when facing disruptions in the primary forwarding path (PP) and the more complex multipath subgraph (MPSG).

In the realm of PP disruption, as illustrated in Figure 7.15, the captivating tale unfolds. Figure 7.15(a) reveals that the EEFR gracefully maintains a steadfast 0%. Meanwhile, Figure 7.15(b) showcases the maximum EEDT, gracefully dancing within the bounds of [0.319, 13.817] ms. A twist in the plot begins at 12.895 ms, with the LLR reaching 97%, only to gracefully descend to 1.351 ms at the pinnacle of 100% LLR. This intriguing journey is guided by the residual penalty under the enigmatic condition #2, affirming the recursive nature of the performance.

Additionally, Figure 7.15(c) brings another revelation – the OOR steadfastly remains at 0%, a testament to the maximum EEDT gracefully tiptoeing within the PMU sending period.

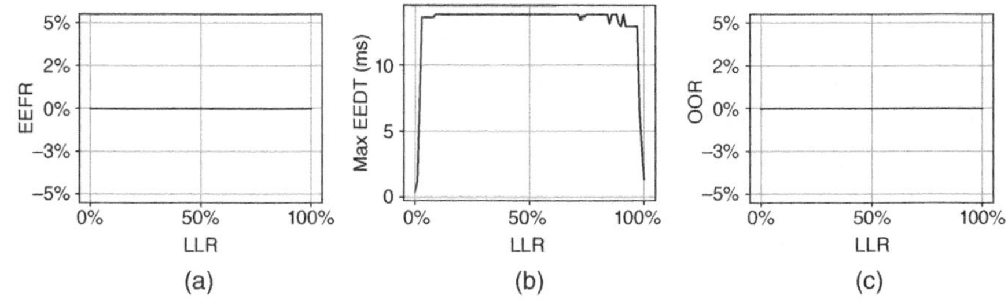

Figure 7.15 Performance of DRTP in the typical MPSG under the full PP disruption with different LLR. (a) EEFR, (b) the maximum EEDT, and (c) OOR.

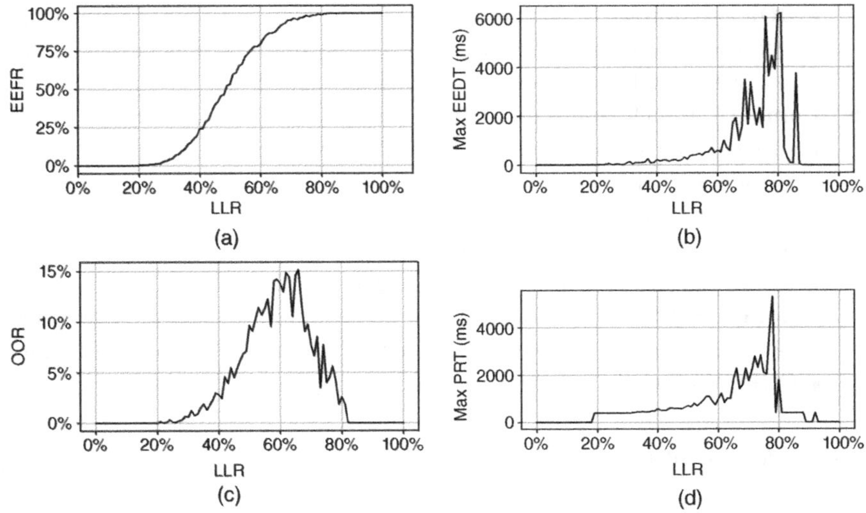

Figure 7.16 Performance of DRTP in the typical MPSG under the complete MPSG disruption with different LLR. (a) EEFR, (b) the maximum EEDT, (c) OOR, and (d) the maximum PRT.

Now, brace yourself for the saga within the MPSG disruption, unveiled in Figure 7.16. Figure 7.16(a) paints the canvas with all EEFR set at a tranquil 0%, gradually succumbing to deterioration in the realms of [0.125, 100]%. The corresponding LLR gracefully guides this narrative through [0, 18]% and [19, 100]%. Simultaneously, Figure 7.15(b) unravels the maximum EEDT saga, beginning in the serene [0, 14.403] ms territory, only to transition into the more tumultuous [14.608, 6221.623] ms realm. At the climax of this story, the maximum EEDT reaches a grand 3741.677 ms when the LLR ventures into the daring [81, 100]% territory. A dramatic turn of events, as the corresponding EEFR experiences significant deterioration in [99.733, 100]%, confirming the maximum EEDT's ability to be bound when LLR behaves modestly, staying under the 18% threshold.

Additionally, Figures 7.16(c) and 7.16(d) depicts the steadfastness of OOR and the maximum PRT at 0% and 0, respectively, when the LLR dwells in the realms of [0, 19]%. This tranquility is due to the corresponding EEDT remaining within the confines of the sending period. A shift occurs beyond LLR of 20%, ushering in a gradual deterioration.

Finally, Figure 7.17 unravels the epilogue, showcasing the average MSR for DRTP under the captivating disruptions of PP and MPSG. The constants in these tales affirm the wisdom shared in Section 7.2.3, bringing the captivating journey to a fulfilling conclusion.

Embark on a journey into the networking resilience with DRTP against disruptions. Even in the face of a total primary forwarding path (PP) disruption,

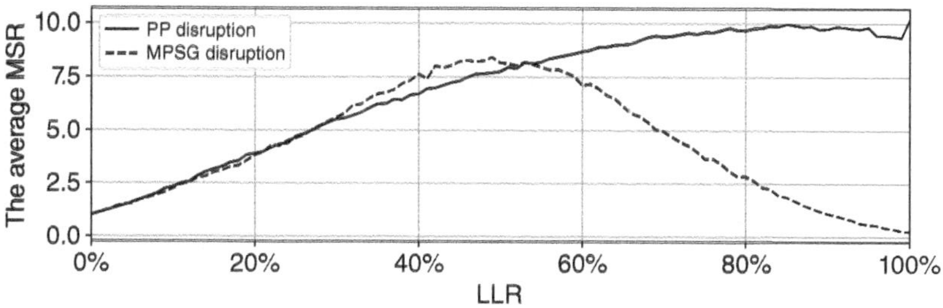

Figure 7.17 The average MSR of DRTP in the typical MPSG.

DRTP stands strong, maintaining a flawless 0% EEFR and a controlled EEDT across the entire spectrum of [0, 100]% LLR. But that's not all. Dive into the MPSG disruption, where DRTP continues to shine, keeping the magical 0% EEFR and a tight grip on EEDT within the bounds of [0, 18]% LLR. As our narrative unfolds, a remarkable discovery takes center stage. These results not only showcase resilience but also spotlight top-tier packet delivery. Think about this: EEDT standing tall, never dropping below 20 ms, and a steadfast 0% OOR – meeting the high standards set by IEEE C37.118.2 [2011]. It's more than a solution; it's a networking marvel in simple terms!

7.4.3.2 DRTP Performance in General MPSG

In the realm of general MPSG, DRTP's prowess unfolds under the full disruption of the PP. Venture into the visual tapestry of Figure 7.18, showcasing the performance of DRTP within the IEEE 300-bus and SC 500-bus topologies.

Specifically, the journey begins with Figures 7.18(a) and 7.18(b), capturing the fluctuations in EEFR under varying LLR for the two topologies. Observe closely: at LLR values of 20%, 60%, and 100%, the mean EEFR for both topologies dances to a rhythm of (2.1, 2.4)%, (37.5, 42.18)%, and (70, 86)%, respectively.

The plot thickens with Figures 7.18(c) and 7.18(d), unraveling the transformations in EEDT under diverse LLR scenarios for the two topologies. A closer look reveals that at LLR values of 20%, 60%, and 100%, the mean EEDT for both topologies orchestrates a symphony of (3.23, 3.2), (34.21, 62.91), and (2.76, 3.46) ms, respectively.

The grand finale arrives with Figure 7.18(e), unveiling the SRE under varying LLR – a tale of satisfaction told through percentages. Behold: at LLR values of 20%, 60%, and 100%, the SRE for the two topologies enchants with (98.88, 99.07)%, (75.53, 60.97)%, and (100, 100)%, respectively.

Under the encompassing disruption of the complete MPSG, witness the prowess of DRTP in Figure 7.19 as it unfolds its performance tale. Delve into the intricate dance portrayed by Figures 7.19(a) and 7.19(b), illustrating the metamorphosis in EEFR under varying LLR for the two topologies. Observe closely: at LLR values of

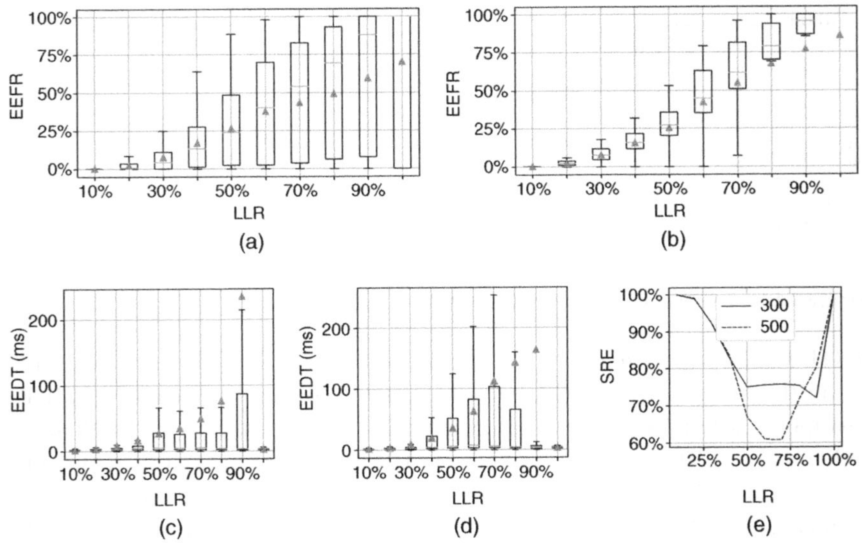

Figure 7.18 Performance of DRTP in general MPSG under the full PP disruption with different LLR. (a) EEFR in IEEE 300 buses, (b) EEFR in SC 500 buses, (c) EEDT in IEEE 300 buses, (d) EEDT in SC 500 buses, and (e) SRE.

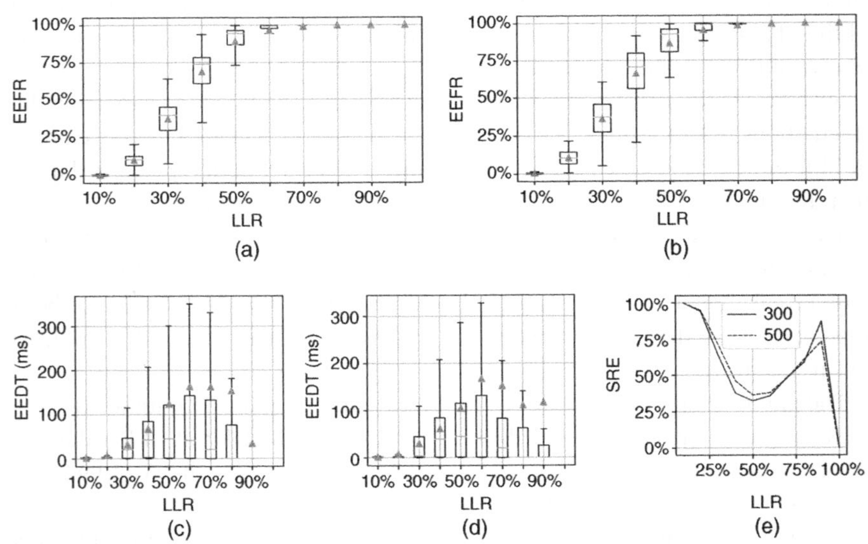

Figure 7.19 Performance of DRTP in the general MPSG under the complete MPSG disruption with different LLR. (a) EEFR in IEEE 300 buses, (b) EEFR in South Caro. 500 buses, (c) EEDT in IEEE 300 buses, (d) EEDT in South Caro. 500 buses, and (e) SRE.

20%, 60%, and 100%, the mean EEFR for both topologies orchestrates a symphony of (10.13, 10.69)%, (96.42, 95.34)%, and (100, 100)%, respectively.

The plot thickens with Figures 7.19(c) and 7.19(d), unveiling the transformations in EEDT under diverse LLR scenarios for the two topologies. A closer look reveals that at LLR values of 20%, 60%, and 100%, the mean EEDT for both topologies orchestrates a symphony of (7.32, 7.16) ms, (162.72, 167.56) ms, and (∞, ∞) ms, respectively.

The grand finale arrives with Figure 7.19(e), where SRE unfurls its narrative under varying LLR – a tale of satisfaction told through percentages. Behold: at LLR values of 20%, 60%, and 100%, the SRE for the two topologies enchants with (94.02, 94.88)%, (35.66, 37.89)%, and (0, 0)%, respectively.

In summary, these results illuminate the prowess of DRTP, showcasing its ability to curtail EEFR while maintaining a certain EEDT not exceeding the bound. This holds true under the most formidable PP disruption scenarios and specific complete MPSG disruption cases.

7.4.4 Comparative Studies Under Full PP Disruptions

Figure 7.20(a) unfolds a visual narrative of EEFR for MFTP, R^2T, and SDATP as they grapple with increasing LLR. Their dance on this disruptive stage reflects a collective deterioration, a consequence of inadequate leveraging of RSPs during retransmissions to counteract the escalating PP losses. Notably, SDATP fares worse due to distributed collaboration challenges. In contrast, DRTP shines, showcasing significantly reduced EEFR reaching 0% during PP disruptions.

Moving to Figure 7.20(b), a spotlight on load capacity per router in MAR unveils a riveting tale. When LLR is at or below 0%, 6%, and 16%, the count of DRTP surpasses SDATP, R^2T, and MFTP, respectively. As LLR climbs beyond 17%, the number of DRTP rises, surpassing its counterparts. DRTP tirelessly performs additional retransmissions, effectively recovering packet losses in PP disruptions at 0% EEFR.

Figure 7.20(c) captures the OOR drama. DRTP, MFTP, and R^2T maintain a flawless 0% OOR, while SDATP falters, reaching a peak of 16.6% at 12% LLR due to extended waiting times during retransmission.

Figures 7.20(d)–7.20(h) further unfolds the CDF of EEDT for DRTP, R^2T, MFTP, and SDATP under varied LLR conditions. DRTP stands out with maximum EEDT values of 413 μs, 13.82 ms, 13.82 ms, 13.82 ms, and 1.35 ms at 0%, 25%, 50%, 75%, and 100% LLR, respectively. In comparison, MFTP, R^2T, and SDATP demonstrate EEDT values at or below 413 μs, 13.82 ms, 13.82 ms, 1.35 ms, confirming DRTP's advantage in maintaining significantly low EEFR and EEDT with a commendable message load capacity (see also the extended experiments in Figures 7.21 and 7.22).

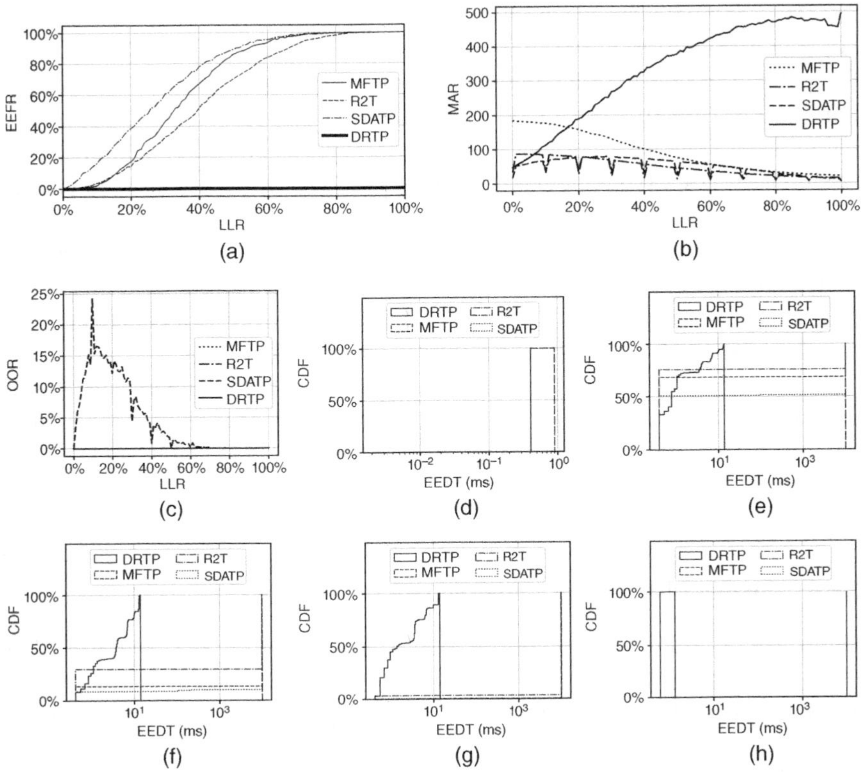

Figure 7.20 Comparative studies in the typical MPSG under the full PP disruption with different LLR. (a) EEFR, (b) MAR, (c) OOR, (d) EEDT under 0% LLR, (e) EEDT under 25% LLR, (f) EEDT under 50% LLR, (g) EEDT under 75% LLR, and (h) EEDT under 100% LLR.

7.5 Summary

In the realm of resilient communication, our DRTP steps forward as a stalwart guardian, harnessing the power of plentiful RSPs for hop-by-hop retransmission control. Picture it as a superhero for your data, ensuring that each step of its journey is sturdy and reliable. But there's more to the story – the magic lies in its recursiveness, a feature often overlooked in existing methods. This superhero doesn't just tackle problems; it addresses the crucial issue of distributed collaboration, bringing a new level of reliability to the table.

In our deep dive into the world of power grids, we took DRTP for a spin, evaluating its performance in a simulated environment inspired by real power transmission grids. Picture it as a journey through the electronic veins of our energy landscape.

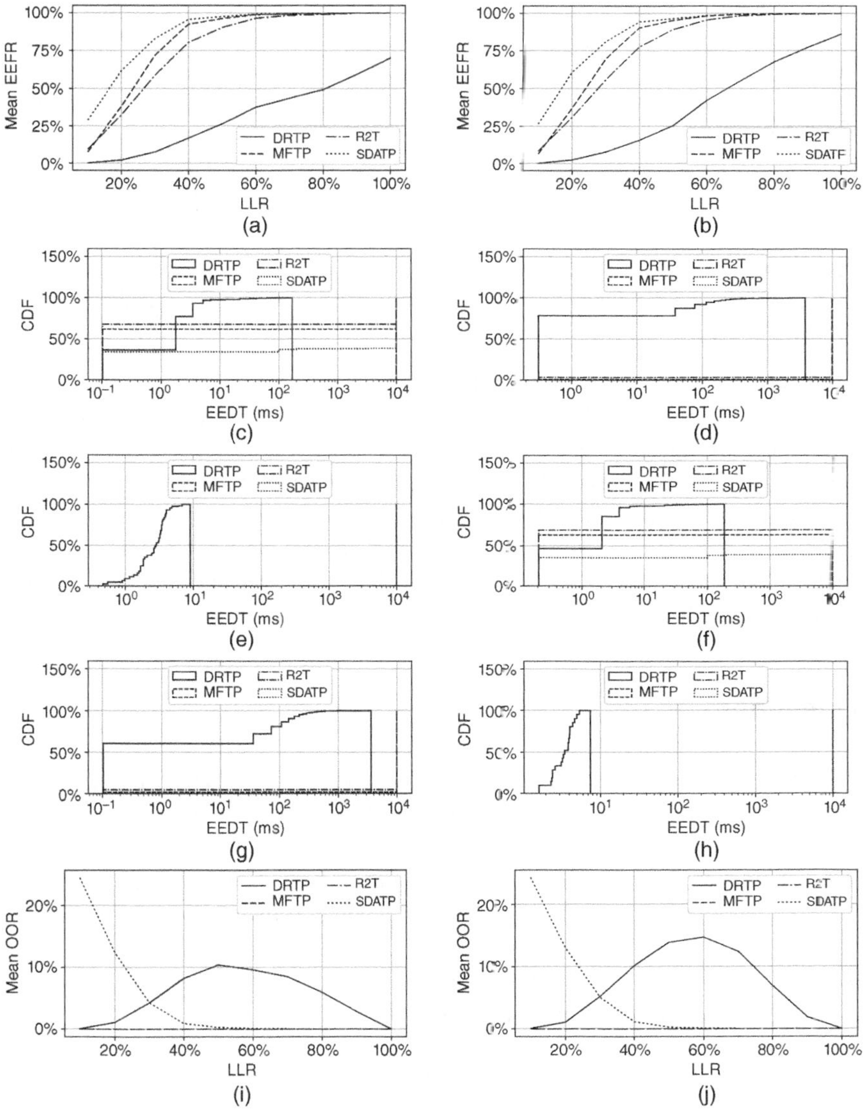

Figure 7.21 Performance comparison in the general MPSG under the full PP disruption with different LLR. (a) Comparison of mean EEFR for IEEE 300 buses, (b) Comparison of mean EEFR for SC 500 buses, (c) EEDT under 20% LLR for IEEE 300, (d) EEDT with 60% LLR for IEEE 300, (e) EEDT with 100% LLR for IEEE 300, (f) EEDT under 20% LLR for SC 500, (g) EEDT with 60% LLR for SC 500, (h) EEDT with 100% LLR for SC 500, (i) Comparison of mean OOR for IEEE 300 buses, and (j) Comparison of mean OOR for SC 500 buses.

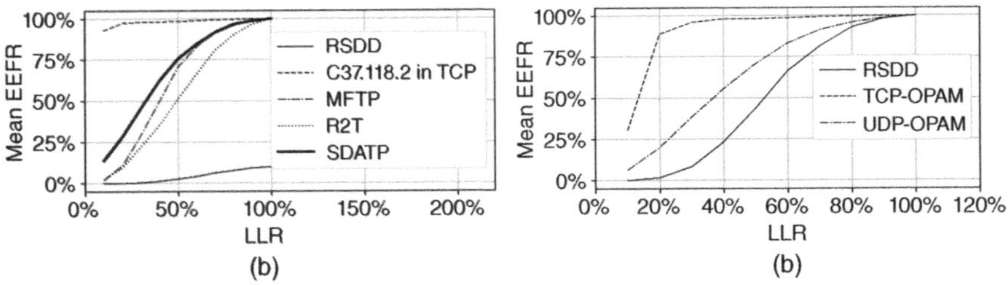

Figure 7.22 Comparative analysis in the reliability of transport protocols in the IEEE 14-bus system. (a) EEFR under full PP disruption versus LLR and (b) EEFR under full MPSG disruption versus LLR.

We scrutinized the capabilities of DRTP, our trusty vehicle, in both the reliable MPSG neighborhood and the more general MPSG territory. These scenarios, crafted by our algorithm and inspired by real-world topologies like the IEEE 300 bus and SC 500 bus, provided the backdrop for our exploration. It's like navigating the twists and turns of real transmission grids. We brought out our measuring tools, comparing DRTP against other transport control methods, judging them on metrics like EEFR and EEDT. The numerical verdict was clear – DRTP stood tall, significantly reducing EEFR even when links faced serious disruptions.

Questions

7.1 What are the key factors influencing the reliability of data transport protocols in the context of power transmission grid state estimation?

7.2 How does the periodic electric variation in bus #1, as depicted in Figures 12.14(a) and 12.14(b), impact the overall power transmission grid dynamics?

7.3 In the scenario of full PP disruptions, why does DRTP consistently outperform other protocols in terms of EEFR?

7.4 What role does the recursiveness of DRTP play in addressing the distributed collaboration challenges, and how does this contribute to its reliability in power grid communication?

7.5 How do different protocols respond to disruptions, and what implications does this have on the observability and accuracy of power grid state estimation?

7.6 What are the distinctive features of DRTP that make it stand out in comparison to TCP, MFTP, R2T, and SDATP under both PP and MPSG disruptions?

7.7 How do the OGR values provide insights into the observability of the power transmission grid, and why is DRTP consistently more effective in maintaining grid observability?

7.8 Elaborate on the role of RSPs in DRTP. How do they contribute to hop-by-hop retransmission control, and why is this feature beneficial for power grid communication?

7.9 Can you discuss potential challenges and benefits of integrating DRTP into existing power grid communication systems?

Bibliography

Saikat Chakrabarti, Elias Kyriakides, Tianshu Bi, Deyu Cai, and Vladimir Terzija. Measurements get together. *IEEE Power and Energy Magazine*, 7(1):41–49, 2009.

Jiayin Chen, Qiang Ye, Wei Quan, Si Yan, Phu Thinh Do, Weihua Zhuang, Xuemin Sherman Shen, Xu Li, and Jaya Rao. SDATP: An SDN-based adaptive transmission protocol for time-critical services. *IEEE Network*, 34(3):154–162, 2020.

G. Dileep. A survey on smart grid technologies and applications. *Renewable Energy*, 146:2589–2625, 2020.

Chavoosh Ghasemi, Hamed Yousefi, Kang G. Shin, and Beichuan Zhang. A fast and memory-efficient trie structure for name-based packet forwarding. In *IEEE ICNP*, pages 302–312, 2018.

GitHub, Inc. Disruption resilient transmission protocol, 2022. URL https://github.com/zhouby-zjl/drtp.

Yih-Fang Huang, Stefan Werner, Jing Huang, Neelabh Kashyap, and Vijay Gupta. State estimation in electric power grids: Meeting new challenges presented by the requirements of the future grid. *IEEE Signal Processing Magazine*, 29(5):33–43, 2012.

IEC 60794-4-10. Optical fibre cables - Part 4-10: Family specification - optical ground wires (OPGW) along electrical power lines, 2014.

IEEE. IEEE 30∼300 bus dataset, 2022. URL http://labs.ece.uw.edu/pstca/.

IEEE C37.118.2. IEEE standard for synchrophasor data transfer for power systems, 2011.

Cédric Josz, Stéphane Fliscounakis, Jean Maeght, and Patrick Panciatici. AC power flow data in MATPOWER and QCQP format: iTesla, RTE snapshots, and PEGASE. *arXiv preprint arXiv:1603.01533*, 2016.

Yanbiao Li, Dafang Zhang, Xian Yu, Wei Liang, Jing Long, and Hong Qiao. Accelerate NDN name lookup using FPGA: Challenges and a scalable approach. In *IEEE FPL*, pages 1–4, 2014.

John D. C. Little. Little's law as viewed on its 50th anniversary. *INFORMS Operations Research*, 59(3):539–549, 2011.

Bin Liu, Huichen Dai, Wenquan Xu, Tong Yun, and Ji Miao. Scalable hardware content router: Architecture, modeling and performance. In *IEEE/ACM IWQOS*, pages 1–7, 2021.

Spyridon Mastorakis, Alexander Afanasyev, and Lixia Zhang. On the evolution of ndnSIM: An open-source simulator for NDN experimentation. *ACM SIGCOMM Computer Communications Review*, 47(3):19–33, 2017.

Rui Miguel, Salvatore Signorello, and Fernando M. V. Ramos. Named data networking with programmable switches. In *IEEE ICNP*, pages 400–405, 2018.

Named Data Networking (NDN). NDN packet format specification, 2022. URL https://named-data.net/doc/NDN-packet-spec/current/.

Named Data Networking (NDN). NFD: Named data networking forwarding daemon¶, 2023. URL https://docs.named-data.net/NFD/current/.

Gelli Ravikumar, Dan Ameme, Satyajayant Misra, Sukumar Brahma, and Reza Tourani. iCASM: An information-centric network architecture for wide area measurement systems. *IEEE Transactions on Smart Grid*, 11(4):3418–3427, 2020.

Redmine. NDNLPv2, 2022. URL https://redmine.named-data.net/projects/nfd/wiki/NDNLPv2.

RTDS Technologies Inc. Real-time digital simulator (RTDS), 2023. URL https://www.rtds.com/.

Junxiao Shi, Davide Pesavento, and Lotfi Benmohamed. NDN-DPDK: NDN forwarding at 100 Gbps on commodity hardware. In *ACM ICN*, pages 30–40, 2020.

Kai Su, Francesco Bronzino, K. K. Ramakrishnan, and Dipankar Raychaudhuri. MFTP: A clean-slate transport protocol for the information centric MobilityFirst network. In *ACM ICN*, pages 127–136, 2015.

Junji Takemasa, Yuki Koizumi, and Toru Hasegawa. Toward an ideal NDN router on a commercial off-the-shelf computer. In *ACM ICN*, pages 43–53, 2017.

Zhaoxu Wang, Hongbin Luo, Huachun Zhou, and Jiawei Li. R2T: A rapid and reliable hop-by-hop transport mechanism for information-centric networking. *IEEE Access*, 6:15311–15325, 2018.

Lixia Zhang, Alexander Afanasyev, Jeffrey Burke, Van Jacobson, Christos Papadopoulos, K. C. Claffy, Lan Wang, Patrick Crowley, and Beichuan Zhang. Named data networking. *ACM SIGCOMM Computer Communication Review*, 44(3):66–73, 2014.

Boyang Zhou, Qiang Yang, Yansong Wang, and Chunming Wu. Reliable communication in transmission grids based on nondisjoint path aggregation using software-defined networking. *IEEE Transactions on Industrial Informatics*, 15(2):842–855, 2019.

Boyang Zhou, Chunming Wu, Qiang Yang, and Xiang Chen. DRTP: A disruption resilient hop-by-hop transport protocol for synchrophasors measurement in electric transmission grids. *IEEE Access*, 10:133898–133914, 2022.

Ray Daniel Zimmerman, Carlos Edmundo Murillo-Sánchez, and Robert John Thomas. MATPOWER: Steady-state operations, planning, and analysis tools for power systems research and education. *IEEE Transactions on Power Systems*, 26(1):12–19, 2011.

8

Implementing DRTP: Orchestrating State Transition Machines

In retrospect to Disruption Resilient Transport Protocol (Disruption Resilient Transport Protocol (DRTP)) early mentioned in Chapter 7, maintaining the recursive nature during retransmission control with timeouts poses challenges.

Yet, realizing the retransmission process of DRTP with recursiveness proves to be a formidable challenge. The crux of the matter lies in sustaining this recursiveness, especially when a hop in the Primary Forwarding Path (PP) triggers its local arrival timeouts, initiating retransmission instances that unfold in an unforeseen sequence. This sequence is dictated by shared retransmission states – variables encapsulating the times and Identifications (IDs) of arrival messages, carrying crucial contents such as the latest arrived payloads and current retransmission requests. Unfortunately, within this orchestration, these states are susceptible to an unprotected parallel race condition. This condition allows instances to execute simultaneously, following a non-deterministic sequence.

In such a race condition, instances may intertwine at the outset of timeout counting, resulting in an execution trajectory that becomes overly aggressive. This aggression can lead to cache mismatch and an augmented End-to-End Failure Rate in Packet Delivery (EEFR), as the sequence generates an excess of unexpected retransmission messages. As we grapple with these intricacies, the pursuit of a resilient and efficient power grid communication system unveils the intricate dance between precision and unpredictability.

In response to this challenge, we introduce Resilient Sensor Data Dissemination (RSDD), a purpose-built system meticulously crafted to implement the retransmission process of DRTP while preserving its inherent recursive nature. Leveraging the classical State Transition Machine (STM) model [Börger, 2010; Sunshine et al., 1982], RSDD strategically employs 8 STMs to adeptly manage incoming DRTP messages, seamlessly aligning with the recursive collaboration steps of DRTP.

Communication Networks in Smart Power Grids, First Edition. Boyang Zhou.

Companion Website: www.wiley.com/go/SmartPowerGrids1e

These STMs play a crucial role in enabling stateful message processing, facilitating asynchronous scheduling, and executing concurrent retransmission instances. Each execution is exclusively dedicated to a specific lost DRTP payload message, ensuring precise arrival timeouts based on the latest arrival time of the corresponding incoming DRTP message.

To fortify the robustness of RSDD, a meticulous theoretical verification process is undertaken. This involves a comprehensive examination of all potential execution trajectories of RSDD, confirming its status as a correct DRTP implementation. The outcomes of this verification affirm that RSDD adheres to the desired recursive nature while maintaining bounded message processing costs. As we unfold the intricacies of RSDD, a beacon of assurance shines on the path toward a resilient and meticulously controlled power grid communication system.

The evaluation extended to the data dissemination process from Phasor Measurement Unit (PMU)s [IEEE, 2011] and critical sensors within a simulated cyber–physical environment given in Section 7.4.1. Our experiments generated valuable insights, summarized as follows:

The evaluation of RSDD unfolded through rigorous assessments on a challenging South Carolina 500-bus system [Josz et al., 2016], showcasing its resilience. Our tests affirmed RSDD's correctness by consistently aligning with DRTP, ensuring exclusive message processing and maintaining End-to-End Payload Packet Delivery Time (EEDT) within expected constraints. In scenarios with a full PP disruption and a 20% Link Loss Rate in Packets (LLR), RSDD demonstrated notable mean EEFRs (2.44%), mean EEDTs (2.7 ms), and EEDT satisfactory ratios (99.57%).

In a comparative analysis with state-of-the-art transport protocols on the 500-bus system, RSDD emerged as a frontrunner. Under full PP disruption with a 10% LLR, RSDD exhibited significantly lower mean EEFR and mean equivalent EEDT (0.13%, 0.02 seconds) compared to other protocols, such as IEEE C37.118.2 in TCP (98.16%, 14.91 seconds), MFTP (6.46%, 0.97 seconds), Rapid and Reliable Transport Mechanism (R2T) (8.36%, 1.26 seconds), and Adaptive Transmission Protocol based on Software-Defined Networking (SDATP) (26.54%, 3.99 seconds).

8.1 Overall Design

RSDD is designed to provide a dependable data dissemination service within optical Internet-of-things (IoT) networks for power transmission grids. Illustrated in Figure 8.1, RSDD is built upon the existing Named Data Network (NDN) forwarding daemon [NDN, 2023] operating on the Linux OS. It functions as a pluggable strategy, tied to a specified name prefix in NDN, allowing the daemon

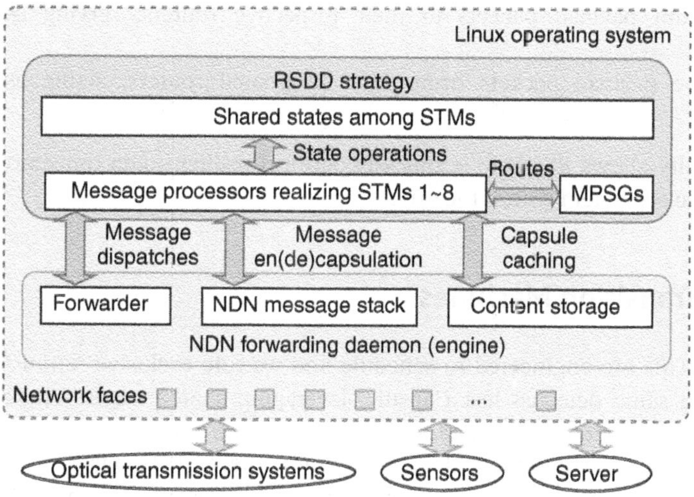

Figure 8.1 RSDD architecture based on NDN.

Table 8.1 Shared states among STMs in RSDD.

State	Description
E	A map from a data ID to the ID of an STM activating M_5 or M_6
T_{req}	A map from a data ID to a retransmission request time
T_{cap}	A map from a data ID to a Capsule sending time
Δ_{losts}	A set containing all data IDs in Reports sent from upstream routers
Δ_{tran}	A set containing all data IDs having been transmitted
δ_{cur}	the data ID of the currently arrived Capsule

to route messages matching the given prefix to the RSDD strategy. The RSDD strategy consists of three key components:

- Message processors represented as STMs 1 to 8.
- Multipath Subgraph (MPSG)s for all pairs that guide message processors in the accurate forwarding of messages.
- Shared states associated with each pair among STMs (details in Table 8.1), facilitating retransmission control.

Parameters of RSDD, such as η, ζ, the MPSGs, and the size of Content Store (CS), are adjustable through a configuration file.

RSDD is tailored to collaborate with different device types:

- Local optical transmission systems engage at the network layer for communication with neighboring routers.

- Sensors transmit payload packets to their respective routers, serving as producers.
- Servers receive payload packets from their designated routers, acting as consumers.

This adaptability allows RSDD to seamlessly perform resilient data transport across the diverse ecosystem within the grid.

8.2 State Transition Machines

In RSDD, the STMs are engineered to schedule and execute exclusive retransmissions for the same detected lost Capsule, leveraging shared states. These STMs operate asynchronously, denoted as M_0 to M_7, within a PP router r_{d_k}, as illustrated in Figure 8.2. Throughout their functioning, these STMs engage with both upstream routers r_{d_s} and the adjacent downstream router $r_{d_{k+1}}$ through messaging. Moreover, a parent STM transitions to a child STM under a specified predicate (shown in a brace), potentially in a non-blocking manner (indicated by hollow arrows). The execution of these STMs enables RSDD to operate through several key phases:

i) **Manage states on demand:** As depicted in Figure 8.3, when an Interest with a specific prefix is received, STM M_0 initializes its associated shared states (refer to Table 8.1) and allows them to expire. Subsequently, it terminates the

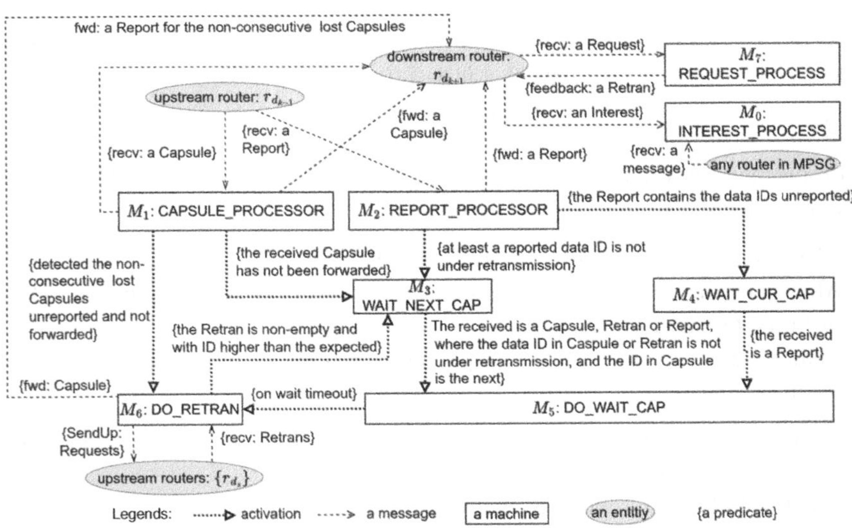

Figure 8.2 Interactions among STMs and entities in RSDD.

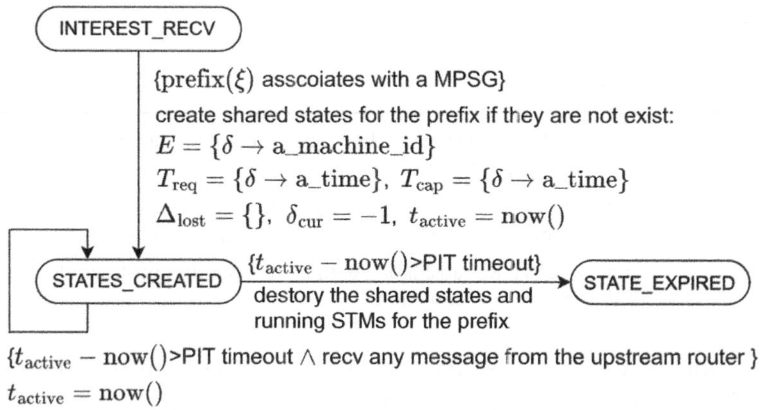

Figure 8.3 STM diagram of M_0: INTEREST_PROCESSOR.

session by destroying the states and all contained STMs in E. This process occurs if there is no message arrival in the router's Pending Interest Table (PIT) timeout [Zhang et al., 2014].

ii) **Exclusively schedule retransmission in dissemination:** To achieve data dissemination, M_1 forwards each received Capsule from $r_{d_{k-1}}$ to r_{d_k} or the consumer if available. Additionally, to ensure exclusiveness, for each data ID in a received message, the current router cancels concurrent schedules for the ID and makes its own schedule for the ID. The scheduling trajectories for the ID fall under three conditions:
   - The Capsule ID is non-consecutive to δ_{cur}, not forwarded, and excluded from the received Reports (not in Δ_{losts}). Exclusion implies that the ID must be under the retransmission of at least one upstream router. For this, M_1 first schedules the retransmission for the detected lost Capsules by activating M_6 and then waits for the next Capsule arrival by activating M_3.
   - The Report ID is not forwarded. In this case, M_2 schedules a backup retransmission for the ID in case of the failure of its upstream retransmission by activating M_4.
   - The Report ID is under the retransmission of at least one upstream router (in Δ_{losts}). M_2 updates the next Capsule it waits for by activating M_3 to avoid retransmission duplication.

   Thus, during dissemination, these trajectories ensure retransmission exclusiveness.

iii) **Exclusively execute the scheduled retransmissions:** For the scheduled IDs, the execution follows a conditional trajectory from M_6, M_3, M_5 to M_6 in a circular manner. In the cases of M_1 and M_2, the activation of M_5 can originate from either M_3 or M_4. M_6 retrieves Retrans by sending Requests for at

Table 8.2 Basic predicate actions.

Action	Description
Forward($\xi, u = 1$)	It forwards ξ to the next PP hop for u times, if the hop is available, or otherwise to the consumer.
SendUp(ξ)	It broadcasts ξ to all upstream routers of r_{d_k} along with all its Redundant Sub-Path (RSP)s and the upstream PP link.
Feedback(ξ)	It feedbacks a Retran ξ along with all its RSPs and upstream PP link in the reverse direction.

most η rounds with the request timeout, navigating through message losses and cache mismatch challenges. To ensure exclusiveness, M_6 updates which Capsule to await by activating M_3 if the retrieved Retran is non-empty and carries a new ID. Subsequently, M_3 calculates the next Capsule ID, which is the smallest unreported and not yet forwarded. It then waits for this ID along with its Capsule Inter-Arrival Time (CIAT) by activating M_5. If the ID surpasses the timeout, M_5 activates M_6 in return. Meanwhile, to retransmit the failed IDs of M_2, M_4 computes Capsule Delivery Time (CDT) and activates M_5.

To respond to a Request initiated by M_6, M_7 at the upstream router sends back a non-empty Retran if it is matched in its local CS; otherwise, it sends the empty Retran. This circular process enables the retransmission of both the current and next Capsule losses.

With the above STMs, RSDD can realize the DRTP collaboration process with the exclusiveness outlined in Sections 7.2 and 7.3. Before delving into the details, for convenience, we define the basic predicate actions listed in Table 8.2 and denote the data ID of a message ξ as $\delta(\xi)$. The functionalities of M_1 and M_2 are specified below. Next, we elucidate how the scheduling and execution mechanisms operate.

8.2.1 Dissemination and Retransmission Schedules

CAPSULE_PROCESSOR M_1. M_1 promises that the schedule upon receiving a Capsule is exclusive in the dissemination through its transitions presented in Figure 8.4 as follows. When receiving a Capsule of ξ, M_1 cancels all unnecessary ongoing retransmissions for $\delta(\xi)$ in E, records its arrival time in T_{cap} to help M_7 to compute Δ_{drift}, and saves ξ to the local CS. Then, according to Δ_{losts}, it distinguishes if $\delta(\xi)$ has been under the upstream retransmission or not. If yes, by removing $\delta(\xi)$ from Δ_{losts}, it avoids duplicated retransmission of ξ at other upstream routers. Then, it forwards ξ to the next and activates M_3. Otherwise, it finds the lost Capsule IDs between $\delta_{cur} + 1$ to $\delta(\xi)$ that are not reported in upstream as Δ_{fou}. If found, to avoid duplication, its forwards a Report indicating Δ_{fou} under retransmission of r_{d_k}. Then, to retransmit them, it immediately

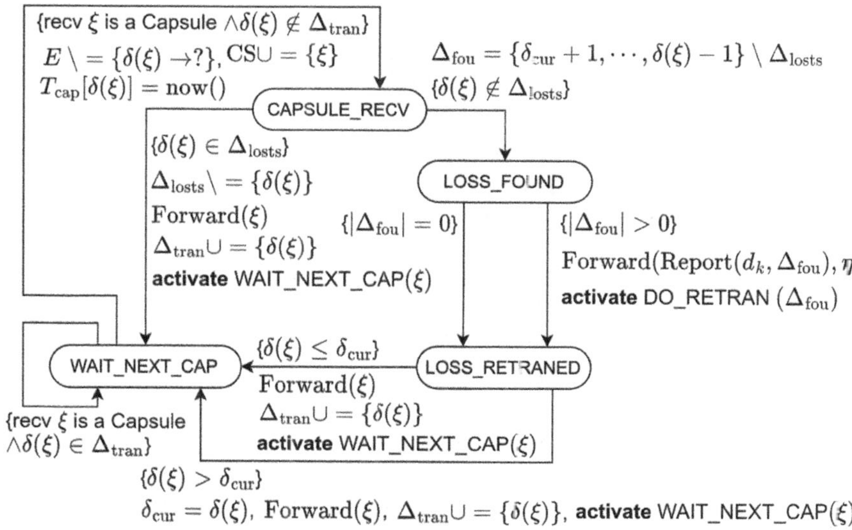

Figure 8.4 STM diagram of M_1: CAPSULE_PROCESSOR.

activates M_6. Lastly, it decides if ξ is with a known ID under upstream retransmission or not. For the latter, M_1 needs to update δ_{cur} to $\delta(\xi)$. Meanwhile, it forwards ξ and activates M_3. Thus, the transitions keep the exclusiveness.

M_1 ensures exclusive scheduling upon receiving a Capsule, as outlined in Figure 8.4. Upon receiving a Capsule ξ, M_1 takes the following steps:

1) Cancels unnecessary ongoing retransmissions for $\delta(\xi)$ in E.
2) Records the arrival time of ξ in T_{cap} to aid M_7 in computing Δ_{drift}.
3) Saves ξ to the local CS.

Next, based on Δ_{losts}:

- If $\delta(\xi)$ has undergone upstream retransmission, M_1 removes $\delta(\xi)$ from Δ_{losts} to avoid duplicating retransmissions at other upstream routers. It then forwards ξ to the next router and activates M_3.
- If $\delta(\xi)$ has not undergone upstream retransmission, M_1 identifies lost Capsule IDs between $\delta_{\text{cur}} + 1$ and $\delta(\xi)$ that are unreported upstream (denoted as Δ_{fou}). If found, to avoid duplication, it forwards a Report indicating Δ_{fou} under retransmission of r_{d_k}. It then activates M_6 for the retransmission process.

Finally, M_1 decides if ξ has a known ID under upstream retransmission. If yes, it updates δ_{cur} to $\delta(\xi)$, forwards ξ, and activates M_3. These transitions maintain exclusiveness in scheduling.

REPORT_PROCESSOR M_2. M_2 ensures that the schedule upon receiving a Report is exclusive through its transitions provided in Figure 8.5. When receiving a Report ξ with Δ_{new} from r_{d_s}, M_2 notifies downstream router Δ_{new} under

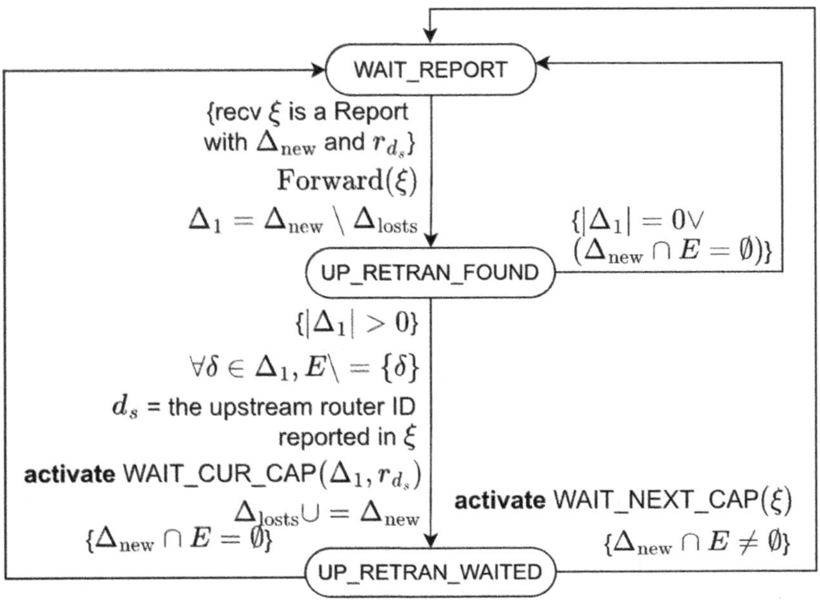

Figure 8.5 STM diagram of M_2: REPORT_PROCESSOR.

retransmission by forwarding ξ, and identifies a set of new data IDs as Δ_1 that are excluded in Δ_{losts}. For each ID in Δ_1, for the failed upstream retransmission, M_2 avoids duplication by canceling its current one in E, and activates M_4 to wait for Δ_1. In addition, $\Delta_1 \cup E \neq \emptyset$ means that the local retransmission for a ID in Δ_1 is duplicated for upstream retransmission. Therefore, it updates the next Capsule ID to wait by activating M_3. Otherwise, if the two sides are equal, the update is futile. Thus, the transitions maintain the exclusiveness.

M_2 ensures exclusive scheduling upon receiving a Report, as depicted in Figure 8.5. When M_2 receives a Report ξ with Δ_{new} from r_{d_s}, it takes the following steps:

1) Notifies downstream router Δ_{new} under retransmission by forwarding ξ.
2) Identifies a set of new data IDs, denoted as Δ_1, that are excluded in Δ_{losts}.

For each ID in Δ_1:

- In the case of failed upstream retransmission, M_2 avoids duplication by canceling its current one in E and activates M_4 to wait for Δ_1.
- Checks if $\Delta_1 \cup E \neq \emptyset$, indicating that local retransmission for an ID in Δ_1 is duplicated for upstream retransmission. In such cases, M_2 updates the next Capsule ID to wait by activating M_3.

Finally, if $\Delta_1 \cup E = \emptyset$, the update is deemed futile. The transitions maintain exclusiveness in scheduling.

8.2.2 Executing Scheduled Retransmissions

DO_RETRAN M_6. M_6 guarantees that the above Retran retrieval is loss tolerant and exclusive with its transitions provided in Figure 8.9 in aspects below.

First, to cope with the losses, M_6 sends Requests with Δ to all faces toward the upstream PP link and Redundant Sub-Path (RSP)s at most η times, to wait the first feedback of their Retrans. Each round prolongs the request timeout to ensure its completion. Upon receiving a non-empty Retran ξ, M_6 forwards it to the next and records its time in T_{cap} to help compute Δ_{drift} in M_7.

Second, to deal with the cache mismatch, for its latest sent Request, M_6 waits for the corresponding Retran with the longest CDT by assuming that the upstream router is r_{d_1}, where the CDT is computed by Eq. (7.7).

Third, to ensure the exclusiveness of M_6, M_6 cancels the concurrent IDs of M_5 and M_6 by removing them from E at the beginning of M_6 execution and upon the Retran receiving.

Eventually, if the non-empty Retran is with its ID greater than δ_{cur}, M_6, M_6 updates δ_{cur} to the ID and activates M_3. Those transitions ensure the loss tolerance and exclusiveness.

M_6 ensures loss-tolerant and exclusive Retran retrieval through transitions provided in Figure 8.9:

1) **Handling losses:**
   1) Sends Requests with Δ to all faces toward the upstream PP link and RSPs for at most η rounds, prolonging the request timeout each round to ensure completion.
   2) Upon receiving a non-empty Retran ξ, M_6 forwards it to the next and records its time in T_{cap} to help compute Δ_{drift} in M_7.
2) **Cache mismatch handling:** For the latest sent Request, M_6 waits for the corresponding Retran with the longest CDT, assuming the upstream router is r_{d_1}, where CDT is computed by Eq. (7.7).
3) **Exclusiveness:** Cancels concurrent IDs of M_5 and M_6 by removing them from E at the beginning of M_6 execution and upon Retran receiving.
4) **Update and activation:** If the non-empty Retran has an ID greater than δ_{cur}, M_6 updates δ_{cur} to the ID and activates M_3.

These transitions ensure loss tolerance and exclusiveness.

WAIT_NEXT_CAP M_3. M_3 ensures the correct waiting for the next Capsule through transitions in Figure 8.6 with the following steps:

1) **Efficient waiting:**
   - Upon receiving ξ in Capsule or Retran with Δ_{new}, M_3 efficiently waits for the next Capsule after $CIAT_1$ or $CIAT_2$, respectively.
   - If a timeout occurs, M_3 retransmits the next Capsule.

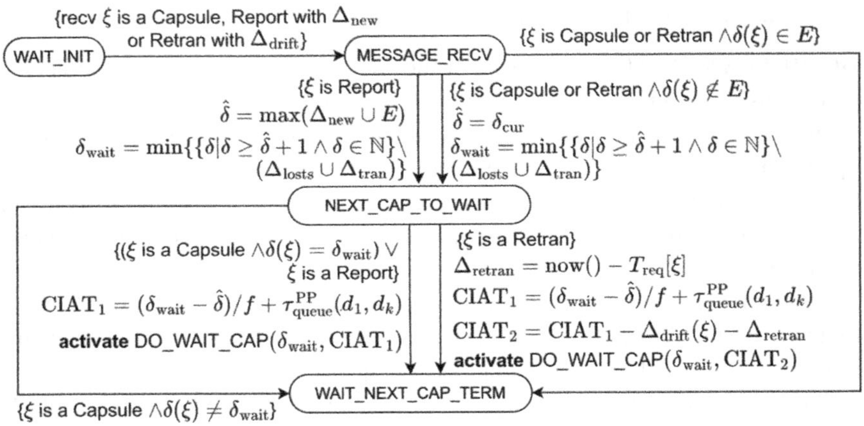

Figure 8.6 STM diagram of M_3: WAIT_NEXT_CAP.

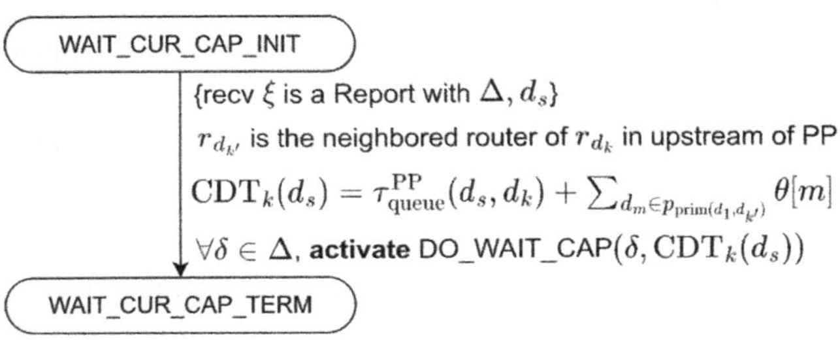

Figure 8.7 STM diagram of M_4: WAIT_CUR_CAP.

2) **Inference of next Capsule ID:**
   - Based on the latest received Capsule ID δ_{cur}, M_3 infers the next Capsule ID δ_{wait}, which is the minimum one not under retransmission upstream.
   - If there's at least one upstream retransmission and the current router is waiting for the next Capsule ID, then δ_{wait} remains unchanged.
   - Otherwise, the Capsule and Retran are associated with CIAT_1 and CIAT_2 computed in Eqs. (7.1) and (7.6), respectively.
3) **CIAT timeout:**
   - To wait for the δ_{wait} for a CIAT timeout, M_3 activates M_5 for recovering the next Capsule loss.

   These transitions determine the correct next ID with its matched CIAT (Figure 8.7).

DO_WAIT_CAP M_5. Figure 8.8 illustrates M_5 waiting for a delay τ to retransmit the IDs in Δ by activating M_6. The waiting period is interruptible by an event deletion in E.

WAIT_CUR_CAP M_4. M_4 ensures the accuracy of the schedule for failed retransmissions, utilizing its CDT. The transitions, depicted in Figure 8.7, outline

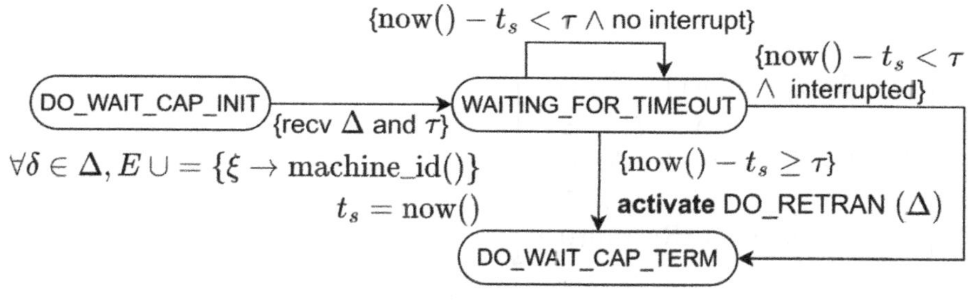

Figure 8.8 STM diagram of M_5: DO_WAIT_CAP.

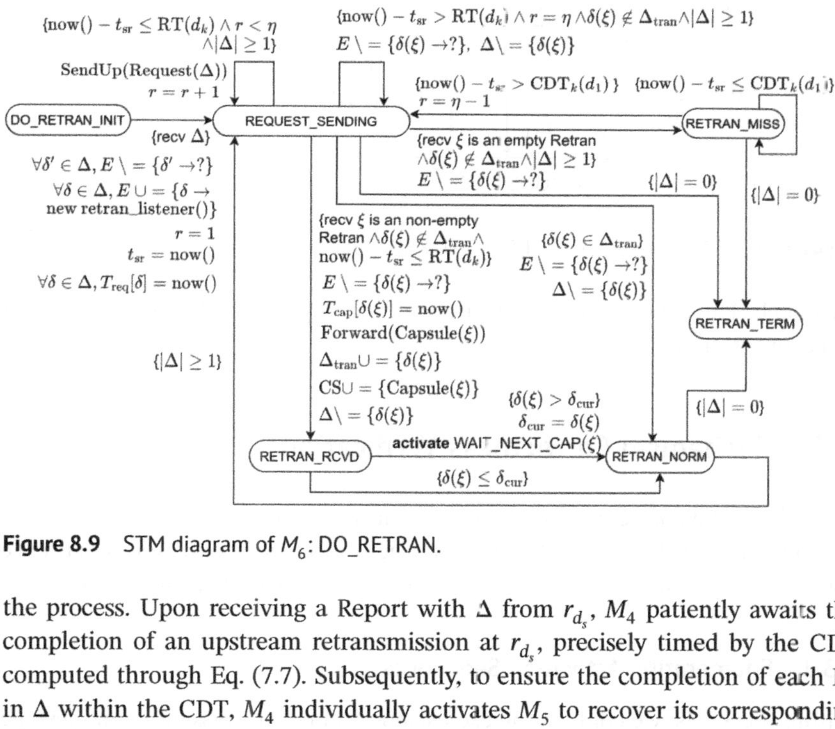

Figure 8.9 STM diagram of M_6: DO_RETRAN.

the process. Upon receiving a Report with Δ from r_{d_s}, M_4 patiently awaits the completion of an upstream retransmission at r_{d_s}, precisely timed by the CDT computed through Eq. (7.7). Subsequently, to ensure the completion of each ID in Δ within the CDT, M_4 individually activates M_5 to recover its corresponding Capsule loss. This guarantees the correctness of the schedule for losses associated with their respective IDs in Δ (Figure 8.9).

REQUEST_PROCESSOR M_7. M_7 ensures the correctness of the response to a Request, detailing its transitions in Figure 8.10. When presented with a Request containing IDs Δ, M_7 responds by providing a corresponding Retran, searching for a matching Capsule copy within its local CS. Upon locating the copy, M_7 calculates the drift (Δ_{drift}) between the current time and the arrival time of the requested Capsule, facilitating CIAT_2 computation for M_3. Accompanied by the copy and Δ_{drift}, M_7 sends the Retran back to the requested router along the reversed path. In cases where no matching copy is found, it sends an empty Retran, signifying a

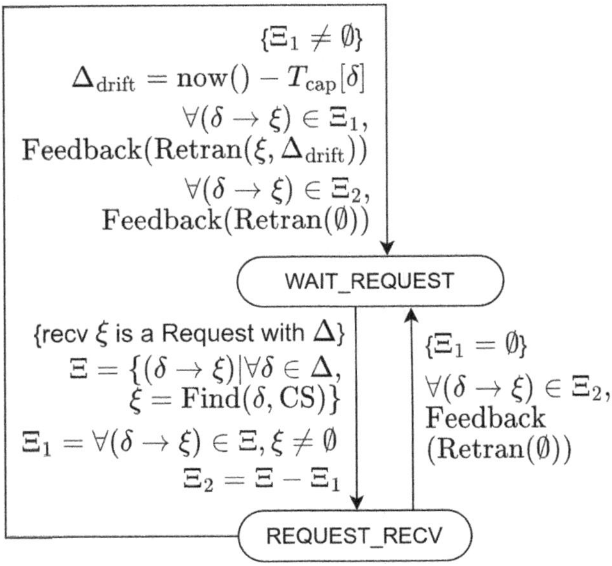

Figure 8.10 STM diagram of M_7: REQUEST_PROCESSOR.

mismatch. Thus, the transitions ensure accurate responses to Requests with their respective Retran.

8.3 Mechanism Correctness Verification

We validate the correctness of RSDD in terms of exclusiveness and arrival timeouts by exhaustively enumerating potential activation sequences. Additionally, we conduct an analysis of the message processing costs associated with RSDD.

8.3.1 Enumerating Activation Sequences

Within the STMs, we confirm the exclusiveness of retransmission upon message arrival by demonstrating that a detected lost Capsule is addressed in only one activation sequence. Illustrated in Figure 8.2, all conceivable activation sequences encompass σ_1 to σ_4 as outlined below.

- $\sigma_1 = \sigma_1'(\rightarrow \delta_s)*$, where $\sigma_1' = M_1$
- $\sigma_2 = \sigma_2'(\rightarrow \delta_s)*$, where $\sigma_2' = M_1 \rightarrow M_3 \rightarrow M_5$
- $\sigma_3 = \sigma_3'(\rightarrow \delta_s)*$, where $\sigma_3' = M_2 \rightarrow M_3 \rightarrow M_5$
- $\sigma_4 = \sigma_4'(\rightarrow \delta_s)*$, where $\sigma_4' = M_2 \rightarrow M_4 \rightarrow M_5$
- $\delta_s = M_6 \rightarrow M_3 \rightarrow M_5$

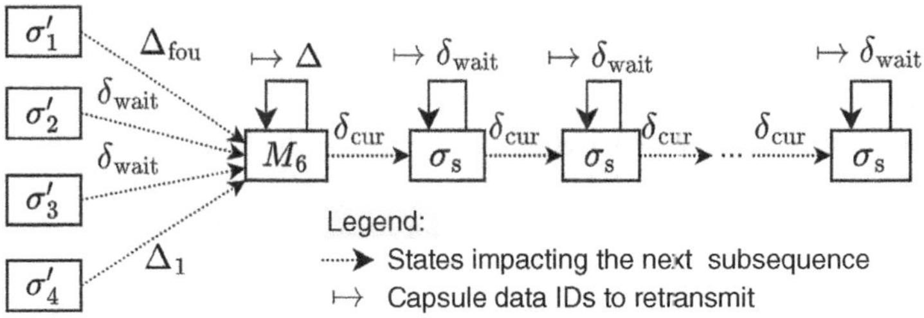

Figure 8.11 State transition sequence in RSDD.

In the aforementioned sequences, each one can be further divided into multiple subsequences by the action of activating M_6, which is responsible for executing retransmissions. These subsequences are denoted as δ'_1 to δ'_4 and δ_s, where δ_s, initiated by M_6, may repeat zero or more times. Regarding retransmission, the relationship between subsequences is illustrated in Figure 8.11. Here, σ_1 to σ_4 may trigger the activation of δ_s with individually computed Capsule IDs: Δfou, δ_{wait}, δ_{wait}, and Δ_1, respectively. Subsequently, the q-th δ_s could activate the $q+1$th δ_s with its computed Capsule ID being $\delta_{wait}^{(q+1)}$, where $q \leq 1$. This process continues until no further activations are feasible. During this repetition, the 1st and $\delta_s^{q'}$ retransmit Δ and $\{\delta_{wait}^{(q')}\}$, respectively, where $q' \leq 2$.

In any given sequence, at the initiation of a subsequence δ_s, the activated M_6 retransmits the lost Capsule IDs that differ from the lost IDs in other sequences. This arises because, prior to the activation of δ_s, its satisfied preceding predicates must be exclusive compared to those in other sequences. To validate this, we scrutinize the associated transition steps during the activation process. For simplicity, irrelevant steps including predicates and actions related to CS, Forward, unreferenced states, and other activations have been omitted from the sequences. The simplified transitions for all subsequences are presented in subsequences 1, 2, 3, 4, and 5, respectively. Each subsequence is associated with its specific input parameters, as indicated in the caption, and carries out its steps until completion, potentially activating subsequence 5 asynchronously. Notably, each parameter is passed by reference (not by value). The subsequences are specified as follows.

- Subsequence 1 undergoes transitions upon receiving ξ, dictated by Δ_{tran} and Δ_{losts}. Initially, δ'_1 transitions from receiving an untransmitted and unreported Capsule ξ to potentially activating δ_s with a non-empty Δ_{fou}.
- Subsequence 2 executes transitions when receiving ξ with E, Δ_{tran}, and Δ_{losts}. δ'_2 shifts from receiving an untransmitted Capsule not under retransmission in E to activating δ_s with δ_{wait}, which is the Capsule ID itself. The transitions cover various trajectories of M_1, considering whether the predicate $|\Delta_{fou}|$ is empty, whether $\delta(\xi)$ is greater than δ_{cur}, and more.

- Subsequence 3 progresses through transitions upon receiving ξ with E and Δ_{losts}. σ_3' transitions from receiving a Report with Δ_{new} containing unreported IDs under retransmission in E to activating δ_s with δ_{wait}.
- Subsequence 4 involves transitions upon receiving ξ with shared E and Δ_{losts}. δ_4' transitions from receiving a Report with Δ_{new} containing IDs not under retransmission by upstream routers (each ID not in Δ_{losts}) to activating δ_s with a non-empty Δ_1.
- Subsequence 5 covers transitions upon receiving data IDs of Δ with E and Δ_{tran}. δ_s transitions from receiving the lost IDs in Δ to activating σ_s with δ_{wait} when a received Retran ID is higher than δ_{cur}. The transitions include various branches of M_6 from REQUEST_SENDING state to RETRAN_RECV, RETRAN_MISS, or RETRAN_NONE state. Line 4 pauses for a specific duration until a Retran is received or reaches the request timeout.

Subsequence 8.1 Simplified $\sigma_1'(\xi, \Delta_{\text{tran}}, \Delta_{\text{losts}})$

1: **if** $\{\xi$ is Capsule $\wedge\ \delta(\xi) \notin \Delta_{\text{tran}} \wedge \delta(\xi) \notin \Delta_{\text{losts}}\}$ **then**
2:     $E \setminus= \{\delta(\xi) \rightarrow ?\}$
3:     $\Delta_{\text{fou}} = \{\delta_{\text{cur}} + 1, \dots, \delta(\xi) - 1\} \setminus \Delta_{\text{losts}}$
4:     **if** $\{|\Delta_{\text{fou}}| > 0\}$ **then**
5:         **activate** $\delta_s(\Delta_{\text{fou}}, E, \Delta_{\text{tran}})$
6:     **end if**
7: **end if**

Subsequence 8.2 Simplified $\sigma_2'(\xi, E, \Delta_{\text{tran}}, \Delta_{\text{losts}})$

1: **if** $\{\xi$ is Capsule $\wedge\ \delta(\xi) \notin \Delta_{\text{tran}}\}$ **then**
2:     $E \setminus= \{\delta(\xi) \rightarrow ?\}$
3:     **if** $\{\delta(\xi) \in \Delta_{\text{losts}}\}$ **then**
4:         $\Delta_{\text{losts}} \setminus= \{\delta(\xi)\}$
5:     **end if**
6:     **if** $\{\delta(\xi) \notin \Delta_{\text{losts}} \wedge \delta(\xi) > \delta_{\text{cur}}\}$ **then**
7:         $\delta_{\text{cur}} = \delta(\xi)$
8:     **end if**
9:     $\Delta_{\text{tran}} \cup= \{\delta(\xi)\}$
10:     **if** $\{\delta(\xi) \notin E\}$ **then** $\delta_{\text{wait}} = \min\{\{\delta | \delta \geq \delta_{\text{cur}} + 1 \wedge \delta \in \mathbb{N}\} \setminus (\Delta_{\text{losts}} \cup \Delta_{\text{tran}})\}$
11:         **if** $\{\{\delta(\xi) = \delta_{\text{wait}}\}\}$ **then** $E \cup= \{\delta_{\text{wait}} \rightarrow \text{machine\_id}()\}$
12:             **activate** $\delta_s(\{\delta_{\text{wait}}\}, E, \Delta_{\text{tran}})$
13:         **end if**
14:     **end if**
15: **end if**

Subsequence 8.3 Simplified $\sigma_3'(\xi, E, \Delta_{\text{losts}})$

1: **if** $\{\xi$ is Report with Δ_{new} and $r_{d_s}\}$ **then**
2: $\quad \Delta_1 = \Delta_{\text{new}} \setminus \Delta_{\text{losts}}$
3: $\quad$ **if** $\{|\Delta_1| > 0\}$ **then**
4: $\quad\quad \forall\delta \in \Delta_1, E\setminus = \{\delta\}$
5: $\quad\quad \Delta_{\text{losts}} \cup = \Delta_{\text{new}}$
6: $\quad\quad$ **if** $\{\Delta_{\text{new}} \cap E \neq \emptyset\}$ **then**
7: $\quad\quad\quad \hat{\delta} = \max(\Delta_{\text{new}} \cup E)$
8: $\quad\quad\quad \delta_{\text{wait}} = \min\{\{\delta | \delta \geq \hat{\delta} + 1 \wedge \delta \in \mathbb{N}\} \setminus (\Delta_{\text{losts}} \cup \Delta_{\text{tran}})\}$
9: $\quad\quad\quad E \cup = \{\delta_{\text{wait}} \rightarrow \text{machine\_id}()\}$
10: $\quad\quad\quad$ **activate** $\delta_s(\{\delta_{\text{wait}}\}, E, \Delta_{\text{tran}})$
11: $\quad\quad$ **end if**
12: $\quad$ **end if**
13: **end if**

Subsequence 8.4 Simplified $\sigma_4'(\xi, E, \Delta_{\text{losts}})$

1: **if** $\{\xi$ is Report with Δ_{new} and $r_{d_s}\}$ **then**
2: $\quad \Delta_1 = \Delta_{\text{new}} \setminus \Delta_{\text{losts}}$
3: $\quad$ **if** $\{|\Delta_1| > 0\}$ **then**
4: $\quad\quad \forall\delta \in \Delta_1, E\setminus = \{\delta\}$
5: $\quad\quad \Delta_{\text{losts}} \cup = \Delta_{\text{new}}$
6: $\quad\quad \forall\delta \in \Delta_1, E \cup = \{\xi \rightarrow \text{machine\_id}()\}$
7: $\quad\quad$ **activate** $\delta_s(\Delta_1, E, \Delta_{\text{tran}})$
8: $\quad$ **end if**
9: **end if**

8.3.2 Verifying Exclusiveness

Based on the five subsequences, we verify that the predicate exclusiveness is always maintained whenever any sequence executes ahead of others in time. The verification fully covers the four cases as follows:

- When σ_1 runs ahead, in Δ_{fou}, each lost Capsule ID cannot be in other sequences because the ID must be in E (see line 3 of subsequence 5), which we discuss below. In subsequence 2, σ_2' activates δ_s only if its received Capsule ID $\delta(\xi)$ is equal to δ_{wait} and not in E. If $\delta(\xi)$ is in Δ_{fou}, then σ_1 cannot retransmit δ_{wait} anymore due to δ_1 ending in line 2 of σ_2'. In subsequence 3, in line 7, the maximum ID of E is greater than $\hat{\delta}$, so each Δ_{fou} is no higher than δ_{wait}. In subsequence 4, for IDs in Δ_1, their retransmission listeners in σ_1 end in line 4. Hence, the lost ID is not in other sequences.

Subsequence 8.5 Simplified $\sigma_s(\Delta, E, \Delta_{tran})$

1: **if** $\{\Delta$ is received$\}$ **then**
2: $\quad \forall \delta' \in \Delta, E \setminus = \{\delta' \rightarrow ?\}$
3: $\quad \forall \delta \in \Delta, E \cup = \{\delta \rightarrow$ new retran_listener()$\}$
4: $\quad$ send Requests and wait for their Retran arrivals
5: $\quad \Delta_{rcv} =$ received Retrans with IDs in Δ
6: $\quad$ **for** $\{\xi \in \Delta_{rcv}\}$ **do**
7: $\qquad E \setminus = \{\delta(\xi) \rightarrow ?\}$
8: $\qquad \Delta \setminus = \{\delta(\xi)\}$
9: $\qquad$ **if** $\{\xi$ is non-empty Retran $\wedge\, \delta(\xi) \notin \Delta_{tran}\}$ **then**
10: $\qquad\quad \Delta_{tran} \cup = \{\delta(\xi)\}$
11: $\qquad\quad$ **if** $\{\delta(\xi) > \delta_{cur}\}$ **then**
12: $\qquad\qquad \delta_{cur} = \delta(\xi)$
13: $\qquad\qquad \delta_{wait} = \min\{\{\delta | \delta \geq \delta_{cur} + 1 \wedge \delta \in \mathbb{N}\} \setminus (\Delta_{losts} \cup \Delta_{tran})\}$
14: $\qquad\qquad E \cup = \{\delta_{wait} \rightarrow$ machine_id()$\}$
15: $\qquad\qquad$ **activate** $\delta_s(\{\delta_{wait}, E, \Delta_{tran}\})$
16: $\qquad\quad$ **end if**
17: $\qquad$ **end if**
18: $\quad$ **end for**
19: $\quad E \setminus = \{\delta(\xi) \rightarrow ? | \delta \in \Delta)\}$
20: $\quad \Delta = \emptyset$
21: **end if**

- When σ_2 runs ahead, δ_{wait} is not in other sequences. In σ_2, when σ_s lingers in line 4, δ_{wait} is in E. Meanwhile, for σ_1, when receiving δ_{wait}, σ_2 ends in line 2 of σ_1'. For σ_3, in line 7, due to the same reason in the 1st case above, δ_{wait} is different from that of σ_2'. In σ_4, line 4 ends the retransmission listeners that have been created by line 3 of δ_s in σ_1. As a result, σ_s is only in a sequence.
- When σ_3 runs ahead, δ_{wait} is different from lost Capsule IDs under retransmission by other sequences. This is because, for σ_3, in line 10 of σ_3', when δ_s is activated with δ_{wait}, δ_{wait} must be higher than each lost Capsule ID in E that contains all IDs of σ_1, σ_2, and σ_4.
- When σ_4 runs ahead, each ID in Δ_1 is exclusive to other sequences. For σ_4, when lingering in line 4, σ_s adds all lost Capsule IDs in Δ_1 to E. For each ID in Δ_1, its retransmission listener ends in line 2 of σ_1 and σ_2. For σ_3, because Δ_1 is in Δ_{new}, the maximum ID in Δ_{new} is no greater than $\hat{\delta}$ in line 7. Each ID in Δ_1 is less than δ_{wait} in line 8, which makes σ_3 cannot retransmit the same ID in Δ_1. As a result, each ID in Δ_1 is either in σ_4, σ_1, and σ_2, and cannot be in σ_3, which is exclusive.

Therefore, from the analysis, during retransmitting each lost Capsule, σ_1, σ_2, σ_3, and σ_4 are exclusive to each other.

8.3.3 Correctness in Arrival Timeouts

To achieve the retransmission recursiveness under the conditions in Table 7.1, with different choices of δ_s starting from σ'_1 to σ'_4, the counting in CIAT and CDT is with its correct constraints below.

- To satisfy CIAT_1, in σ'_2 and σ'_3, M_3's conditions include the receiving of the following messages: (1) the non-duplicated Capsule in the transitions from WAIT_NEXT_CAP to CAPSULE_RECV of M_1, and (2) the Report with new and non-empty data IDs in the transitions from WAIT_REPORT, UP_RETRAN_FOUND to UP_RETRAN_WAITED of M_2.
- To satisfy CIAT_2, in σ_s from δ'_1, δ'_2, δ'_3, or δ'_4, M_3's conditions involve the in-time receiving of the non-empty and non-duplicated Retran, in the transitions from REQUEST_SENDING to RETRAN_RCVD of M_3.
- To satisfy $\text{CDT}_k(d_s)$, in δ'_4, M_4's conditions cover on receiving a non-empty Report with new IDs, in the transitions from WAIT_REPORT, UP_RETRAN_FOUND to UP_RETRAN_WAITED of M_2.
- To satisfy $\text{CDT}_k(d_1)$, in σ_s from δ'_1, δ'_2, δ'_3, or δ'_4, M_6's conditions embody in-time the receiving of the empty Retran with some non-duplicated IDs, in the transitions from REQUEST_SENDING to RETRAN_MISSS of M_6.

Therefore, the conditions in Table 7.1 and the ones in STMs above are consistent with each other in a correct manner.

8.3.4 Retransmission Message Number Under Exclusiveness

Due to the above exclusiveness, for the same lost data ID, we examine the maximum thresholds of the number of messages per node in PP The Number of Messages Sent Per Node of PP (NMN) in Request and Retran below.

- The NMN of Request is no greater than the threshold under different conditions as follows: (1) $2 \cdot \eta$, when σ_4 can cancel σ_1 and σ_2, meaning that a jump from M_5 to σ_s is caused by a failed upstream retransmission, and (2) η, when no canceling exists, meaning that no such a jump is caused by a successful upstream retransmission.
- Due to M_7, the NMN of Request constraints on the NMN of Retran. However, a Request with Δ does not mean there is only one Retran for each data ID in Δ, since for these data IDs, the closest caching finding policy means that there can be duplicated Retrans in feedback.

Furthermore, between Retran and Request, the NMN difference must be not greater than $|\Delta|$. Therefore, under the above different retransmission consequences, the exclusiveness is verifiable via experiments under different link fault intensities.

8.3.5 Message Processing Costs

Due to the exclusiveness, upon receiving of a message, δ_1', δ_2', and δ_3' all take the finite number of steps in $O(1)$ to activate δ_s. Then, δ_s deals with each ID in Δ_{rcv} by activating its next δ_s in parallel, which also takes $O(1)$. Therefore, the process time cost of RSDD is in $O(1)$, and the space cost is the maximum size of shared states in $O(1)$, indicating efficient processing.

8.4 Exploring the Robustness of the Mechanism

In our journey through the intricacies of communication protocols, RSDD emerges as a beacon of reliability within the realm of DRTP. Let's unravel the practical prowess of RSDD as it fortifies the resilience against link faults in disseminating sensor data.

Understanding the essence: At its core, RSDD impeccably embodies the principles of DRTP, providing a concrete and correct algorithmic framework. It's not just a theoretical concept; it's a practical remedy to bolster the robustness of sensor data dissemination, especially when facing multiple link faults.

Key features that set RSDD apart: What sets RSDD apart is its ingenious hop-by-hop retransmission control strategy. This approach incorporates exclusivity and recursiveness, harnessing the power of set dissemination frequency. Moreover, it strategically leverages multiple RSPs, introducing a novel dimension to enhance overall reliability.

Real-world application: Power transmission grids: Let's dive into a real-world scenario where RSDD shines – the dynamic landscape of power transmission grids. Maintaining a low EEFR is pivotal for State Estimation (SE) to uphold topological observability and accuracy in the grid. Here, RSDD strategically navigates data transmission complexities by utilizing MPSG routes intricately woven with RSPs within the NDN architecture.

Navigating challenges with RSDD's STMs: Imagine a scenario where a payload packet goes astray during transmission. This is where RSDD's STMs come to the forefront. Executing a predetermined trajectory and adhering strictly to arrival timeouts as specified in DRTP, these STMs initiate a robust and recursive retransmission process. This strategic approach seizes every opportunity at each hop along PP to recover lost Capsules while minimizing the likelihood of cache mismatches.

Conclusion: Unveiling RSDD's resilience: In conclusion, RSDD emerges as an efficient solution, seamlessly blending resilience and reliability in sensor data dissemination. Its unique features and real-world applications, particularly in power transmission grids, underscore its significance in ensuring the uninterrupted flow of critical information. As we delve into experiments under varying link fault intensities, RSDD stands tall as a testament to the reliability of communication protocols in dynamic environments.

8.5 Performance Evaluation

The performance of RSDD was evaluated under the same cyber–physical environment given in Section 7.4.1. The results confirm its implementation correctness and superior advantage compared to other existing solutions. The concrete statistical results are elaborated as follows.

8.5.1 Ensuring Implementation Correctness

i) **Retransmission exclusiveness findings:** Figures 8.12 and 8.13 showcase box plots of NMNs under PP and MPSG disruptions with LLRs ranging from 10% to 100%.

For PP disruption, refer to Figure 8.12(a) and 8.12(b), revealing that the maximum NMNs of RequestS and RequestM are all 3 less than η, signifying successful retransmissions. Figure 8.12(c) and 8.12(d) shows that the maximum NMN

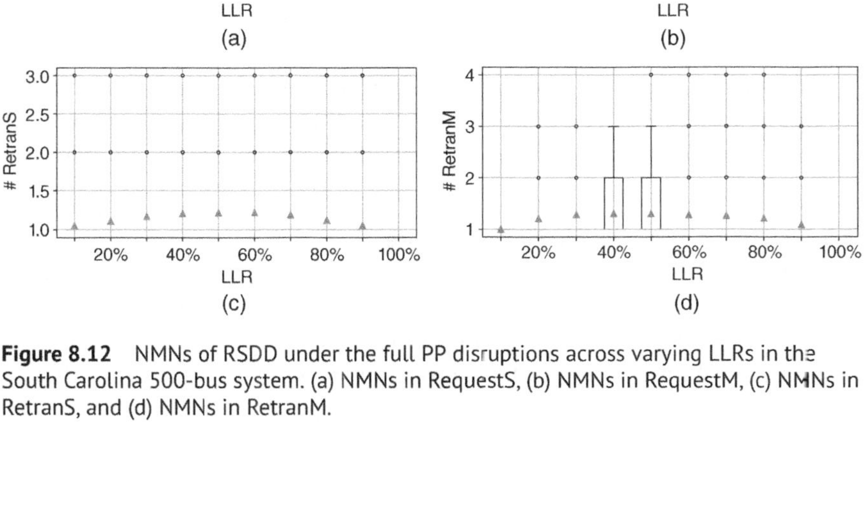

Figure 8.12 NMNs of RSDD under the full PP disruptions across varying LLRs in the South Carolina 500-bus system. (a) NMNs in RequestS, (b) NMNs in RequestM, (c) NMNs in RetranS, and (d) NMNs in RetranM.

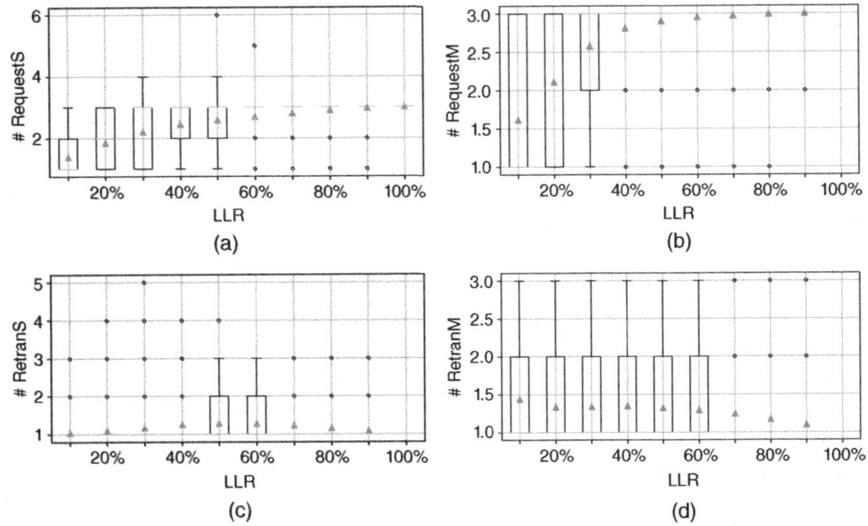

Figure 8.13 NMNs of RSDD under the full MPSG disruptions across varying LLRs in the South Carolina 500-bus system. (a) NMNs in RequestS, (b) NMNs in RequestM, (c) NMNs in RetranS, and (d) NMNs in RetranM.

of RetranM is 3 and 4, respectively, both less than $2 \times \eta$, courtesy of the closest cache matching policy of M_7.

For MPSG disruption, observe Figure 8.13(a) and 8.13(b), displaying box plots indicating the maximum NMNs of RequestS and RequestM as 6 and 3, respectively. Both are less than $2 \times \eta$. Message overhead can be effortlessly absorbed by ample optical link bandwidth, reaching 10~400 Gbps [IEC, 2014; Bedi et al., 2018; Faheem et al., 2018]. Figure 8.13(c) and 8.13(d) reveals that the maximum NMNs of RetranS and RetranM are 5 and 3, respectively. These NMNs are all less than $2 \cdot \eta$, attributed to the closest policy.

These results affirm the exclusiveness of retransmissions.

ii) **Exploring Quality-of-Service (QoS) performance:** Figure 8.14 unfolds the tale of RSDD's EEFR and EEDT under the influence of disruptions from both PP and MPSG, traversing the diverse landscape of LLRs from 10% to 100%.

8.5.1.1 Dance of Disruption: PP Performance

Gaze upon Figure 8.14(a) and 8.14(b), where, amid 20%, 40%, 60%, 80%, and 100% LLRs, the mean EEFRs and EEDTs perform a graceful symphony: (2.44%, 2.7 ms), (18.32%, 12.68 ms), (34.67%, 25.38 ms), (59.31%, 52.05 ms), and (75%, 2.94 ms), respectively. Witness their EEDT satisfactory ratios: 99.57%, 89.97%, 81.45%, 83.27%, and 100%. This narrative unravels a tapestry of high recursive pair

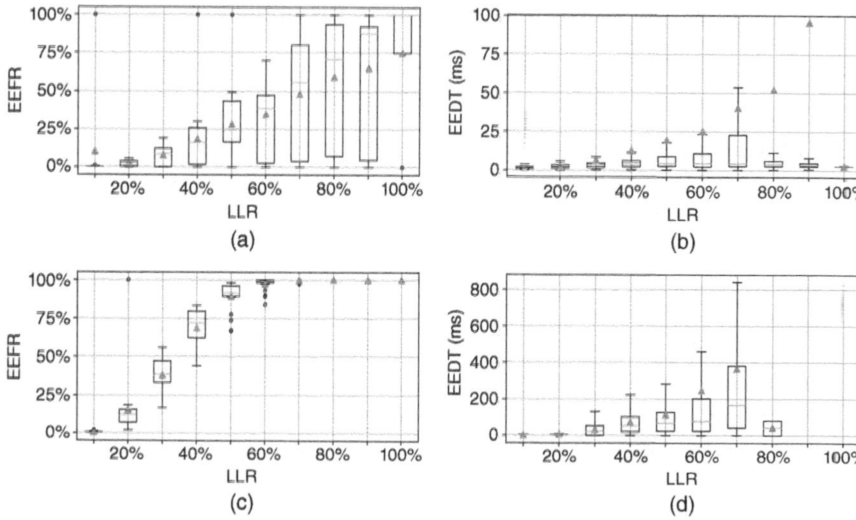

Figure 8.14 Performance of RSDD under different disruption cases across varying LLRs in the South Carolina 500-bus system. (a) EEFRs under the PP case, (b) EEDTs under the PP case, (c) EEFRs under the MPSG case, and (d) EEDTs under the MPSG case.

ratios, with non-recursiveness a rarity, born of assumption violations amid disruption-induced message loss.

8.5.1.2 Theatrics of MPSG Disruption

Delve into Figure 8.14(c) and 8.14(d), where the drama unfolds at 20%, 40%, 60%, 80%, and 100% LLRs, with mean EEFRs and EEDTs taking center stage: (15.08%, 6.85 ms), (68.71%, 73.36 ms), (97.16%, 245.7 ms), (99.98%, 41.82 ms), and (100%, $+\infty$), respectively. Observe their EEDT satisfactory ratios: 94.77%, 39.02%, 29.93%, 50%, and 100%. A plot twist emerges at 40%, 60%, and 80% LLRs, where low EEDT satisfactory ratios reveal the toll of heavy losses in RSPs, challenging the sanctity of recursiveness assumption ii, as detailed in Section 7.3.

In this theatrical journey, the sudden surge in mean EEDT at 70% LLR, followed by a decline beyond 80% LLR, paints a common scene under both disruptions. This phenomenon is born of an excessively high LLR, rendering Reports unreachable to the next hop, triggering a cascade of premature retransmission requests and recovering fewer lost Capsules. At 80% LLR, the stage is set: most retransmissions spanning more than two hops falter, leaving successful retransmissions confined to neighboring hops. This script unfolds the essence of the phenomenon.

Thus, the performance ballet of EEDT and EEFR echoes the melodic notes of DRTP evaluation in Zhou et al. [2022], harmonizing the correctness of RSDD.

8.5.2 Comparative Studies Under Full MPSG Disruptions

The comparison of RSDD performance was an exploration into the South Carolina (SC) 500-bus topologies, unraveling metrics that tell the tale.

i) **EEFR comparison**. In the captivating performance showdown depicted in Figure 7.21, DRTP takes center stage alongside MFTP, R^2T, and SDATP. Figure 7.21(a) reveals their mean EEFR dance at 20%, 60%, and 100% in the IEEE 300-bus topology: $(2.12, 37.92, 33.22, 61.63)\%$, $(37.5, 98.88, 94.62, 99.23)\%$, and $(70, 100, 100, 100)\%$, respectively. Similarly, Figure 7.21(b) showcases their moves in the SC 500-bus topology: $(2.39, 36.91, 30.92, 60.65)\%$, $(42.18, 98.25, 95.57, 98.55)\%$, and $(86, 100, 100, 100)\%$, respectively.
ii) **OOR comparison**. As the plot thickens, Figures 7.21(f), 7.21(g), and 7.21(h) unveils the Cumulative Probability Distribution Function (CDF) of EEDT under 20%, 60%, and 80% in the IEEE 300-bus. Meanwhile, Figures 7.21(f), 7.21(g), and 7.21(h) brings forth the CDF of EEDT under the same LLRs in the SC 500-bus. A closer look reveals their stories, showcasing DRTP's prowess in maintaining low EEDT with an acceptable message load capacity.
iii) **OOR comparison**. In a finale that echoes through Figures 7.21(i) and 7.21(j), the mean Out-of-Order Rate at the Receiver Side (OOR) tale unfolds. MFTP and R2T stay true to a 0% OOR script, while SDATP and DRTP take center stage with OORs of $(12.4, 0.11, 0)\%$ and $(1.08, 9.59, 0)\%$, respectively, at 20%, 60%, and 100% LLR. However, DRTP exhibits a lower EEFR, hinting at a more dedicated effort in packet recovery.

Continuing the saga in Figure 8.15, the spotlight shifts to RSDD's unmatched performance under both PP and MPSG disruptions in the SC 500-bus topology.

i) **EEFR**. Figure 8.15(a) unveils the EEFR brilliance of RSDD against IEEE C37.118.2 in TCP, MFTP, R2T, and SDATP across varying LLRs. RSDD steals the show with mean EEFRs of 0.13%, 15.57%, 55.02%, and 86% at LLRs of 10%, 40%, 70%, and 100%, respectively. A stark contrast unfolds as TCP's EEFR crumbles with rising LLR, while MFTP, R2T, and SDATP grapple with higher EEFRs than RSDD.
ii) **EEDT**. Figure 8.15(b) turns the spotlight on equivalent EEDTs, showcasing RSDD's efficiency with mean EEDTs of 0.02, 2.34, 8.25, and 12.9 seconds for LLRs of 10%, 40%, 70%, and 100%, respectively. The competition falls short with TCP, MFTP, R2T, and SDATP displaying higher EEDTs.
iii) **OOR**. Figure 8.15(c) delves into OOR, revealing RSDD's commendable mean OORs of 0.08%, 10.15%, 12.43%, and 0% at LLRs of 10%, 40%, 70%, and 100%,

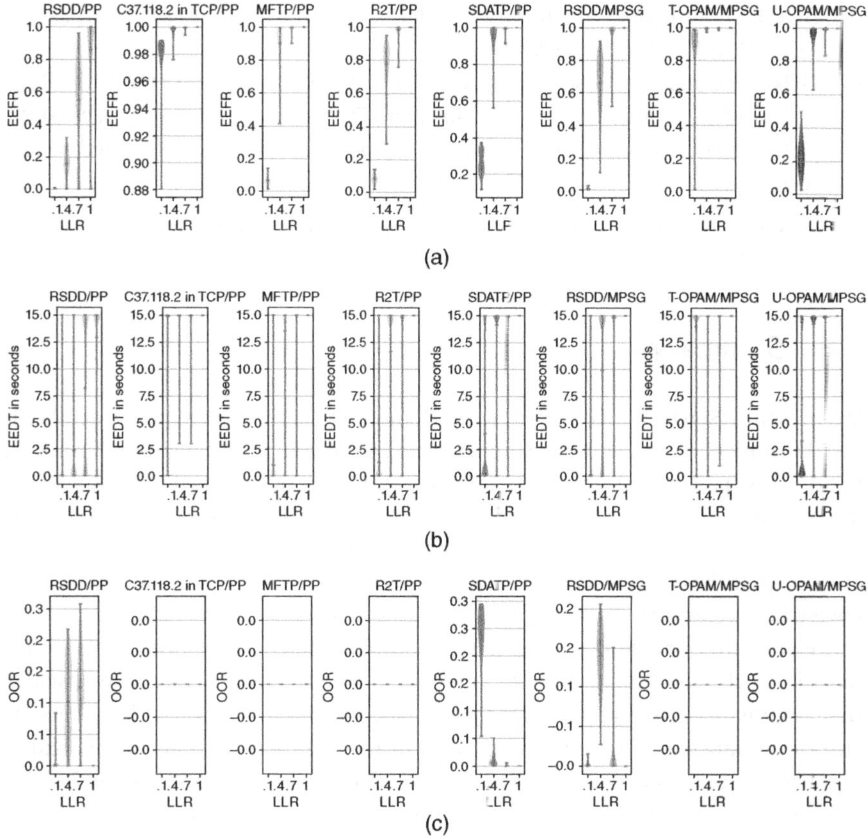

Figure 8.15 Comparative performance studies of transport protocols in the South Carolina 500-Bus System: Violin plots under PP and MPSG disruption scopes. (a) EEFRs across varying LLRs of transport protocols, (b) EEDTs across varying LLRs of transport protocols, (c) OORs across varying LLRs of transport protocols (sub-figure titles are in the format "protocol name/scope").

respectively. In contrast, SDATP grapples with higher OOR and EEFR, hinting at RSDD's dedication to packet recovery.

In a grand finale, the performance saga extends to MPSG disruptions in Figure 8.15(a).

i) **EEFR**. Figure 8.15(a) showcases RSDD's triumph with mean EEFRs of 0.53%, 66.15%, and 98.37% at 10%, 40%, and 70% LLRs, respectively, outshining TCP-OpenFlow-based Non-Disjoint Path Aggregation Mechanism (OPAM) and UDP-OPAM.

ii) **EEDT**. Figure 8.15(b) amplifies RSDD's superiority in EEDT, surpassing TCP-OPAM and UDP-OPAM.

iii) **OOR**. The tale concludes in Figure 8.15(c), showcasing RSDD's mean OORs of 0.17%, 14.04%, and 0.81% at 10%, 40%, and 70% LLRs, respectively, against 0% OOR for TCP-OPAM and UDP-OPAM, emphasizing RSDD's dedicated packet recovery.

In essence, RSDD emerges as a performance champion, blending low EEFR, efficient EEDT, and commendable OOR metrics.

8.6 Summary

We present RSDD – a mechanism to implement DRTP in a correct manner. With RSDD, think of it as the maestro orchestrating a symphony among in-path routers, turning lost payload packet retransmissions into a harmonious dance. How does it work? Each router becomes a proactive seeker, asking its upstream routers for cached lost packets when needed. And here's the twist – a calculated timeout ensures that these requests are made at just the right moment, avoiding interference with ongoing retransmissions.

Imagine this collaboration as a well-choreographed ballet. State transition machines, the choreographers in our story, play a crucial role. They ensure that retransmissions happen asynchronously, with perfect timing. It's like synchronized dancers moving to the rhythm of message arrivals. But in this ballet, correctness takes center stage. We enforce exclusivity for retransmissions of the same lost packet – no stepping on each other's toes. And those timeouts? They are the metronome, keeping everyone in perfect harmony and making sure lost packets are detected accurately.

Enter RSDD, a beacon of hope in the realm of power transmission grids. It's not just a solution; it's a promise to tame the unruly disruptions caused by link faults in the optical IoT networks.

For the curious minds out there, our DRTP prototype, along with its ingenious algorithms, is now open for exploration online [GitHub, Inc., 2023]. Yes, you can peek under the hood and see the magic unfold!

As we gaze into the future, we toss two exciting ideas into the research arena. First up, imagine the RSDD prototype seamlessly integrating into existing high-performance data planes. Think of it as fitting a superhero into an already-powerful team, enhancing efficiency with streamlined pipelines and a dash of efficient CS. And that's not all – there's a second act. Picture RSDD stepping into the limelight for data dissemination applications in the vast landscape of machine-to-machine communications. This time, it's not just about reducing RSDD; it's about predicting paths, navigating the data highways with finesse.

Questions

8.1 How does RSDD address the challenge of link disruptions in optical IoT networks, and what key features make it stand out in comparison to existing solutions?

8.2 In the context of retransmission exclusiveness, what are the conditions under which the maximum NMN of Request is determined to be no greater than a certain threshold, and how does this relate to successful upstream retransmissions?

8.3 What is the role of state transition machines in the RSDD mechanism, and how do they contribute to the efficiency and correctness of the retransmission process?

8.4 How does the concept of exclusivity for retransmissions contribute to the correctness and reliability of the RSDD mechanism, and what experiments or evidence support this claim?

8.5 How does RSDD strategically leverage multiple RSPs, and what role do these RSPs play in enhancing the overall reliability of sensor data dissemination?

8.6 What are the implications of the observed performance of RSDD in power transmission grids, and how might its efficiency contribute to the resilience of critical information flow in such infrastructures?

8.7 What insights can be drawn from the performance comparison of RSDD with other protocols like MFTP, R^2T, and SDATP under different disruption scenarios in the South Carolina 500-bus system?

8.8 In what ways could RSDD be integrated into existing high-performance data planes, and what potential benefits might this integration bring to the efficiency of data dissemination in optical IoT networks?

Bibliography

Guneet Bedi, Ganesh Kumar Venayagamoorthy, Rajendra Singh, Richard R. Brooks, and Kuang-Ching Wang. Review of Internet of things (IoT) in electric power and energy systems. *IEEE Internet of Things Journal*, 5(2):847–870, 2018.

Egon Börger. *The Abstract State Machines Method for High-Level System Design and Analysis*, pages 79–116. Springer, London, 2010.

Muhammad Faheem, Syed Bilal Hussain Shah, Rizwan Aslam Butt, Basit Raza, Muhammad Anwar, Muhammad Waqar Ashraf, Md Asri Bin Ngadi, and Vehbi Cagri Gungor. Smart grid communication and information technologies in the perspective of Industry 4.0: Opportunities and challenges. *Computer Science Review*, 30:1–30, 2018.

GitHub, Inc. The source code of the resilient sensor data dissemination mechanism (RSDD) implementing the disruption resilient transport protocol (DRTP) for power transmission grids, 2023. URL https://github.com/zhouby-zjl/drtp.

IEC 60794-4-10. Optical fibre cables - Part 4-10: Family specification - optical ground wires (OPGW) along electrical power lines, 2014.

IEEE C37.118.2. IEEE standard for synchrophasor data transfer for power systems, 2011.

Cédric Josz, Stéphane Fliscounakis, Jean Maeght, and Patrick Panciatici. AC power flow data in MATPOWER and QCQP format: iTesla, RTE snapshots, and PEGASE. *arXiv preprint arXiv:1603.01533*, 2016.

Named Data Networking (NDN). NFD: Named data networking forwarding daemon¶, 2023. URL https://docs.named-data.net/NFD/current/.

Carl A. Sunshine, David H. Thompson, Roddy W. Erickson, Susan L. Gerhart, and Daniel Schwabe. Specification and verification of communication protocols in AFFIRM using state transition models. *IEEE Transactions on Software Engineering*, 5:460–489, 1982.

Lixia Zhang, Alexander Afanasyev, Jeffrey Burke, Van Jacobson, Christos Papadopoulos, K. C. Claffy, Lan Wang, Patrick Crowley, and Beichuan Zhang. Named data networking. *ACM SIGCOMM Computer Communication Review*, 44(3):66–73, 2014.

Boyang Zhou, Chunming Wu, Qiang Yang, and Xiang Chen. DRTP: A disruption resilient hop-by-hop transport protocol for synchrophasors measurement in electric transmission grids. *IEEE Access*, 10:133898–133914, 2022.

9

Enhancing DRTP Reliability: Cultivating Robust Routes

To enhance Resilient Sensor Data Dissemination (RSDD)'s resilience against fault links, we consider on optimizing the structure of MPSG. The harsh fault links in the MPSG can trigger fragility and make DRTP's retransmission process susceptible to failure cases, seriously impacting DRTP's resilience in EEFR [Zhou et al., 2022]. The two primary failure cases are listed as follows [Zhou et al., 2022]: (1) Collaboration recursiveness between two hops is disrupted due to report losses, resulting in unexpected early requests for cache mismatches at upstream hops. (2) Retransmission requests from each hop receive no response due to packet losses. Thus, to mitigate these cases, less fragility necessitates that the MPSG possesses two specific features [Zhou et al., 2022]: (1) the PP must be highly reliable, and (2) each hop should be connected to as many RSPs as possible.

To bolster the resilience of RSDD against potential fault links, we delve into optimizing the structure of the Multipath Subgraph (MPSG). Within the MPSG, harsh fault links can induce fragility, rendering Disruption Resilient Transport Protocol (DRTP)'s retransmission process susceptible to failure cases, thereby significantly impacting its resilience in terms of End-to-End Failure Rate in Packet Delivery (EEFR) [Zhou et al., 2022]. The primary failure scenarios are outlined as follows [Zhou et al., 2022]: (1) Disruption of collaboration recursiveness between two hops, arising from report losses. This disruption leads to unexpected early requests for cache mismatches at upstream hops. (2) Non-responsive retransmission requests from each hop due to packet losses. To address these challenges and minimize fragility, it is imperative that the MPSG incorporates two specific features [Zhou et al., 2022]: (1) Ensuring high reliability of the Primary Forwarding Path (PP). (2) Connecting each hop to as many Redundant Sub-Paths (RSPs) as possible. "Less fragility" in this context becomes the key to mitigating the outlined failure cases. By fortifying the PP's reliability and maximizing the connections to RSPs at each hop, we strive to enhance the overall robustness of the MPSG.

Despite the enticing features, current approaches fall short in tackling the challenge of crafting MPSGs with reduced fragility to reduce Probability

Communication Networks in Smart Power Grids, First Edition. Boyang Zhou.

Companion Website: www.wiley.com/go/SmartPowerGrids1e

of Retransmission Failure Cases (PRF) and fortify DRTP resilience with a significantly low EEFR. The recursive process must also adhere to the constraint on the end-to-end packet delivery time (End-to-End Payload Packet Delivery Time (EEDT)) at its maximum [Zhou et al., 2022]. The EEDT represents the time difference between when a Phasor Measurement Unit (PMU) transmits a payload packet and when its Phasor Data Concentrator (PDC) receives the packet [Zhou et al., 2022]. Generating MPSGs with minimal fragility proves to be inherently challenging due to their NP-complete computational complexity. Beyond DRTP, other hop-by-hop controls predominantly rely on single paths [Su et al., 2015; Wang et al., 2018; Chen et al., 2020; Garcia-Luna-Aceves and Albalawi, 2020]. As for DRTP, the sole available algorithm for MPSG generation is GenMPSG [Zhou et al., 2022], lacking considerations for fragility and EEDT. GenMPSG constructs the shortest PP and independently identifies the shortest RSPs that are disjoint from both each other and the PP. Fortuitously, the network is interlinked with high-capacity optical lines [IEC, 2014; Bedi et al., 2018] and structured with redundant topological meshes and rings [Zhou et al., 2022]. These factors have inspired the development of a new algorithm to generate MPSGs with reduced fragility while strictly adhering to the maximum EEDT constraint.

To tackle the aforementioned challenge, this chapter introduces a groundbreaking HeuMPSG algorithm – an innovative heuristic approach designed to generate MPSGs and enhance the resilience of DRTP. The primary objective is to minimize the Fragility Index (FI) metric for each MPSG, considering constraints related to the recursive collaboration process and the maximum EEDT, as computable via a known equation [Zhou et al., 2022]. FI measures the highest PRF achieved under the worst conditions, particularly when the first failure case involves complete payload packet loss. HeuMPSG adopts a step-by-step methodology, navigating through the selection of a candidate PP hop by hop while actively pursuing a maximum of β maximally disjoint RSPs starting from the current hop. Each selection is made by comparing its calculated FI with its previously recorded value, while simultaneously meeting the EEDT constraint. This incremental approach streamlines the RSP search process and progressively diminishes the FI for each MPSG. The outcome is not only improved computational efficiency but also heightened DRTP resilience, surpassing the capabilities of the existing GenMPSG algorithm.

We conducted a comprehensive evaluation, pitting HeuMPSG against GenMPSG to showcase its prowess in improving both the efficiency of MPSG generation and the resilience of DRTP within real grid networks. The evaluations were carried out through simulations using ndnSIM [Mastorakis et al., 2017]. Our thorough assessment yielded promising numerical results. In a 500-bus network, HeuMPSG exhibited efficient MPSG generation, significantly reducing the FI by discovering, on average, 171.51% more RSPs at an impressive 7.07 times the maximum speed. Leveraging the inherent robustness of the generated MPSGs, HeuMPSG notably strengthened resilience in a 300-bus network. It achieved substantial reductions in mean EEFRs and EEDTs by up to 29.11% and 2.41 seconds, respectively, under challenging conditions marked by heavy random fault links with 50% and 100% Link Loss Rate in Packetss (LLRs). Consequently, HeuMPSG emerges as a potent solution for enhancing DRTP resilience, delivering significantly lower EEFR and EEDT compared to alternative methods. These statistics mark a significant stride in resilience advancement.

9.1 Problem on Generating MPSG Less Fragile

The fragility of the MPSG is assessed through the Fault Index (FI), a metric measuring PRF. In simpler terms, FI gauges the likelihood of failure in rerouting scenarios within the Disruption-Resilient Transport Protocol (DRTP), considering both recursive violation and no response in Requests. Let's break down the mathematical derivation of FI. In the context of DRTP Step ii (refer to Chapter sec:retran-control-process), we focus on the probability of successful notification. This occurs when the first Primary Forwarding Path (PP) hop effectively transmits a Report to the last hop, denoted as $\alpha(d_b)$, as defined in Eq. (9.1).

Assuming scenario i, where there are no exceptions during the sequential execution of the aforementioned steps, this case represents a successful operation under ideal conditions.

$$\alpha(d_b) = \prod_{2\leq c\leq b} 1 - \epsilon\left(d_{c-1}, d_c\right)^{\zeta} \tag{9.1}$$

Recursive Retransmission Process and CPRF

Moving on to the second scenario, let's explore the condition where the retransmission process is recursive, as detailed in Chapter 7. In this context, the Conditional Probability of Reroute Failure (Conditional Probability of Retransmission Failure Cases (CPRF)), denoted as $F(d_b)$, arises when each PP hop from d_2 to r_{d_b} fails to recover a lost Capsule. Let's delve into the specifics: Firstly, the packet loss rate from r_{d_m} to r_{d_k} in a path ($\mathbf{p}$) is defined by Eq. (9.2). Next, the failure rate of η times retransmission for a Capsule from r_{d_m} to r_{d_k} in $\mathbf{p}$ is calculated using Eq. (9.3). Moving on, the failure rate of r_{d_k} in delivering a Capsule from $r_{d_{k-1}}$ to r_{d_k} with η retransmissions is computed as $\phi(d_k)$, as expressed in Eq. (9.4). Finally, by setting $k = b$, we compute $F(d_b)$ in Eq. (9.5) as the sum of the lost probabilities in Capsule for each router, where $\psi(d_k)$ becomes a recursive function, defined in Eq. (9.6).

$$\epsilon_{\mathbf{p}}(d_m, d_k) = 1 - \prod_{m+1\leq c\leq z} \left(1 - \epsilon\left(d_{c-1}, d_c\right)\right) \tag{9.2}$$

$$\epsilon_{\mathbf{p}}^{\eta}(d_m, d_k) = (1 - (1 - \epsilon_{\mathbf{p}}(d_m, d_k))(1 - \epsilon_{\mathbf{p}}(d_k, d_m)))^{\eta} \tag{9.3}$$

$$\begin{aligned} \phi(d_k) &= \epsilon(d_{k-1}, d_k)(1 - \epsilon(d_{k-1}, d_k)) \\ (1 - \epsilon(d_k, d_{k-1}))^{\eta} & \prod_{\mathbf{p} \in P_{\mathrm{rsp}}[k] \cup \{(d_k, d_{k-1})\}} \epsilon_{\mathbf{p}}^{\eta}(d_k, d_s) \end{aligned} \tag{9.4}$$

$$\begin{aligned} F(d_k) &= \sum_{2 \le m \le k} \left(\prod_{1 \le w \le m-1} \left(1 - \phi\left(d_w\right)\right) \right) \phi(d_m) \\ &= F\left(d_{k-1}\right) + \psi\left(d_k\right) \phi\left(d_k\right) \end{aligned} \tag{9.5}$$

$$\psi(d_k) = \begin{cases} \psi(d_{k-1}) \left(1 - \phi(d_{k-1})\right) & z \ge 2 \\ 1 & z = 1 \end{cases} \tag{9.6}$$

To gauge the fragility of the MPSG, we define $\alpha(d_b)$ as the probability when the retransmission is recursive. Combining this with $F(d_b)$, the FI represents the probability under the worst conditions, where at least one of the two retransmission failure cases occurs (refer to the beginning of Chapter 9). This is captured by Eq. (9.7), denoted as $\gamma(d_b)$, where its value falls within the range $[0, 1]$.

For FI, we adopt a conservative strategy. In the extreme exception case where the Report notification fails with a probability of $1 - \alpha(d_b)$, the retransmission is deemed to fail entirely (100%). Consequently, a lower FI value indicates a higher fragility of MPSG within the context of DRTP.

$$\gamma(d_b) = \alpha(d_b) \cdot F(d_b) + \left(1 - \alpha(d_b)\right) \cdot 1 \tag{9.7}$$

The challenge at hand is to generate Multipath Subgraph (MPSG) configurations for each pair, aiming to minimize the FI as defined in Eq. (9.7). This optimization is conducted under the constraint of the maximum EEDT of D, considering the recursiveness outlined in Eq. (7.9). This requirement aligns with the specific needs of the grid application [IEEE, 2011].

The problem at its core is reducible to the Steiner tree issue, where the constraint is additive and concave, while the FI is both additive and multiplicative. Consequently, the computational complexity of this problem falls under NP-completeness [Wang and Hou, 2000].

9.2 Overall Design

In the interconnected network G equipped with Named Data Network (NDN) routers featuring the DRTP [Zhou et al., 2022]—as illustrated in Figure 9.1 – HeuMPSG emerges as an ingenious solution. Operating seamlessly within any router, HeuMPSG leverages the NDN forwarding daemon [NDN, 2021], tightly integrated with DRTP [Zhou et al., 2022].

Figure 9.1 HeuMPSG design.

The NDN forwarding daemon, enriched by DRTP, not only captures the network's inherent topology [Zhang et al., 2014] but also serves as the foundation for HeuMPSG's algorithmic design. HeuMPSG's approach involves the systematic generation of Multipath Subgraphs (MPSGs) from each producer to its respective consumers. Subsequently, these optimized MPSGs are intelligently configured within each daemon, aligning with the principles of DRTP. The entire orchestration unfolds in a centralized manner [Zhang et al., 2014].

In the journey toward an optimized MPSG, Figure 9.2 unveils the sequential steps of HeuMPSG – a methodical approach that refines the selection of candidate MPSG hop by hop along the Primary Forwarding Path (PP). The objective is to minimize the Fault Index (FI) while adhering to the constraint D and limiting the RSPs for each PP hop to at most β.

Step i initiates a candidate PP, establishing a trajectory from the producer to the current node (the rightmost router). This trajectory undergoes continuous optimization, contributing to a reduction in fragility.

Steps ii–v introduce a backtracking technique to craft candidate RSPs. It temporarily disrupts links between the current router and all upstream routers (step ii), introduces a virtual router with 0% LLR and zero total delay (step iii), computes candidate RSPs with maximum link disjointness (step iv), and then restores the graph to its original state for ongoing optimization (step v).

Step vi concludes with the selection of an MPSG boasting a lower FI.

By iterating through these steps, HeuMPSG systematically refines the MPSG toward the consumer, achieving the minimum FI under the specified constraint.

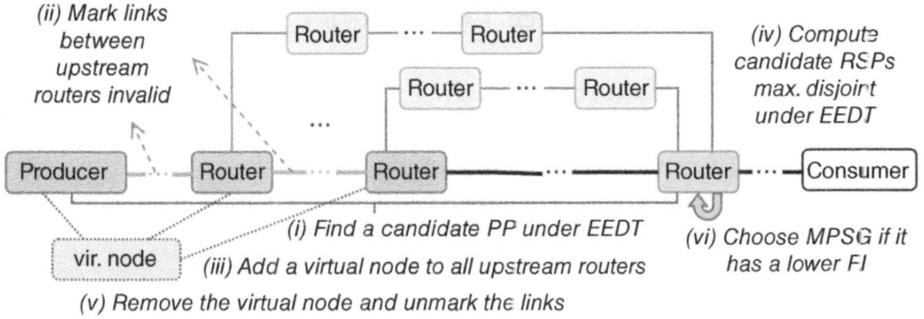

Figure 9.2 MPSG generation process.

9.3 HeuMPSG Algorithms

The HeuMPSG algorithm is designed to ensure both correctness and convergence while minimizing the FI under the maximum EEDT constraint, denoted as D, and the parameters η and β. To achieve this, HeuMPSG utilizes two key algorithms: the GenRRV algorithm, outlined in Alg. 9.1, and its subroutine, the ComRSP algorithm, detailed in Alg. 9.2. The workflow of both algorithms is shown in Figure 9.3.

i) **GenRRV algorithm**: Within the network topology represented as $G(R, L)$ – where R refers to the set of routers, and L to the set of links – GenRRV generates PPs and RSPs as part of Resilient Route Vectors (RRVs). The RRV originates from a given node r_i, and the algorithm heuristically explores each non-first hop of the PP. During this exploration, it considers paths with lower FI values

---

Algorithm 9.1 GenRRV

---

function GenRRV(R, L, n, i, η, β, D)
  $\omega = [-1]_n, X = \{\}, T_{\text{all}} = [0]_n, \gamma = [1]_n, \gamma[i] = 0$
  $q_1 = \text{FiboHeap}(\gamma, \text{des})$, **for all** $r_k \in R$ **do** $q_1 \leftarrow k$
  while $|q_1| > 0$ **do**
    $u = q_1.\text{poll}()$
    for all $r_v \in \text{EgressNodes}(u)$ **do**
      if $v \notin q_1 \vee \tau_{\text{all}}^{\text{PP}}(i, v) > D$ **then continue**
      $\mathbf{p}' = \text{ConvertPath}(\omega, i, u)$
      $\tau' = T_{\text{all}}[u] + \tau_{\text{all}}^{\text{PP}}(u, v), \tau_m = D - \tau'$
      $P_{\text{rsp}} = \{\}, \tilde{L} = \{\}, \hat{L} = \text{links}(\mathbf{p}')$
      for $(i = 0; i < \beta; i = i + 1)$ **do**
        $\mathbf{p}_{\text{r}} = \text{ComRSP}(R, L, n, \hat{L}, \tilde{L}, v, \mathbf{p}', \eta, \tau_m)$
        if $\mathbf{p}_{\text{r}} = \text{null} \vee \mathbf{p}_{\text{r}} \in P_{\text{rsp}}$ **then break**
        $P_{\text{rsp}} = P_{\text{rsp}} \cup \mathbf{p}_{\text{r}}$
        if $i < \beta - 1$ **then** $\tilde{L} = \tilde{L} \cup \text{links}(\mathbf{p}_{\text{r}})$
      end for
      compute $\gamma'(v)$ using Eq. (9.7)
      if $\gamma'(v) \geq \gamma[v]$ **then continue**
      $\gamma[v] = \gamma'(v), q_1.\text{updateKey}(v), \omega[v] = u$
      $T_{\text{all}}[v] = T_{\text{all}}(v)$ computed in Eq. (7.9)
      if $|P_{\text{rsp}}| > 0$ **then** $X[v] = P_{\text{rsp}}$
    end for
  end while
  return (ω, X)
end function

Algorithm 9.2 Compute a RSP

function ComRSP($R, L, n, \hat{L}, \tilde{L}, v, \mathbf{p}', \eta, \tau_m$)
  $\kappa = [-\infty]_{n+1}, \kappa[v] = 0, \theta' = [0]_{n+1}, \omega_2 = [-1]_{n+1}$
  $R = R \cup \{r_{\text{vir}}\}, L = L \cup \{l = (r_s, r_{\text{vir}}) | r_s \in \mathbf{p}'\}$
  $q_2 = \text{FiboHeap}(\kappa, \text{des})$, **for all** $r_h \in R$ **do** $q_2 \leftarrow h$
  while $|q_2| > 0$ **do**
    $w = q_2.\text{poll}()$
    for all $x \in \text{EgressNodes}(w)$ **do**
      if $(r_w, r_x) \in \hat{L} \vee x \notin q_2$ **then continue**
      $\tau_r = \theta'[w] + (\tau(w, x) + \tau(x, w))\eta$
      if $\tau_r > \tau_m$ **continue**
      compute $\epsilon_{\mathbf{p}}^{\eta}(v, x)$ in Eq. (9.3), where $\forall\, l \in \tilde{L}$, $\epsilon(l)$ is replaced with ϵ_{ex}
      if $\epsilon_{\mathbf{p}}^{\eta}(v, x) \geq \kappa[x]$ **then continue**
      $\omega_2[x] = w, \kappa[x] = \epsilon_{\mathbf{p}}^{\eta}(v, x), \theta'[x] = \tau_r$
      $q_2.\text{UpdateKey}(r_x)$
    end for
  end while
  $R = R \setminus \{r_{\text{vir}}\}, L \setminus = \{(r_s, r_{\text{vir}}) | r_s \in \mathbf{p}'\}$
  return $\text{ConvertPath}(\omega_2, r_v, r_{\text{vir}}) \setminus \{r_{\text{vir}}\}$
end function

Algorithm 9.3 Construct an MPSG

function ConMPSG(L, ω, X, i, j)
  $\mathbf{p}_{\text{prim}} = \text{ConvertPath}(\omega, i, j)$
  $X_{\text{rsp}} = \{r_k \rightarrow X[r_k] | r_k \in \mathbf{p}_{\text{prim}}\})$
  return $(\mathbf{p}_{\text{prim}}, X_{\text{rsp}})$
end function

relative to the parameters η and β, ensuring compliance with the maximum delay constraint D, where $n = |R|$ is the total number of routers.

In this algorithm, the construction of each RSP (lines 9–21) is an iterative process, ensuring that each RSP is maximally disjoint from previously computed paths in $\tilde{L}$, while adhering to the delay constraint τ_m. Notably, links in $\mathbf{p}'$ are excluded from the path computation to ensure resilience.

ii) **ComRSP algorithm**: The ComRSP algorithm, which operates under the delay constraint τ_m, computes an RSP from a router r_v to an upstream node along the path $\mathbf{p}'$. This computation occurs in the subgraph $(R, L \backslash \hat{L})$, where $\hat{L}$ is the set of links already considered in earlier computations. ComRSP selects an RSP that is maximally disjoint from each link in $\tilde{L}$, a set of previously identified critical links.

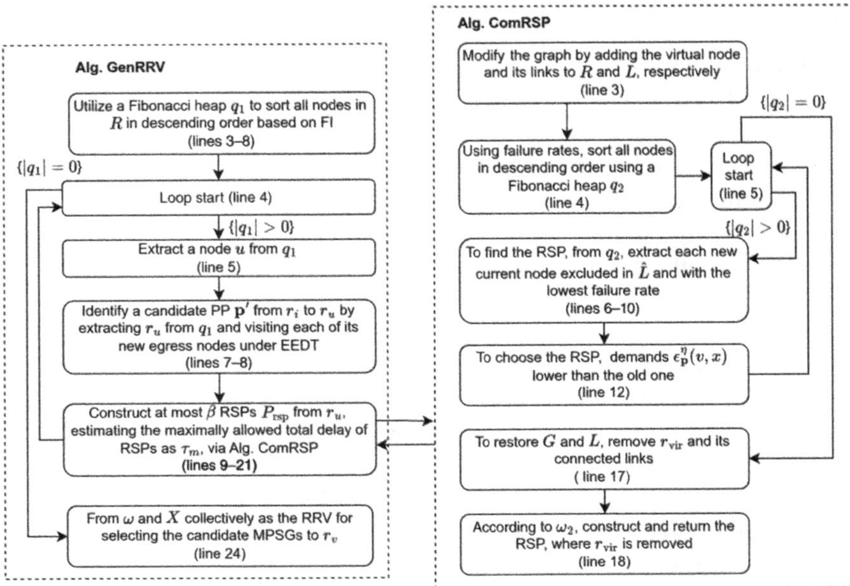

Figure 9.3 Workflow of the GenRRV and ComRSP algorithms.

For each hop, ComRSP ensures that up to η retransmissions occur, aiming to reduce the failure rate as defined in Eq. (9.3). The maximum time for these retransmissions is computed as per Eq. (7.2), ensuring the delay remains within the τ_m constraint.

In ComRSP, the node with the lowest failure rate is selected to ensure that the total delay of the RSP, τ_{r}, does not exceed τ_m (see line 9). The failure rate for each RSP is calculated as $\epsilon_{\mathbf{p}}^{\eta}(v, x)$, where ϵ_{ex} represents the worst-case link loss rate, a hypothetical value greater than the worst link loss in the network.

Finally, the GenRRV algorithm ultimately produces an RRV, which contains critical information for further constructing Multipath Survivable Groups (MPSGs) in subsequent stages of the network resilience strategy.

Based on the RRV, Algorithm 9.1 includes ConMPSG, which constructs an MPSG by finding a Primary Forwarding Path (PP) from r_j to r_k in line 2. For each router on the PP, it extracts its RSPs as X_{rsp} in line 3.

9.4 Computational Complexity

The scalability of the algorithm is analyzed as follows. In ComRSP, its execution time is as follows: (1) the lines 2–4 in $O(|R|)$, (2) the line 6 in $O(\log|R|)$, (3) the lines 7–14 in $O(|L|/|R|)$, and (4) the lines 15–16 in $O(\log|R|)$. Here, $|L| = |R| \cdot$

$\log|R|$ [Soltan and Zussman, 2016]. Hence, the total execution time of `ComRSP` is in $\nu = O(|R| \cdot \log|R| + |R| + \log|R|)$.

Based on `ComRSP`, `GenRRV` has the following time complexity: (1) the lines 1–3 in $O(\log|R|)$, (2) the line 5 in $O(\log|R|)$, and (3) the lines 6–20 in $O(|L|/|R| \cdot (\log|R| + \beta \cdot \nu))$. Therefore, the overall time is in $O((\beta+1) \cdot |R|^2 + (\beta+2) \cdot |R| \cdot \log|R| + (\beta+1) \cdot |R| + (\beta+1) \cdot \log^2|R| + (\beta+2) \cdot |R|)$, which is simplified to $O(|R|^2)$, lower than that of `GenMPSG` in $O(|R|^2 \cdot \log|R|)$ [Zhou et al., 2022].

The execution time of `ConMPSG` is in $O(\log|R|)$.

The execution space of `ComputeRSP` is analyzed as follows: (1) $O(|R|)$ in lines 2–15, and (2) $O(\log|R|)$ in line 16. Hence, the `ComputeRSP` space is in $O(|R| + \log|R|)$. `GenRRV` is analyzed below: (1) $O(|R|)$ in line 2–3, (2) $\log|R|$ in lines 8, and (3) $O(|R| + \log|R|)$ in lines 10–15. Hence, the `GenRRV` space is in $O(|R| + (\beta+1) \cdot \log|R|)$. Meanwhile, the space of `ConstructRR` is in $O(\beta \cdot \log|R| + |R|)$. These show HeuMPSG exhibits scalability in time and space.

9.5 Performance Evaluation

9.5.1 Experiment Settings

We have implemented HeuMPSG with its source code released in Zhou [2023] based on DRTP [Zhou, 2022], where HeuMPSG can generate MPSGs for each DRTP-enabled NDN forwarding daemon. We carried out the evaluations in the performance of HeuMPSG with DRTP by comparing it to the existing GenMPSG algorithm [Zhou et al., 2022; Zhou, 2022] via simulations in ndnSIM [Mastorakis et al., 2017]. The simulations target at the phasor data transport scenarios in the IoT networks built according to the real power transmission grids. The parameters of HeuMPSG are listed below: (1) D is 200 ms for non-critical applications, according to IEEE C37.118.2 [2011]; (2) β is 5 or 10 abundant enough; (3) η and ζ are 3 for sufficient reliability, being the same as GenMPSG in DRTP [Zhou et al., 2022]. The only parameter of GenMPSG β is also 5 or 10. All experiments were on a server with 2 Intel Xeon Gold 6248R CPUs and 376 GB memory.

The performance tests of HeuMPSG aimed to examine RSP redundancy, MPSG fragility, and generation efficiency using metrics, respectively, below: (1) the mean RSP number of an MPSG is the division between the total number of its RSPs and the number of its PP hops minus 1, (2) FI and CPRF, and (3) the mean algorithmic execution time. These tests target 300 randomly chosen pairs in the IEEE 300-bus system large enough, under the different numbers of random link faults (NFL). We also tested EEFR, EEDT, PIAT, and the EEDT satisfactory ratio. EEDT of a lost Capsule is assumed to be impossible 10 s. The ratio is the division between the number of Capsules received with EEDTs not greater than the maximum EEDT in Eq. (7.9) and the total number of Capsules, revealing how well the recursiveness can hold.

To show the feasibility, we evaluate HeuMPSG with DRTP in EEFR and EEDT under harsh fault links via comparison to other solutions in three categories below: (1) the current IEEE C37.118.2 adopts TCP allows to recover loss packets [IEEE, 2011]; (2) the recent hop-by-hop transport protocols based on the single path include R2T [Wang et al., 2018], MFTP [Su et al., 2015], and SDATP [Chen et al., 2020]; and (3) DRTP adopts the MPSG routes [Zhou et al., 2022; Zhou, 2022].

9.5.2 MPSG Generation Performance

The performance results of HeuMPSG and GenMPSG are shown as increases of buses. Figure 9.4(a) depicts HeuMPSG, which has a better RSP redundancy. When β is 5 and 10, the mean RSP number of HeuMPSG is higher than that of GenMPSG by (66.74, 69.38), (85.88, 105.59), (80.77, 94.61), (119.84, 153.11), and (115.7, 171.51)%, respectively. Figure 9.4(b) shows that for β of 5 and 10, in the 30 and 57 buses, their FIs are almost the same for each other. For the 118, 300, and 500 buses, the corresponding FIs of HeuMPSG are

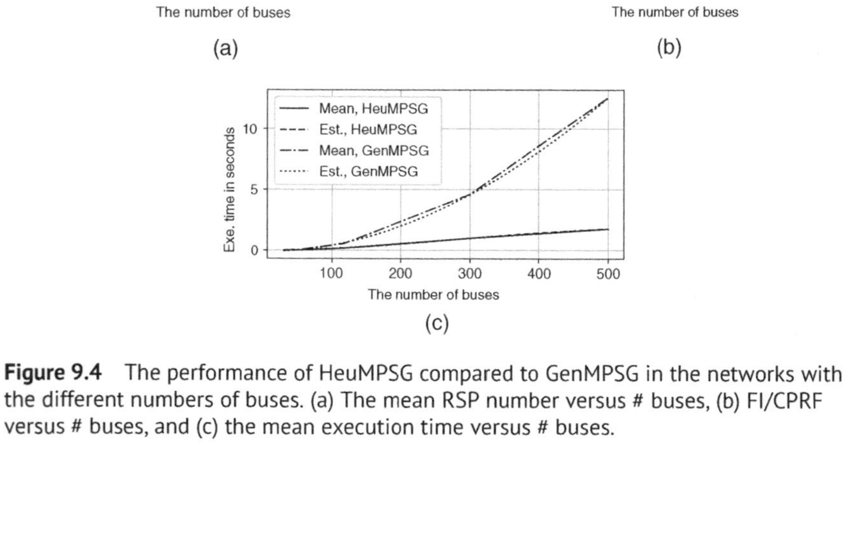

Figure 9.4 The performance of HeuMPSG compared to GenMPSG in the networks with the different numbers of buses. (a) The mean RSP number versus # buses, (b) FI/CPRF versus # buses, and (c) the mean execution time versus # buses.

lower by (0.14, 0.15), (0.24, 0.31), and (0.24, 0.32)%, respectively. For all buses, the CPRFs of HeuMPSG are lower by (0.22, 0.23), (0.32, 0.34), (0.48, 0.49), (0.61, 0.68), and (0.59, 0.68)%, respectively. They indicate that HeuMPSG gains reduced MPSG fragility. Figure 9.4(c) depicts HeuMPSG is 1.65, 2.2, 2.92, 4.58, and 7.07 times faster than GenMPSG for 30, 57, 118, 300, and 500 buses, respectively. The execution times of HeuMPSG and GenMPSG are consistent with their estimates, respectively, below: (1) $O(-1.6710^{-5}|R|^2 + 1.0810^{-2}|R| \log|R| - 6.0110^{-2}|R| + 6.5110^{-2}\log^2|R| - 7.6110^{-3}|\mathrm{R}|)$, and (2) $6.5910^{-7}|R^{2} \log^2|R| + 8.7110^{-4}|R|\log^2|R| - 3.4110^{-3}|R| \log|R|$ [Zhou et al., 2022]. The results indicate efficiency.

9.5.3 Resilience Improvement Performance

The resilience improvement brought by HeuMPSG to DRTP is shown with comparison to that by GenMPSG in Figure 9.5. Under 50% LLR, Figure 9.5(a) depicts that the mean EEFRs of HeuMPSG are lower in 5.5%, 26.08%, 23.49%, and 6.63%, when NFLs are 10, 30, 70, and 110, respectively. Under 100% LLR, Figure 9.5(b) indicates the mean EEFRs of HeuMPSG are lower by 27.22% and 29.11%, when NFLs are 10 and 20, respectively, where for no less than 30 NFLs, their EEFRs are 100%. Under 50% LLR, Figure 9.5(c) depicts that HeuMPSG's EEDTs are lower by 416.82, 2173.65, 1954.08, and 543 08 ms, for NFLs of 10, 30, 70 and 110, respectively. Figure 9.5(d) depicts that HeuMPSG's EEDTs under 100% LLR are lower by 2248.65 and 2410.89 ms, for NFLs of 10 and 20, respectively, where for no more than 30 NFLs, their EEDTs are absent. For 30, 70, and 110 NFLs, Figure 9.5(e) shows that the EEDT satisfaction ratios of HeuMPSG under 50% LLR are higher by 28.45, 50.44, and 32.75%, respectively. For 10 NFLs, the ratios of HeuMPSG and GenMPSG under 100% LLR are 99.18% and 99.58%, respectively. For [20, 60] and [20, 30] NFLs, the ratios are all 100%. Such ratio differences confirm the better recursiveness of HeuMPSG. Figure 9.5(f) shows that the mean PIATs of HeuMPSG under 50% LLR are reduced by 28.45, 50.44, and 32.75%, for NFLs of 10, 30, 70, and 110, respectively. Under 100% LLR, their PIATs are similar to each other. The results manifest that in EEFR, EEDT, and PIAT, the improvements of HeuMPSG to DRTP are all remarkable.

9.5.4 Comparative Studies with Existing Solutions

We show that with HeuMPSG, the resilience of DRTP stands out above other existing solutions with significantly lower EEFR and EEDT for 30 pairs selected randomly for generality in the IEEE 300-bus system, despite the harsh link faults. In particular, HeuMPSG and GenMPSG generate their own MPSGs under the severe condition of 10 NFL and 50% LLR. Under the same condition, TCF, R2T,

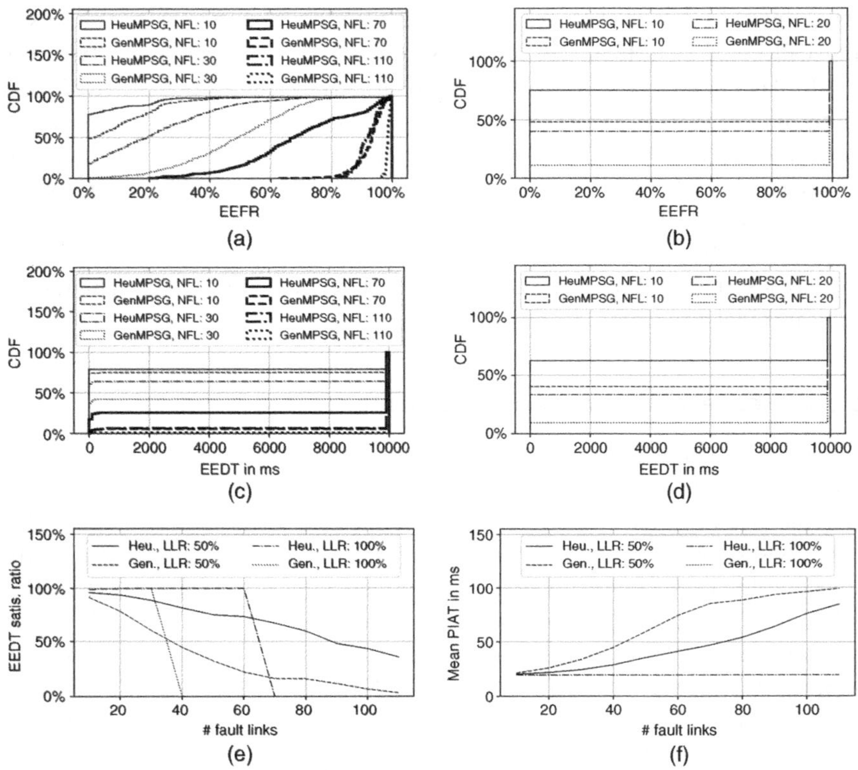

Figure 9.5 The resilience improvements of HeuMPSG to DRTP compared to GenMPSG in the network of the IEEE 300-bus system. (a) EEFR CDF under 50% LLR, (b) EEFR CDF under 100% LLR, (c) EEDT CDF under 50% LLR, (d) EEDT CDF under 100% LLR, (e) EEDT satisfactory ratio, and (f) mean PIAT.

MFTP, and SDATP are reliant on the shortest PP of the MPSG generated by HeuMPSG, where PP experiences certain faulty links with 50% LLR for severe tests. Additionally, each test lasts for 13 s where faults occurred between 3 s and 13 s, which is enough to capture performance dynamics. Figure 9.6 shows that the mean EEFRs of IEEE C37.118.2 in TCP, R2T, MFTP, SDATP, DRTP with HeuMPSG, and DRTP with GenMPSG are 98.2%, 42.12%, 59.11%, 58.55%, 4.61%, and 10.11%, respectively. At the same time, Figure 9.7 depicts that the mean EEDTs of them are 9.97s, 5.18s, 6.59s, 6.55s, 2.1s, and 2.51s, where we treat EEDT of a lost payload packet to an impossible value of 10s. With the generated MPSGs being less fragile (see Section 9.5.2), these above results show that DRTP with HeuMPSG is superior in EEFR and EEDT.

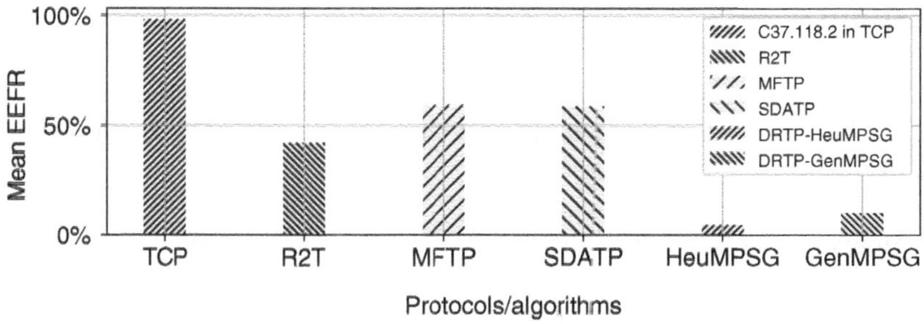

Figure 9.6 EEFR comparison in 300-bus system for 10 NFL and 50% LLR.

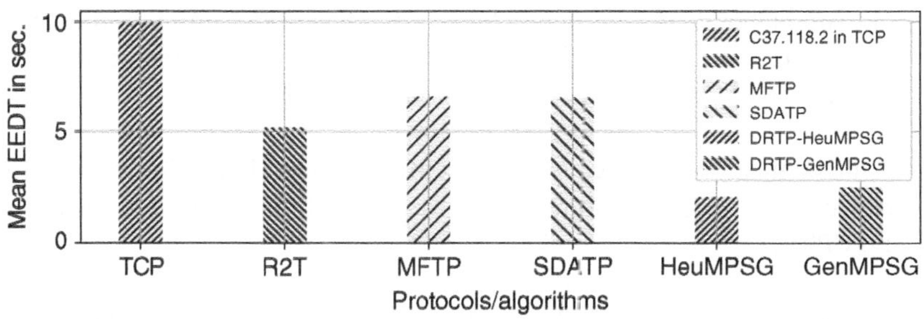

Figure 9.7 EEDT comparison in 300-bus system for 10 NFL and 50% LLR.

9.6 Summary

In our quest to bolster the resilience of DRTP for phasor data, we presented a heuristic algorithm designed to efficiently generate less fragile MPSGs. The algorithm mitigates failures during the hop-by-hop retransmissions carried out by DRTP. The evaluation results, obtained through simulations on real power grid networks, clearly illustrate the remarkable enhancements brought about by HeuMPSG. Under severe random link faults, HeuMPSG not only significantly improves the resilience of DRTP by reducing EEFR and EEDT but also offers the advantage of faster execution times. These improvements far surpass the existing solutions, including the currently adopted IEEE C37.118.2 in TCP, the original GenMPSG algorithm of DRTP, and various other hop-by-hop transport protocols. Our study underscores the importance of generating suitable routes to augment the resilience of in-network retransmissions, a critical advancement with vast potential for future data transport in redundant IoT networks.

Questions

9.1 Regarding the resilience improvements in EEFR and EEDT, what challenges do the various existing solutions face, and how does HeuMPSG address or overcome these challenges?

9.2 How does the heuristic algorithm HeuMPSG contribute to addressing the challenges of in-network retransmissions, and what potential does it hold for the future of data transport in redundant IoT networks based on the study's findings?

9.3 Can you elaborate on the significance of the mean RSP number in MPSG generation performance and how HeuMPSG's RSP redundancy compares to that of GenMPSG across different numbers of buses?

9.4 In the context of resilience improvement, how does HeuMPSG manage to reduce MPSG fragility, and what are the practical implications of lower FI and CPRF?

9.5 How does the resilience of DRTP, equipped with HeuMPSG, compare to existing solutions like TCP, R2T, MFTP, and SDATP, particularly in terms of EEFR and EEDT under harsh link fault conditions?

9.6 Can you explain the rationale behind choosing specific parameters such as D, β, η, and ζ in the experiment settings, and how sensitive are the performance metrics to changes in these parameters?

9.7 Considering the execution time comparison between HeuMPSG and GenMPSG, can you discuss the implications of HeuMPSG being several times faster for various network sizes, and how might this impact real-world applications?

Bibliography

Guneet Bedi, Ganesh Kumar Venayagamoorthy, Rajendra Singh, Richard R. Brooks, and Kuang-Ching Wang. Review of Internet of things (IoT) in electric power and energy systems. *IEEE Internet of Things Journal*, 5(2):847–870, 2018.

Jiayin Chen, Qiang Ye, Wei Quan, Si Yan, Phu Thinh Do, Weihua Zhuang, Xuemin Sherman Shen, Xu Li, and Jaya Rao. SDATP: An SDN-based adaptive transmission protocol for time-critical services. *IEEE Network*, 34(3):154–162, 2020.

J. J. Garcia-Luna-Aceves and Abdulazaz Ali Albalawi. Connection-free reliable and efficient transport services in the IP Internet. In *IEEE CNSM*, pages 1–7, 2020.

IEC 60794-4-10. Optical fibre cables - Part 4-10: Family specification - optical ground wires (OPGW) along electrical power lines, 2014.

IEEE C37.118.2. IEEE standard for synchrophasor data transfer for power systems, 2011.

Spyridon Mastorakis, Alexander Afanasyev, and Lixia Zhang. On the evolution of ndnSIM: An open-source simulator for NDN experimentation. *ACM SIGCOMM Computer Communications Review*, 47(3):19–33, 2017.

NDN. NDN-0021, NFD Developer's Guide. Technical report, NDN, 2021. URL https://named-data.net/.

S. Soltan and G. Zussman. Generation of synthetic spatially embedded power grid networks. In *IEEE PESGM*, pages 1–5, 2016.

Kai Su, Francesco Bronzino, K. K. Ramakrishnan, and Dipankar Raychaudhuri. MFTP: A clean-slate transport protocol for the information centric MobilityFirst network. In *ACM ICN*, pages 127–136, 2015.

Bin Wang and J. C. Hou. Multicast routing and its QoS extension: Problems, algorithms, and protocols. *IEEE Network*, 14(1):22–36, 2000.

Zhaoxu Wang, Hongbin Luo, Huachun Zhou, and Jiawei Li. R2T: A rapid and reliable hop-by-hop transport mechanism for information-centric networking. *IEEE Access*, 6:15311–15325, 2018.

Lixia Zhang, Alexander Afanasyev, Jeffrey Burke, Van Jacobson, Christos Papadopoulos, K. C. Claffy, Lan Wang, Patrick Crowley, and Beichuan Zhang. Named data networking. *ACM SIGCOMM Computer Communication Review*, 44(3):66–73, 2014.

Boyang Zhou. The source codes of the disruption-resilient transport protocol (DRTP) and the GenMPSG algorithm, 2022. URL https://github.com/zhouby-zjl/drtp.

Boyang Zhou. The source code of the HeuMPSG algorithm, 2023. URL https://github.com/zhouby-zjl/heumpsg.

Boyang Zhou, Chunming Wu, Qiang Yang, and Xiang Chen. DRTP: A disruption resilient hop-by-hop transport protocol for synchrophasors measurement in electric transmission grids. *IEEE Access*, 10:133898–133914, 2022.

10

Crafting Application Data: Unleashing Adaptive RaptorQ Coding

In the 4G LTE private network, wireless interference amplifies the channel packet loss rate and introduces delays in transmitting control commands to terminals within the distribution power grid. This, in turn, undermines control reliability. While recovering lost packets is achievable through RaptorQ (RQ) coding and real-time optimization of RQ coding parameters, the challenge lies in the non-linearity of the RQ failure probability model.

Considering the scenario of power grid control, this chapter takes advantage of the abundant bandwidth of the wireless channel while recognizing the low data transmission rate of three-remote data. It proposes an Adaptive RaptorQ Encoder and Decoder (ARED) with variable-length message redundancy codes to enhance the reliability of LTE wireless communication access for power grid control terminals. By utilizing idle transmission bandwidth, the proposed approach improves transmission reliability by increasing the code length, strengthening the Forward Error Correction (FEC) capability of the Binary Erasure Channel (BEC) in distribution terminals. Additionally, it establishes an IP tunnel layer with RQ redundancy adjustability between the sender and receiver. In situations with a high packet loss rate, the system dynamically increases the RQ redundancy code length using the remaining LTE bandwidth, enhancing automatic recovery of lost data packets and resisting noise interference. Conversely, in scenarios with a low packet loss rate, the system reduces the RQ code length, thereby lowering the power consumption of ARED devices.

To address the challenge of dynamically adjusting RQ encoding parameters and overcoming the nonlinearity of the RQ decoding error rate model, a suboptimal RQ encoding error model needs to be established. This model aims to dynamically adjust RQ encoding parameters under the constraints of latency and bandwidth, based on the varying packet loss rate of the channel, with the goal of minimizing the throughput of the transmission. In this regard, the paper proposes the design of a Raptor encoder and decoder that can adapt to wireless channel quality, featuring adjustable redundancy code lengths. These Raptor encoders and decoders

Communication Networks in Smart Power Grids, First Edition. Boyang Zhou.

Companion Website: www.wiley.com/go/SmartPowerGrids1e

are deployed at the front end of the terminal network interface and between the master station and the wireless core network, respectively, forming an adjustable redundancy IP tunnel layer. Telemetry, telecontrol, and remote signaling data from both the terminal and the master station are transmitted through this IP tunnel layer. In the presence of noise interference, the IP tunnel layer senses changes in data transmission quality, swiftly adjusts the encoding length, selects the minimum corrective encoding length, reduces the packet loss rate of IP data packets, ensures that packet arrival times fall within the maximum delay range, thereby enhancing the transmission success rate of upper-layer IP services. This approach aims to seamlessly integrate with existing distribution network devices and LTE communication equipment.

Based on a typical wireless topology structure of a distribution network, this chapter constructs a simulation experiment environment using the NS-3 LTE simulator [Piro et al., 2011]. The simulation models communication between a base station and wireless terminals in the distribution network with a distance of 5 km. The results indicate that in the uplink, with a Noise Figure (NF) in the range of [18.6, 20.5] dB, the packet loss rates for ARED and the original LTE are 7.8% and 21.82%, respectively. ARED reduces the packet loss rate by 64.25%. In the downlink, with NF in the range of [25.6, 28.7], the packet loss rates for ARED and the original LTE are 1.7% and 16.86%, respectively. ARED decreases the packet loss rate by 89.92%. Additionally, as NF increases, the average utilization of both uplink and downlink channels approaches 79.58%, meeting the 99% confidence interval values for the Poisson distribution of distribution network business flows. Leveraging the characteristics of low throughput and ample redundant bandwidth in power distribution network business flows, ARED significantly improves LTE's resistance to interference.

10.1 Problem Modeling and Analysis

As Figure 10.1 shown, distribution network terminals connect to base stations (LTE eNodeB) through wireless communication terminals (LTE UE) via the air interface. The eNodeB processes the baseband signals of accessing terminals to generate business data, which is then transmitted to the EPC for further processing. The EPC connects to the distribution network master station to send and receive business data through business ports. In a distribution network BEC, IP business data flows with a length of k are transmitted from the sender. Due to the relatively fixed transmission process of distribution network three-remote data, the arrival intervals of adjacent *IP* data packets follow a Poisson distribution, as shown in Figure 10.2, which illustrates the data flow characteristics of distribution network terminal business in a specific city. An analysis of the cumulative

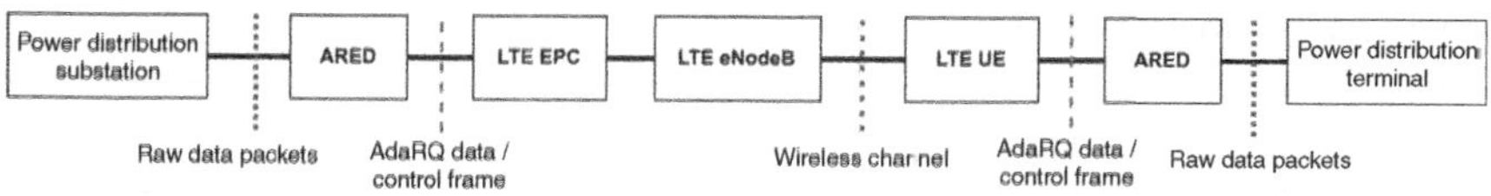

Figure 10.1 Illustration of LTE terminal deployment in power distribution networks.

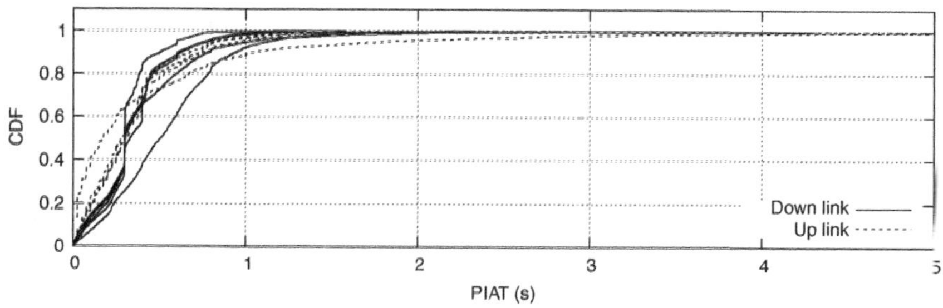

Figure 10.2 Packet Inter-Arrival Time (PIAT) CDF of typical measurement and control service flows in power distribution networks.

probability density distribution (CDF) of arrival times between the distribution network master station and terminals is derived.

Assuming the transmission rate of X is λ packets per second (pps), the available channel bandwidth is B pps, and the maximum channel bandwidth is $B_{\max}$ pps, with a channel packet loss rate of L. The upper limit of the remaining bandwidth on the channel, when interference occurs, is $B - \lambda$. In the presence of interference, L exhibits intermittent and bursty packet losses, displaying random packet loss when interference occurs. Table 10.1 lists the commonly used data symbols and explanations relevant to the remainder of this chapter.

In a wireless BEC channel, ARED encodes multiple business data packets from the sender into an RQ encoding block X_p, which can be encoded by RQ into a coding symbol block $Y_p = \text{RQ}(X_p)$. The receiving end decodes and corrects Y_p, generating an estimate $\hat{X}_p$ of X_p. The occurrence of $\hat{X}_p \neq X_p$ is defined as the decoding error probability when the message redundancy code length is o, denoted as $\bar{f}_k(o)$. In the presence of noise interference, the error rate of signal demodulation in the uplink and downlink increases, resulting in packet loss in X_p and an increase in $f_k(o)$. When δ has a constant value, the magnitude of $f_k(o)$ is inversely proportional to o and approximately linearly related to L, similar to the decoding error rate of k-symbol-random fountain codes. Therefore, $f_k(o) = O(\delta/k^{(1-\delta)\cdot o+\log_2 k})$. Additionally, the time complexity of RQ encoding and decoding is $O((k+o)\cdot \cot\log(k/o))$.

As the link state changes are unpredictable and L measurement results are difficult to achieve in real time, ARED's key issue is described as follows: given a Poisson business flow X_p, under bandwidth constraints, dynamically change the

Table 10.1 Mathematical notations.

Notation	Description
X_p	RQ source symbol block, $X_p = \{x_1, \cdot, x_k\}$, where x_i represents the i-th source coding symbol
Y_p	RQ coding symbol block, $Y_p = (x_1, \cdot, x_k, z_1, \cdot, z_o)$, where x_i is the source code symbol and z_i is the repair code symbol. x_i and z_i are all values in the Galois domain. The length of Y_p can be calculated as $n = o + k$
p	Identification (ID) of starting symbol of an RQ block
k	The length of an AdaRQ data code block
f	The data flow Identification (ID)
i	ESI that refers to the encoding symbol Identification (ID) in IETF RFC 6330, which can be represented by bit clock in any length
o	The length of an AdaRQ repair code block
L	The drop ratio of AdaRQ data frames
$\hat{L}$	The estimation of L
L_{max}	The packet loss rate of AdaRQ data blocks, maximally tolerant
λ	The transmission speed of AdaRQ data frames
$\hat{\lambda}$	The estimation value of λ
α_λ	The Confidence Ratio (CR) of a random variable λ
τ_{max}	The Half Round-Trip Time (HRRT) in a single direction, maximally tolerant for service flows
$\tau_p(k, n)$	The processing delay of ARED coding under its parameters of k and n
S	The symbol data used for coding in RQ
C, D	Constants
W_0	Lambert W function with 0 value
k_{min}	The minimum length of RQ coding, referred in IETF RFC 6330
Poisson	The cumulative distribution function in Poisson
$\text{Poisson}_{\alpha_\lambda}^{-1}(x)$	The inverse function of the cumulative distribution function in Poisson with the Confidence Ratio (CR) of α_λ and the parameter of x
Norm	The norm cumulative distribution function
B_{max}	The maximum bandwidth

encoding rate of RQ, k/n, based on the dynamic changes of L. This is done to minimize both $f_k(o)$ and the encoding processing delay, avoiding sending redundant data that may cause network congestion. Considering the constant forwarding delay, this problem can be formulated as Formula 10.1.

$$\min\{\mathrm{argmin}_{n,k}(f_k(o), L_{\max})\}, \text{subject to :}$$
$$\mathrm{Poisson}^{-1}_{\alpha_\lambda}(\lambda \cdot n/k) \le B_{\max} \wedge \tag{10.1}$$
$$n/\mathrm{Poisson}^{-1}_{\alpha_\lambda}(\lambda \cdot n/k) + \tau_p(k,n) \le \tau_{\max}$$

10.2 Adaptive RaptorQ Encoder and Decoder

10.2.1 Architecture

The structure of the Adaptive RQ (ARED) coder and decoder, as shown in Figure 10.3, consists of the Adaptive RQ (AdaRQ) module for self-adjusting RQ encoding, RQ encoder, RQ decoder, packet sender, and packet receiver. ARED operates on a general computing platform with two duplex physical network interfaces provided by the computing platform: the host-side physical network port and the wireless-side physical network port. Both network ports connect to the packet receiver and packet sender. The packet receiver and sender use standardized IP network interface software or hardware, while the RQ encoder and decoder leverage existing software and hardware systems compliant with IETF RFC 6330 standards for implementation. The AdaRQ module achieves RQ encoding parameter adjustments, comprising RQ channel feedback unit and encoding parameter optimization unit.

ARED establishes AdaRQ Dynamic Coding Tunnels (ADCT) on the sender and receiver sides, enabling dynamic RQ encoding parameter transmission of

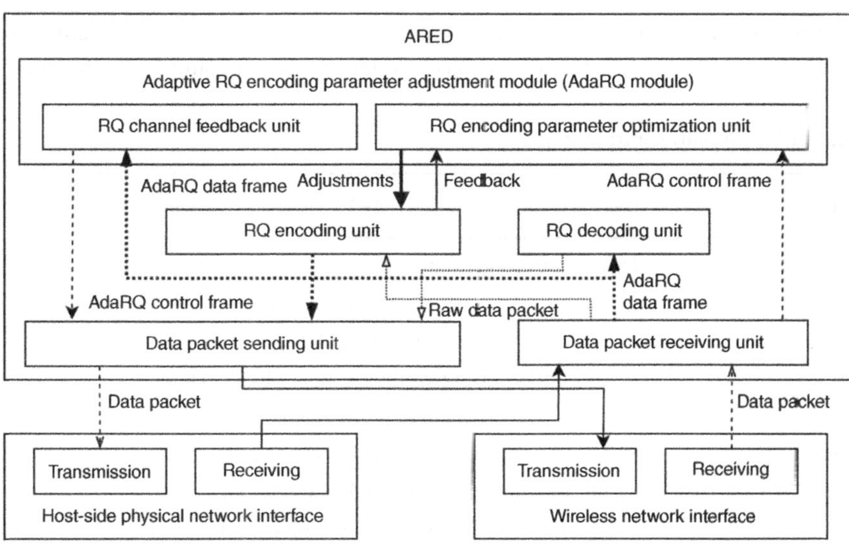

Figure 10.3 Architecture of ARED.

RQ encoding symbol blocks. ADCT comprises data frames and control frames. The RQ channel feedback unit generates control frames based on received data frames and sends them to the host-side physical network interface through the packet sender unit. The RQ encoding parameter optimization unit receives control frames and data frames from the packet receiver unit, along with feedback information from the RQ encoding unit. It then generates encoding adjustment parameters for the RQ encoder, enhancing interference resistance by producing RQ adjustment parameters matching channel capacity and quality when channel quality deteriorates. To ensure real-time forwarding of source data, the modules use a pipeline execution approach, and the receiving end employs a coding packet scheduling strategy based on business priority and low latency.

10.2.2 Dynamic Coding Tunnel Protocol

ADCT is based on the IPv4 protocol, with the assigned IP protocol number defined as 0x9F. When the IPv4 protocol field value is 0xFD, ARED activates the decoder to restore transmitted data; otherwise, it directly forwards the data, ensuring compatibility with the original IP protocol. ADCT consists of data frames and control frames, implementing RQ symbol payload and channel quality feedback functions. The first two bits of both frame types are the frame type, taking values 01 or 11 and 10. Bits 2–16 represent the business flow Identification (ID) to distinguish different services. When the frame type of a data frame is 11, it indicates both a data frame and a control frame, with a control frame field added at offsets 10–14.

1) Data frame structure, as shown in Table 10.2, can be formally represented as $Y_p(f, i, p, k, o, \tau_{\max}, S)$. Each field has a length of 16 bits, and the data frame

Table 10.2 AdaRQ control format.

offset	0–1	2	3–15
0	frame type	urgent bit	stream ID
2	priority		the maximum delay
4	the symbol ID		
6	the starting symbol ID in the RQ block		
8	the length of the RQ block		
10	the length of the RQ repair code		
12	the number of RQ symbol losses (optional)		
14	the inter-arrival time of consecutive symbols in RQ		
12/16~MTU-44/48	RQ symbol data		

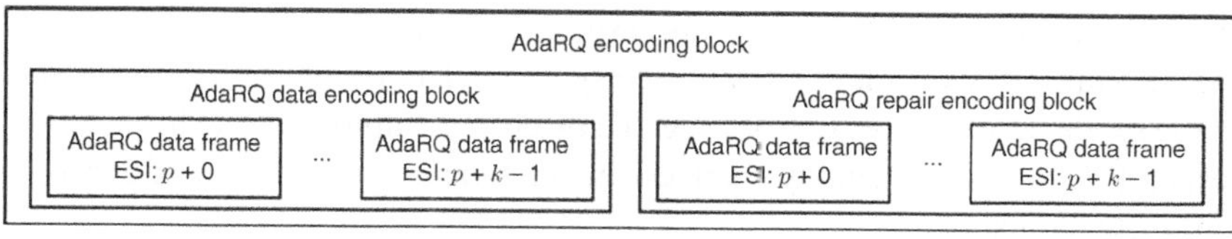

Figure 10.4 Architecture of ARED.

Table 10.3 AdaRQ data format.

offset	0–1	2	3–15
0	frame type	urgent bit	stream ID
2	the number of losses in RQ symbols		
4	the throughput of RQ symbols		

allows a maximum of 2^16-sized encoding symbol blocks. As Y_p is transmitted, the value of ESI (p+i) increments from 0 to $2^16 - 1$ and then resets to 0. The receiver forms an AdaRQ-encoded data block upon receiving symbol Y_p (see Figure 10.4), and after successful decoding, Y_p is sent to the receiving end.

For each transmitted data frame, the IP layer ID (IETF RFC 791) is incremented by 1 to recognize the temporal and unique arrival sequence of data packets. To prevent packet misordering, which could lead to the overwriting of Y_p^X by $Y_p^{X'}$ (X and X' being two temporally close packets), the receiver determines the order of IP data packet arrival by inspecting Y_p and the IP ID. It then places Y_p^X back into the corresponding position in the receiving buffer, avoiding the reception of the same Y_p and potential RQ decoding failure.

2) Control frames. $C(f, \lambda, L)$ are sent at fixed intervals or based on events, as shown in Table 10.3. These frames convey feedback on the number of lost RQ symbols and the interval between consecutive symbol arrivals. The sending of control frames can be initiated independently by the receiver to inform the sender, or the sender can use the ADCT protocol to convey them to the receiver through data frames with frame type 11. Upon receiving these frames, the receiver passes λ and L to the AdaRQ algorithm, updating the encoding strategy of RQ.

10.2.3 RaptorQ Channel Feedback Unit

The RQ channel feedback unit operates on the receiving side, continuously measuring the throughput estimate $\hat{\lambda}$ of the data stream Y_p using methods like sliding averages or time series analysis. It also estimates the packet loss rate $\hat{L}$ by detecting the continuity of AdaRQ data frame symbols. The results $C(f, \hat{\lambda}, \hat{L})$ are sent to the sender at fixed intervals or triggered by events, improving synchronization, responding to channel quality changes, and enhancing the accuracy of encoding parameter adjustments.

10.2.4 RaptorQ Encoding Parameter Adjustments

The RQ encoding parameter adjustment optimization unit operates on the sending side. It receives $C(f, \hat{\lambda}, \hat{L})$ from the RQ Channel Feedback Unit, extracts $\hat{L}$, and uses it to initiate the AdaRQ algorithm, generating AdaRQ encoding structure features (k, o). The AdaRQ algorithm relies on optimal estimates of RQ encoding parameters k and o to minimize the result of formula 1, as given by Theorem 10.1.

Theorem 10.1 Assuming the throughput of the source encoding stream is $\hat{\lambda}$ and RQ encoding parameters satisfy $o \leq k \leq k_{\max}$, with given $\tau_{\max}$ and an upper limit on RQ encoding processing and forwarding delay $\tau_p(k, o)$, the RQ encoding parameter k should satisfy the range specified in Formula 10.2, where W_0 is the Lambert W function with a value of 0, and C and D are constants, and $k_{\min}$ is the minimum value for parameter k, representing the minimum length of RQ encoding.

$$k \leq \frac{1}{1+Z} \exp\left(W_0 \left(\frac{\tau_{\max} - D}{C + 1/\left(\log(1+Z) \cdot \text{Poisson}_{\alpha_\lambda}^{-1}\left((1+Z)\hat{\lambda}\right) k_{\min}\right)} \right)\right) \tag{10.2}$$

Proof: When the RQ encoding redundancy length $\sigma \geq 2\hat{\delta}_{\max} k / \left(1 - 2\hat{\delta}_{\max}\right)$ [Shokrollahi, 2006], the improvement of RQ encoding on channel quality becomes unclear. Simultaneously, RQ encoding allows a maximum channel error rate of 50%. Since $\sigma \leq k \leq k_{\max}$, formula 1 leads to $F \cdot e^t + C \cdot t \cdot e^t \leq \tau_{\max} - D$, where F is defined in Eq. (10.3), $t = \log((1+Z)k)$, and Z is defined in Eq. (10.4). Based on the above derivation, it can be further obtained that $t \cdot e^t \leq \left(\tau_{\max} - D\right) / (C + D)$, where M is defined in Eq. (10.5). Using the Lambert W function $W_0(x) = \sum_{i=1}^{\infty} \left(-i^{i-1}/i!\right) \cdot x^i$, Formula 10.2 is derived, thus proving the statement.

$$F = 1/\text{Poisson}_{\alpha_k}^{-1}\left((1+Z)\hat{\lambda}\right) \tag{10.3}$$

$$Z = 2\hat{L}_{\max} / \left(1 - 2\hat{L}_{\max}\right) \tag{10.4}$$

$$M = 1/\left(\log(1+Z) \cdot \text{Poisson}_{\alpha_\lambda}^{-1}\left((1+Z)\hat{\lambda}\right) \cdot k_{\min}\right) \tag{10.5}$$

The theoretical packet loss rate of RQ(k, o) approximates Eq. (10.6) and is related to the source encoding probability distribution. However, actual measurement results exhibit significant fluctuations. Therefore, a heuristic algorithm is employed to adjust RQ encoding parameters. On the sender side, the AdaRQ algorithm dynamically adjusts RQ parameters k and o based on variations in λ and L. Given Eq. (10.7) (where k and o are trained and fitted during device initialization by measuring RQ encoding and decoding runtime), the specific steps of the AdaRQ algorithm are as follows:

$$d(L, o, k) = L \cdot f_k(o) = C \cdot L^2 / k^{(1-L)\cdot o + \log_2 k} \tag{10.5}$$

$$\tau_p(k, o) = C \cdot (k + o) \log(k + o) + D \tag{10.7}$$

1) Upon establishing an ADCT tunnel request, calculate parameter k using Formula 10.2 based on $\tau_{\max}$ and $\hat{\delta}_{\max} = 1$.
2) Based on the received $C(f, \hat{\lambda}, \hat{L})$ message, gather recent data stream throughput values $M = (\lambda_1, \cdot, \lambda_g)$ and packet loss rate values $L = (\delta_1, \cdot, \delta_g)$. Establish Poisson$(\overline{M})$ and Norm$(\overline{L})$ regression models, enabling the sender to detect trends in λ and δ. By assuming the mean values of these measurements, verify whether channel quality has changed. When the throughput $\overline{M_{\text{cur}}}$ or packet loss rate $\overline{L_{\text{cur}}}$ is outside the confidence intervals of Poisson$(\overline{M})$ and Norm$(\overline{L})$, determine that the channel quality has changed.
3) Upon detecting a change in channel quality, calculate the estimated percentiles of the normal distributions for M and L as $\hat{\delta}_{\max}$ and $\hat{\lambda}_{\max}$ respectively. Use Formula 10.2 to calculate the theoretical value of k. Simultaneously, calculate the theoretical length of the current RQ modified encoding as $o_{\text{th}} = Z \cdot k$. Compute the current theoretical block size allowed by $B_{\max}$ and the encoding rate and test whether it satisfies Formula 10.8. If o_{th} does not satisfy, set o according to Eq. (10.9); otherwise, set o to o_{th}.

$$\text{Poisson}_{\alpha_\lambda}(\lambda)(o + k)/k \le B_{\max} \tag{10.8}$$

$$\sigma = \lfloor B_{\max} k / \text{Poisson}_{\alpha_\lambda}^{-1}(\lambda) - k \rfloor \tag{10.9}$$

4) After the current data frame is transmitted by the RQ encoding unit, set the new value of o to the RQ encoding unit for data frame transmission.

The AdaRQ algorithm, described above, dynamically adjusts RQ parameters to update the encoding strategy in real time, ensuring that the interval between packet arrivals does not exceed $\tau_{\max}$. When the encoding rate n/k exceeds the allowed bandwidth range, AdaRQ reduces the encoding rate to prevent channel congestion. In case of intermittent channel failures, extending the source encoding of length k to an RQ encoding of length n improves the success rate of data packet transmission. When channel quality improves, AdaRQ reduces the encoding rate, freeing up channel resources to avoid congestion.

10.2.5 Wireless Spectrum Optimization Techniques

To complement the RQ presented above, fundamental wireless spectrum optimization techniques can be employed to further improve the reliability of wireless data transmission. The main causes of radio frequency interference in the distribution and utilization of wireless communication network systems include: occupying existing frequency resources, improper configuration of different wireless networks, transmitter setup issues, overlapping cell areas, environmental factors, and electromagnetic compatibility (EMC).

Since public networks use the infrastructure of mobile operators, if the security situation awareness system of the power distribution wireless network detects that the interference is caused by wireless network interference leading to bit errors, it will notify the mobile operator to modify the network's spectrum and configuration, or identify the source of interference. If the interference is caused by an attack due to base station hijacking, it will notify the mobile operator to inspect and clear any fake base stations.

For private networks, which use self-built networks and radio frequency equipment from manufacturers like Huawei, a management system and policy need to be established to standardize the parameters of radio frequency equipment. This includes systematic research on frequency guard bands, filtering accuracy, site spacing, antenna installation optimization, etc., to form an active anti-interference wireless spectrum coordination and allocation mechanism for self-distribution wireless communication networks.

1) **Frequency guard band**: The solution of increasing the frequency guard band is achieved through frequency planning, ensuring a certain level of separation between the transmission band of the interfering system and the reception band of the interfered system in the frequency domain. As the separation increases, the components of the interfering system's transmission signal that fall into the receiving bandwidth of the interfered system decrease, while the attenuation of the interfering system's transmission signal by the receiving system's filter increases, thereby reducing inter-system interference.

   An appropriately sized frequency guard band can effectively mitigate interference issues. At the same time, when considering the use of additional filters to limit interference signals, an ideal linear filter is difficult to achieve. Therefore, a certain guard band should be reserved as a transition zone for the filter. However, due to the scarcity of frequency resources and the different frequency response characteristics of transmission and reception filters, other interference mitigation solutions should also be considered when using a guard band, aiming to minimize the bandwidth of the guard band as much as possible.
2) **Filter precision**: The solution to improving filter precision involves enhancing the filtering characteristics of the transmitter or receiver in the existing wireless

transceiver system by using high-precision filters or additional filters. This aims to achieve the level of isolation required for inter-system coexistence. Increasing filter precision is an effective way to resolve interference issues, but it also comes with the trade-off of higher costs.

3) **Site spacing**: Increasing site spacing can effectively reduce interference, but this method is limited by the scarcity of site resources and the coexistence of multiple operators, making practical implementation in engineering projects more challenging.
4) **Antenna installation optimization**: Optimizing antenna installation involves adjustments to antenna tilt, azimuth, vertical and horizontal isolation, etc. By implementing these optimization measures, coupling loss between antennas can be increased, thereby reducing interference.
5) **Restricting equipment parameters**: Setting sufficient equipment specifications ensures coexistence between adjacent transmission and reception frequencies, primarily through strict limitations on transmission power, among other parameters.
6) **Frequency hopping spread spectrum technology**: Spread spectrum communication technology is widely used in both public and private networks, and it can be divided into two types: direct-sequence spread spectrum and frequency-hopping spread spectrum.

A frequency-hopping system relies on random carrier frequency hopping to avoid interference, keeping interference out of the receiving channel to achieve anti-interference goals. The public 2G and 4G networks used in distribution wireless networks already have frequency-hopping technology, but these are slow-hopping systems (generally with fewer than 1000 hops), which are not robust enough and lose many of the inherent advantages of frequency-hopping technology. As a result, they are very sensitive to narrowband interference. Therefore, an evaluation of frequency-hopping spread spectrum technology for private distribution wireless networks should be conducted, along with research into frequency-hopping spread spectrum technology standards, and experimental studies on fast frequency-hopping technology.

10.3 Performance Evaluation

The goal of the experiments is to verify that ARED significantly improves the performance of IP data transmission for power distribution networks in LTE channels under varying levels of noise interference, with a packet loss rate of less than 50%. The aim is to provide end-to-end service quality assurance within a certain confidence interval, meeting the performance indicators required by the IEC 60870-104 protocol.

The simulation environment for power distribution wireless communication is based on NS-3 LTE [Piro et al., 2011], operating on Ubuntu 16.10 64bit. The underlying hardware platform uses an Intel i7 3667U CPU and 8GB DDR3 1333MHz memory. This environment replicates the typical IP communication process between the power station and terminal illustrated in Figure 10.3. The ARED prototype system is implemented as a module in NS-3, building on the IPv4 protocol stack provided by NS-3 and realizing the ADCT protocol stack. The test data flow carries distribution network three-remote business data based on IP and conforms to a Poisson distribution model.

On the LTE base station side, communication bandwidth between eNodeB and the power distribution network main station is 1 Gbps, MTU is 1500 bytes, and transmission delay is 100 ms. On the LTE terminal side, the IP packet sending interval for power distribution network terminals is 10 ms. The distance between the base station and wireless front-end is 5 km. For the uplink and downlink, AdaRQ algorithm parameters are set as follows: $B_{max} = 1$ Mbps, $\tau_{max} = 500$ ms, $\alpha_\lambda = 0.99$, $k_{min} = 10$. To measure ARED's interference resistance, NF is defined as the ratio of actual LTE receiver noise output power at 290 K temperature to the ideal LTE receiver noise output power at the same gain and bandwidth, measured in dB.

The testing scenarios include activating the ARED module and two wireless communication processes in the power distribution network for both the ARED module and the original LTE system [3GPP, 2016]. The following technical indicators are measured: (1) Packet loss rate: the rate at which the receiving end experiences packet loss due to channel noise causing decoding failures. (2) Half Round-Trip Time (HRRT): one-way delay between the sender and receiver. (3) Channel bandwidth utilization (i.e., n/k): indicating the efficiency of RQ encoding.

Firstly, Figure 10.5 provides a comparison of uplink and downlink channel packet loss rates for typical IP data services in both scenarios. Table 10.4 details the packet loss rate comparison between ARED and LTE under different NF values, demonstrating ARED significantly improves interference resistance compared to LTE. Figures 10.6 and 10.7 show the HRRT of uplink and downlink ARED at different packet loss rates, respectively, along with the variations in the delays incurred by RQ encoding and decoding. The experimental results illustrate stable changes in RQ encoding and decoding delays. As the packet loss rate increases, the adjustment of AdaRQ's parameter o meets formula 1, and the complexity of encoding and decoding satisfies $O\left((k+o)\cdot\log(k/o)\right)$. When the uplink and downlink packet loss rates are not less than 21.82% and 44.14%, respectively, the delay reaches its maximum. At this point, the sum of RQ encoding and decoding times accounts for 51.86% and 54.55% of HRRT, respectively. With the

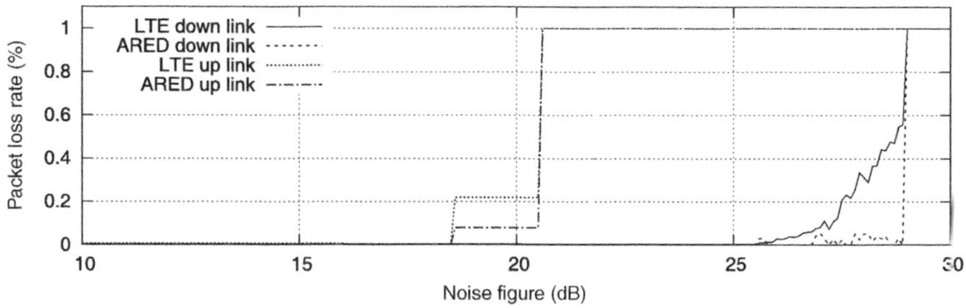

Figure 10.5 Packet loss rates in up and downlinks under different noise interference.

Table 10.4 Comparison in packet losses between ARED and LTE.

Channel	NF (dB)	ARED	LTE
uplink	[20.6, 45]	100%	100%
	[18.6, 20.5]	7.8%	21.82%
	[0, 18.5]	0%	0.469%
downlink	[28.9, 45]	100%	100%
	28.8	3%	54.65%
	[25.6, 28.7]	1.7%	16.86%
	[0, 25.5]	0%	0.2%

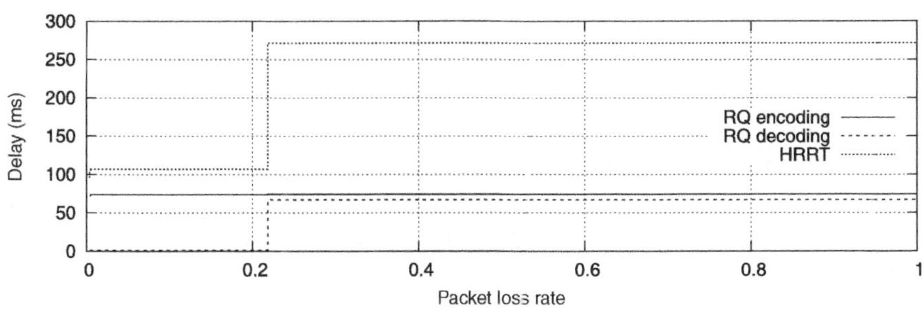

Figure 10.6 Uplink delay under different packet loss rate.

improvement in hardware platform performance, RQ encoding and decoding times are further reduced.

Finally, Figure 10.8 presents the average channel utilization in uplink and downlink for different NF values. The experimental results indicate that as NF increases, the AdaRQ module gradually adjusts RQ encoding parameters k and o, making the channel utilization approach theoretical values. When NF is not less than

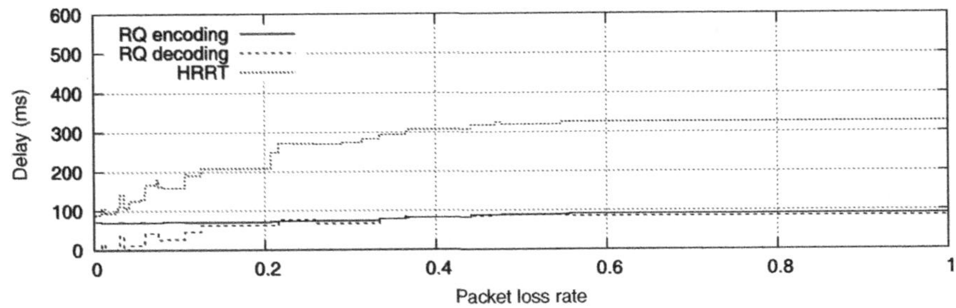

Figure 10.7 Downlink delay under different packet loss rate.

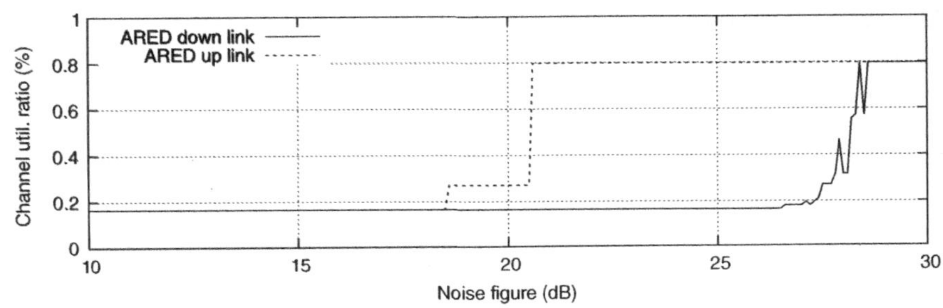

Figure 10.8 Average channel utilization ratio under different noise interference.

20.6 dB and 29 dB, respectively, the average utilization of uplink and downlink channels approaches 79.58%. This is due to considering that the throughput of power business follows a Poisson distribution with a confidence interval parameter of $\alpha_\lambda = 0.99$, avoiding channel interference when throughput exceeds the channel bandwidth limit.

The above experimental results demonstrate that ARED can timely respond to changes in channel quality when disturbed, adjust RQ encoding parameters, reduce packet loss rates under interference, ensure HRRT within an acceptable range for distribution network operations, and optimize channel bandwidth utilization.

10.4 Summary

This chapter presents an adaptive RQ encoder–decoder with variable message redundancy length (ARED) for wireless communication in power distribution networks. ARED effectively utilizes the remaining bandwidth of LTE to enhance communication reliability. Through simulating typical LTE communication processes in power distribution networks on NS-3, experimental results demonstrate that ARED significantly reduces packet loss rates under channel noise interference and meets the one-way transmission delay requirements.

Questions

10.1 How does the Adaptive RaptorQ (ARED) system adjust RQ encoding parameters to respond to changes in channel quality, and what impact does this adjustment have on packet loss rates?

10.2 Can you explain the role of the RQ channel feedback unit in the ARED system and how it contributes to improving synchronization and encoding parameter adjustments?

10.3 In the context of the AdaRQ algorithm, how does the system dynamically adjust parameters such as k and o based on variations in throughput (λ) and packet loss rate (L)?

10.4 How does ARED utilize the remaining bandwidth of LTE to enhance communication reliability, especially in the presence of noise interference?

10.5 How does ARED optimize channel bandwidth utilization, and what role does the AdaRQ algorithm play in achieving efficient RQ encoding?

10.6 In the context of ARED, what is the significance of ADCT, and how does it contribute to the dynamic RQ encoding parameter transmission?

10.7 Explain the role of the Lambert W function in the formula for adjusting RQ encoding parameters, as presented in Theorem 10.1. How does this function contribute to the optimization of encoding parameters?

Bibliography

3GPP TS 36.401. LTE; Universal terrestrial IO access network (E-UTRAN); architecture description version 13.0.0 release 13, 2016.

G. Piro, N. Baldo, and M. Miozzo. An LTE module for the NS-3 network simulator. In *ICST*, pages 415–422, 2011.

Amin Shokrollahi. Raptor codes. *IEEE Transactions on Information Theory*, 52(6):2551–2567, 2006.

11

Mastering Link-Flooding Attack: Route Diversification

As our power grids embrace the digital era, the convergence of OT networks with IT networks brings forth a formidable challenge – the stealthy Link-Flooding Attack (LFA). This attack inundates persistent routes (PRs) with low-rate floods directed at their decoys, disrupting end-to-end connectivity to the target.

In the initial attack stage, PRs quietly undergo probing, paving the way for a more insidious renaissance stage. This latter phase remains largely overlooked, allowing the identification of undefended and vulnerable PRs. To counter this two-stage menace, this chapter delves into a novel approach – Persistent Route Diversification Defense (PRDD). This mechanism adeptly mitigates each identified PR under attack, thwarting flooding and congestion.

The brilliance of PRDD lies in its ability to shroud PRs, making them impervious to probing or speculation during their respective attack stages. By depriving adversaries of these crucial insights, PRDD eliminates their flooding choices, thus preventing attacks. Designed with scalable algorithmic complexity, PRDD balances computational efficiency and minimal overhead.

Here's how it works: Firstly, PRDD identifies all the vulnerable PRs disrupting the target by gauging the congestion levels on the links. Next, PRDD puts the brakes on ongoing floods, alleviating congestion on these identified PRs. Simultaneously, it throws adversaries off the scent by making these PRs elusive – unfindable during probing in the initial attack or speculation in the renaissance stage.

How does PRDD achieve this feat? By transforming each potential decoy and target for identified PRs into multiple proxies, each armed with a randomized network address. This not only discards flooding attempts but also camouflages paths, evading PR probing. Moreover, PRDD strategically handles overloaded traffic headed to victim decoys, preventing disruptive flooding during the renaissance stage. The result? The adversary's attempt to validate PR speculation fails,

Communication Networks in Smart Power Grids, First Edition. Boyang Zhou.

Companion Website: www.wiley.com/go/SmartPowerGridsle

rendering the identified PRs invisible and thwarting further disruptive flooding. Thus, the two-stage attack is effectively repelled.

Moving beyond theory, we assess PRDD's defense prowess and performance using the NS-3 simulator [Riley and Henderson, 2010] and Mininet emulator [Lantz et al., 2010]. The results speak volumes: PRDD effectively mitigates PRs during non-congestion states in two-stage attacks. Moreover, its execution time, space requirements, and overhead prove scalable to large networks, aligning seamlessly with theoretical expectations. Notably, PRDD outperforms existing defenses by handling more congested PRs and responding to a larger number of bots.

11.1 Defending Two-Stage Attacks

To tackle the nuanced challenge of the two-stage LFA, our innovative defense mechanism, PRDD, unfolds a tailored strategy. It employs distinct defense processes for identified PRs based on their stages of engagement, as outlined below.

11.1.1 Defensive Measures in Initial Attack Stage

During the initial attack stage, PRDD deploys a scalable strategy by creating distinct *proxies* for each destination host, strategically placed at the ingress switch of the PR under attack. These proxies, acting as lightweight network-level delegates, handle all traffic to and from their associated destination hosts using randomized addresses. The implementation relies on OpenFlow (OF) [Open Networking Foundation, 2012]-based rules, a proven approach detailed in Section 11.2.3.2. Figure 11.1 illustrates the defense process with the following steps:

1) **Identification of PRs under the initial attack stage**: Identify PRs targeted during the initial attack; for instance, consider a PR between switches 1 and 4.

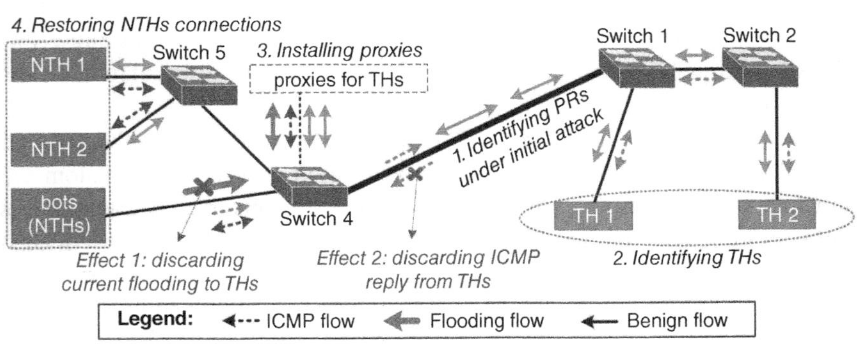

Figure 11.1 Defense process using the proxy creation.

2) **Identification of *threatened hosts* (THs)**: Extract all destination hosts downstream of the attacked PR, labeled as THs (e.g., THs 1 and 2). THs encompass not only the target and decoys but also benign hosts downstream. This inclusive approach anticipates potential adversary exploitation of any TH (excluding the target) as a decoy. The defender remains unaware of the target for the adversary.
3) **Proxy deployment. Installing proxies**: For each TH (e.g., TH 1 or 2), multiple proxies are strategically placed (see dashed squares), with each proxy installed at the ingress switch (e.g., switch 4) of a PR reachable by the TH identified during the initial stage (Step 1). The process of determining the proxy location is explained in Section 11.2.3.2.

   Furthermore, each proxy at a switch (e.g., switch 4) facilitates secure data packet exchange between a TH corresponding to the proxy and any Non-Threatened Host (NTH) using a randomized network address. This is achieved through:

   - **ICMP request handling**: The switch promptly responds to Internet Control Message Protocol (ICMP) requests with the proxy's address (see dashed arrows). Simultaneously, it discards subsequent ICMP replies linked to the request, preventing the PR from being exposed to bots (cross on the dashed arrow). The decision to discard ICMP replies is based on whether each reply originates from an identified TH or not. Additionally, the switch forwards ICMP requests addressed to a TH (dashed arrows) to the next hop (e.g., switch 1). The switch permits the forwarding of ICMP replies unrelated to the request, provided each reply lacks the source address of an identified TH. This ensures that all NTHs can only probe paths excluding the PR.
   - **Non-ICMP data packet handling**: The TH associated with a proxy responds to any non-ICMP data packet sent to/from the proxy's address (see solid thin arrows).
4) **Reconnecting NTHs**: This step reinstates end-to-end connections to the identified THs. The process involves reconnecting their NTHs to the addresses of the proxies established in Step (3).

11.1.1.1 Discussions

In consequence, the switch efficiently discards flooding packets halfway before reaching the PRs (e.g., observe the cross on the thick arrow between NTHs M and switch 4). This occurs because Step (3) alters the addresses of Threatened Hosts (THs) to those of their proxies. Furthermore, in subsequent initial attack stages on the same PRs, bots are unable to probe any path that includes the PRs since the NTHs encompass the bots. Consequently, the adversary is left with no choice for initiating a disruptive flooding. Thus, the PRDD effectively safeguards the initial attack stage. It is important to note that vulnerable PRs may remain, such as the one between switches 1 and 4, susceptible to speculation in later renaissance attack stages. For expansive networks, considerations must be made regarding performance scalability and overhead issues in the proxy generation process for defense design.

11.1.2 Defense Processes in Renaissance Attack Stage

During the renaissance attack stage, PRDD navigates the high-rate traffic directed toward the THs of the PRs under attack to alleviate congestion within the PRs. Simultaneously, it updates the addresses of the proxies associated with the THs at their respective switches. Figures 11.2 and 11.3 illustrate distinct defense processes for the renaissance attack stage in two corresponding scenarios: one where an optional path detours the PR, and the other where no such alternative path is available. Despite the differences in these processes, they share common steps outlined below.

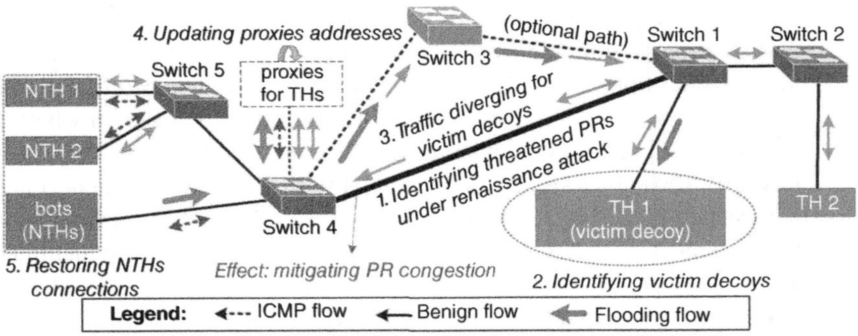

Figure 11.2 Defense process using the traffic diverging.

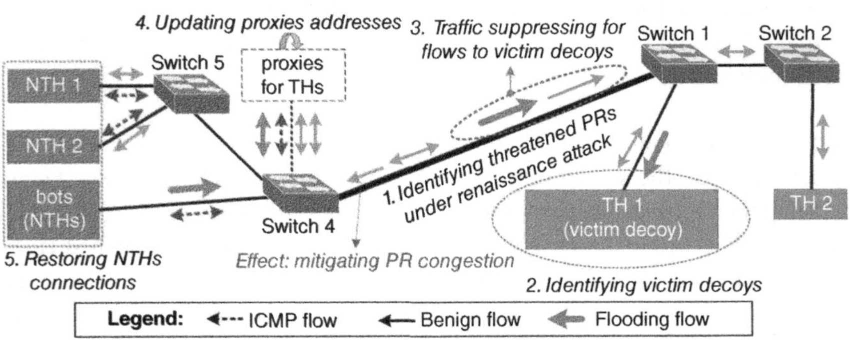

Figure 11.3 Defense process using the traffic suppressing.

i) **Identifying the PRs under the Renaissance Attack Stage**, such as the PR between switches 1 and 4 in Figures 11.2 and 11.3.

ii) **Identifying the victim decoys**: Firstly, extract the Threatened Hosts (THs) associated with the PRs, similar to the initial attack stage mentioned earlier (e.g., THs 1 and 2 in both Figures 11.2 and 11.3). Then, pinpoint the *victim decoys* among these THs based on higher packet receiving rates, attributed to the volume accumulation effect of the flooding.

iii) **Traffic steerings for victim decoys**: These steerings aim to prevent vulnerability speculation on the PRs during this stage. The goal is to thwart the attack from disrupting connectivity to the target through the victim decoys. To achieve this, all data flows from any Non-Threatened Host (NTH) to each victim decoy via a related PR under attack are redirected to noncongested states. This redirection encompasses the flooding flows. The specific steering mechanisms differ depending on whether there is an optional path detouring the PR or not. These conditions are outlined as follows:

- **Traffic diverging**: As illustrated in Figure 11.2, an optional path traverses switches 4, 3, and 1 (depicted by dashed lines). This diversion redirects flows from any Non-Threatened Host (NTH) to each victim decoy over the Persistent Route (PR) to the optional path (indicated by thick and thin arrows).
- **Traffic suppressing**: In Figure 11.3, the mechanism suppresses the forwarding rates of flows to the victim decoys (highlighted by thick and thin arrows within the dashed circle).

Under both conditions, the process effectively mitigates the PRs into a noncongested state when these PRs are under attack by the same bots at any given time. This ensures that each PR is pristine for flooding toward any proxy of the victim decoy connected to the PR. Consequently, the adversary's attempt to speculate on the same decoys associated with these PRs during their subsequent renaissance stage attacks is thwarted.

iv) **Updating proxy addresses**: This step involves re-randomizing the addresses of all previously installed proxies on the switches along their corresponding forwarding paths. For example, observe the light gray arrows in both Figures 11.2 and 11.3. Subsequently, the switches discard flooding packets in the stage, visible in the thick arrows between the bots and switch 4 (in both Figures 11.2 and 11.3). This update compels the ingress switches on the Persistent Routes (PRs) under attack to discard flooding packets, effectively mitigating the PRs into noncongestion states.

v) **Restoring NTH connections**: The restoration process is executed in the same manner as described in Section 11.1.1.

11.1.2.1 Discussions

In essence, step (iv) necessitates the adversary to speculate on the PRs under attack to identify vulnerable ones for subsequent renaissance stage attacks. Simultaneously, step (iii) ensures that no vulnerable PRs are identified, leaving the adversary with no choice but to abandon a disruptive flooding strategy. Consequently, the defense processes effectively thwart both initial and renaissance attack stages.

11.2 Defense Mechanism

To implement the defense processes described earlier, the proposed PRDD is designed with specific algorithms, including an analysis of their computational complexity and overhead.

11.2.1 Overview Design

The PRDD functions as a control service within the Software-Defined Networking (SDN) controller [Floodlight Controller, 2020]. Comprising the Link Congestion Level Analysis (LCLA), Defense Policy Generation (DPG), and Defense Policy Enforcement (DPE) components, the defense collaboration unfolds in distinct phases (see Figure 11.4):

1) The LCLA assesses congestion levels across all links, distinguishing between flooded and legitimate ones.
2) The DPG identifies all PRs under the two-stage attack based on congestion levels and formulates a defense policy with corresponding countermeasures.
3) The DPE enforces the policy on relevant switches, restoring end-to-end communications for impacted NTHs through their proxy addresses.

It is assumed that network addresses are unique and switches automatically decrement the Time-To-Live (TTL) value of outgoing packets by 1 during forwarding. The detailed designs of the LCLA, DPG, and DPE are presented in Sections 11.2.2–11.2.4. Common notations used in the chapter are listed in Table 11.1.

11.2.2 Link Congestion-Level Analysis

The LCLA assesses congestion levels of all network links in two phases:

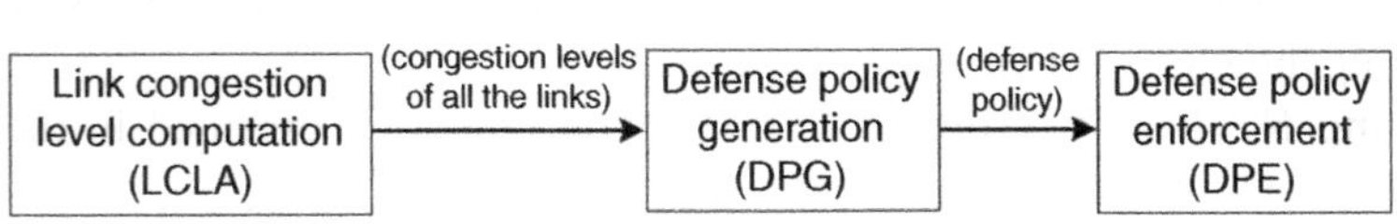

Figure 11.4 Components collaboration of PRDD in defense.

Table 11.1 Mathematical notations.

Notation	Description
G	An undirected graph of data plane, $G = (N, L)$
N	A set of all nodes in the data plane, $N = Y \cup W$
Y	A set of switches, $Y = \{y_1, y_2, \dots, y_{\lvert Y \rvert}\}$
W	A set of end hosts, $W = \{w_1, w_2, \dots, w_{\lvert W \rvert}\}$
L	A set of links connecting switches, $L = \{l_1, l_2, \dots, l_{\lvert L \rvert}\}$
$p(w_r, w_s)$	The forwarding path starting from w_r to w_s that is a sequence of nodes, i.e., $(w_r, y_{j_1}, y_{j_2}, \dots, y_{j_m}, w_s)$
$P^{\text{th}}_{\text{PR}}$	PRs under attack, $P^{\text{th}}_{\text{PR}} \subset L$
W^{th}	A set of THs, $W^{\text{th}} \subset W$
τ_{s}	The period for link statistics sampling
$\tilde{\tau}$	Link throughput smoothing period
ς	The minimum congestion-level threshold value
η	The minimum attack times threshold on a PR identified as in the renaissance attack stage
$v(w_i)$	The proxy of w_i
Υ	A defense policy, denoted as $\Upsilon = (V, R, S)$, where V, R, and S are proxy creation, diverging, and suppressing countermeasures, respectively
$\kappa(l_i, t)$	The congestion level of l_i at the time tick of t
t	The time tick
ρ	The minimum proportional threshold for identifying a threatened host as a victim decoy

i) **Link throughput sampling**: The LCLA samples two statistics for each link in the network at its egress and ingress switches, respectively, within a period of τ_{s}:

1) λ^t_{RX}: Number of receiving packets on the link at time t, obtained via an OFPMP_PORT or OFPMP_PORT_STATS message [Open Networking Foundation, 2012].
2) λ^t_{TX}: Number of packets from data flows forwarded to the same physical ingress port of the link at time t, obtained in a group via an OFPMP_GROUP_STATS message [Open Networking Foundation, 2012].

The sampling efficiency is ensured as follows: Rules for data flows forwarded to the same egress port point to the same group entry [Open Networking Foundation, 2012]. The entry defines the forwarding behavior and provides statistics counting with four fields:

1) **Identifier**: Distinguishes different groups.
2) **Group type**: Initialized to "all."
3) **Counter**: Indicates the number of receiving packets forwarded to the same port.
4) **Action set**: Defines the selection of an output forwarding port via the OFPAT_OUTPUT instruction.

Hence, it collects these statistics in the coarse-grained unit of the physical port of switches, requiring only a round-trip of OF control messages per unit, i.e., in $O(1)$. Thus, the number of control messages has a low complexity of $O(|L|)$.

ii) **Congestion-level analysis**: Using the sampled statistics, Algorithm 11.1 examines the congestion level of each link (l_i) in the following steps shown in Figure 11.5. These steps are detailed below.

1) **Compute in-flight packets at ingress switch**: For a time range of $[t - \tilde{\tau}, t]$ (line 2), Algorithm 11.1 calculates the sequence of in-flight packets forwarded at the ingress switch of l_i as λ_a.
2) **Compute received packets at egress switch**: Similarly, in line 3, it obtains a sequence of the numbers of packets received at the egress switch of l_i as λ_b.
3) **Compute correlation coefficients**: Compute $\mathbf{H}_1$ and $\mathbf{H}_2$ as two series of correlation coefficients between λ_a and λ_b, and the time range in lines 4 and 5, respectively.
4) **Compute overall correlation**: Compute the correlation between $\mathbf{H}_1$ and $\mathbf{H}_2$ as $\tilde{\kappa}$ in line 6, where $\tilde{\tau}$ is the smoothing period.
5) **Compute congestion levels**: Finally, Algorithm 11.1 computes the congestion level for two different conditions as follows. Then, in line 9, it obtains s_{min} by searching among candidate values computed via Eq. (11.1). In line 10, the congestion level (κ) is computed as the product of two parts: (1) the area proportion corresponding to s_{min}, denoted as $A(b, s, \lambda_a)$, and (2) the angle percentage between the corner cut of the slope and a right angle. Finally, it returns κ in line 11.
   - The first condition is $\tilde{\kappa} > C_1$, indicating that λ_a changes significantly in accordance with λ_b, suggesting an affluent link bandwidth. In line 7, the level is set to 0.
   - The second condition occurs when the adversary exhausts the link bandwidth during an attack. To address this, in line 8, Algorithm 11.1

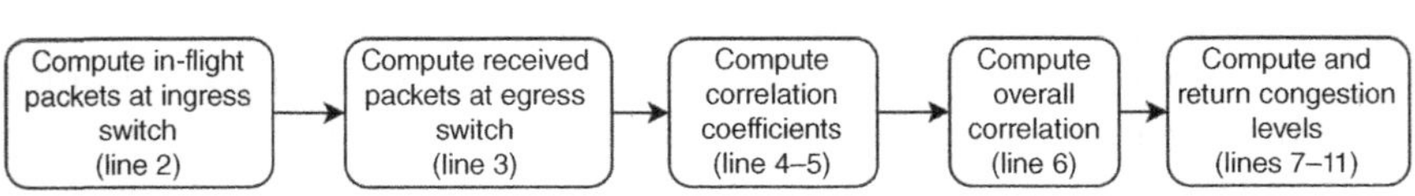

Figure 11.5 Workflows of the LCLA algorithm.

Algorithm 11.1 LCLA

1: **function** LCLA($\lambda_{\mathrm{RX}}^{t}, \lambda_{\mathrm{TX}}^{t}$)
2: $\quad \lambda_{\mathrm{a}} = \left[\left(\lambda_{\mathrm{TX}}^{t'} - \lambda_{\mathrm{RX}}^{t'}\right) / \mathrm{delay}(l_i) | t - \tilde{\tau} \le t' \le t\right]$
3: $\quad \lambda_{\mathrm{b}} = \left[\lambda_{\mathrm{RX}}^{t'} - \lambda_{\mathrm{RX}}^{t'-1} | t - \tilde{\tau} \le t' \le t\right]$
4: $\quad \mathbf{H}_1 = \left[\mathrm{Corrcoef}_{1,2}\left([t' - \tilde{\tau}, t'], \lambda_{\mathrm{a}}\right) | t - \tilde{\tau} \le t' \le t\right]$
5: $\quad \mathbf{H}_2 = \left[\mathrm{Corrcoef}_{1,2}\left([t' - \tilde{\tau}, t'], \lambda_{\mathrm{b}}\right) | t - \tilde{\tau} \le t' \le t\right]$
6: $\quad \tilde{\kappa} = \mathrm{Corrcoef}_{1,2}\left(\mathbf{H}_1, \mathbf{H}_2\right)$
7: $\quad$ **if** $\tilde{\kappa} > \mathrm{C}_1$ **then return** 0
8: $\quad b = \mathrm{argmax}_b b$, subject to: $\bigwedge_{1 \le i \le b}(|\lambda_{\mathrm{a}}[i] - \lambda_{\mathrm{a}}[1]| \le \mathrm{C}_2 \cdot \lambda_{\mathrm{a}}[1]) = \mathrm{True}$
9: $\quad s_{\min} = \mathrm{argmin}_s A(b, s, \lambda_{\mathrm{a}})$, subject to $A \ge C_3$
10: $\quad \kappa = 2 \cdot A(b, s_{\min}, \lambda_{\mathrm{a}}) \cdot \arctan(s_{\min})/\pi$
11: $\quad$ **return** κ
12: **end function**

identifies a starting index b within λ_{a} where its value suddenly changes beyond a constant proportion (C_2) of $\lambda_{\mathrm{a}}[1]$. In line 9, it computes a right triangle with features: (1) Its bottom edge starts from $(b, \lambda_{\mathrm{a}}[b])$ on the left and ends at $(|\lambda_{\mathrm{a}}|, \lambda_{\mathrm{a}}[b])$. (2) Its hypotenuse has the minimum slope ratio ($s_{\min}$), covering at least a constant proportion (C_3) of the area between the bottom edge and the line of λ_{a}.

Algorithm 11.1 effectively distinguishes flooding traffic from legitimate traffic in the current Internet. λ_{b} values in the second condition fluctuate around a constant, while λ_{a} for legitimate traffic are more decurved, resembling a straight line. This is attributed to end-to-end congestion controls adopted in many data transports (e.g., NewReno, LinuxReno, CUBIC, Illinois of TCP [Afanasyev et al., 2010], and QUIC [Langley et al., 2017]). These controls decrease their window size suddenly upon packet loss. In contrast, flooding packets without congestion control continuously overload the link bandwidth, resulting in a constant rate. Therefore, a link under heavier flooding exhibits a higher area proportion and hypotenuse slope. Line 10 of Algorithm 11.1 effectively determines the congestion level.

Operators can optimize performance by adjusting parameters according to these principles:

a) A smaller value of τ_s results in a quicker response time for detecting changes in link congestion.
b) The parameters C_1, C_2, and C_3 are all within the range (0, 1). In this paper, they are set to 0.5, 0.05, and 0.99, respectively, providing tolerance for fluctuations in throughput statistics.
c) $\tilde{\tau}$ is determined to achieve an appropriate resolution in computing correlations.

Additionally, LCLA returns κ to DPG to trigger the execution of GenPolicy for generating the defense policy.

$$A(b, s, \lambda_a) = \frac{\sum_{b+1 \le i \le |\lambda_a|} \min\left((i-b)\cdot s, \lambda_a[i]\right)}{\sum_{b+1 \le i \le |\lambda_a|} \left(\lambda_a[i] - \lambda_a[b]\right)} \tag{11.1}$$

11.2.3 Defense Policy Generation

In this section, we introduce the DPG algorithms, providing insights into computation complexity and overhead analysis.

11.2.3.1 DPG Algorithm

Algorithm 11.2 outlines the DPG algorithm, taking inputs of (1) the network topology G, (2) a set of all forwarding paths within the current data plane P_{all}, and (3) a map from each link to its congestion level at the current time t, denoted as $\Omega(t) = \{l_i \to \kappa^t | \forall l_i \in L\}$, computed by LCLA in Section 11.2.2. The execution steps are as follows:

Firstly, in line 2 of Algorithm 11.2, based on $\Omega(t)$, it identifies the attack intention using the AttackIntention function. Figure 11.6 illustrates the workflow of this function for identifying components involved in the attack intention.

1) Extracting the PRs under attack, denoted as $P_{\mathrm{PR}}^{\mathrm{th}}$, involves two criteria: (1) congestion levels no less than ς, and (2) link density no less than 2. These PRs are then divided into two subsets, Θ and Λ, based on the number of times under attack, whether less than η or not, respectively.
2) The THs in the initial attack stage, W_{Θ}^{th}, are computed as the combination of each destination host of the paths passing through a link in Θ.
3) For the victim decoys in the renaissance attack stage, denoted as $W_{\Lambda}^{\mathrm{th}}$, the computation involves two criteria: (1) the forwarding path of the host passes through a PR in Λ, and (2) the proportion between the data packet receiving

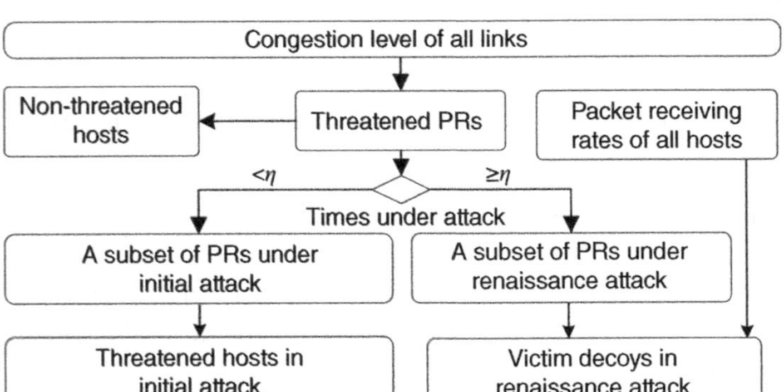

Figure 11.6 Workflow for identifying attack intention.

rate of the host and the bandwidth of its corresponding PR is no less than ρ. This rate is efficiently measured using the sketch hashing count approach provided by the OF controller itself [Yu et al., 2013]. A benign host will suddenly decrease its receiving rate due to congestion control in dealing with continuous packet losses caused by flooding [Afanasyev et al., 2010; Langley et al., 2017].

4) The NTHs, denoted as W^{no}, are computed as all destination hosts of paths disjoint from P^{th}_{PR}.

Secondly, in lines 3–4, Algorithm 11.2 invokes Algorithms 11.3 and 11.4 to create countermeasures for the initial and renaissance stages of the attack, denoted as V and (R, S), respectively. Subsequently, in line 5, Algorithm 11.2 returns the defense policy, denoted as Υ, encompassing the aforementioned countermeasures.

Moreover, operators can fine-tune the parameters as follows: (1) opting for smaller values of η (no less than 2) and ς (more than 0) to enhance sensitivity, and (2) configuring ρ within the range $(0, 1)$ to strike a balance between accuracy (closer to 1) and sensitivity (closer to 0) in victim decoy detection.

11.2.3.2 Countermeasures for Initial Attack Stage

In response to the attack intention identified by Θ and its corresponding threshold W^{th}_{Θ} during the initial attack stage, the GenProxies algorithm (Algorithm 11.3) generates countermeasures aimed at optimizing resource usage while mitigating the attack. This algorithm calculates the optimal locations for creating proxies, specifically designed to minimize switch resource consumption without compromising security. The workflow and steps of the algorithm are illustrated in Figure 11.7.

The algorithm proceeds by initializing key internal states in lines 2–3, which include:

- *Q*: A queue used to perform a Breadth-First Search (BFS) through the network.
- *torch*: A structure that records link Identification (ID) for determining optimal proxy locations.
- *visited*: A set that tracks whether a node has already been processed during the BFS, ensuring that each node is iterated only once in the context of a threatened host.

Algorithm 11.2 DPG algorithm

1: **function** GenPolicy($G, P_{all}, \Omega(t)$)
2: $(\Theta, \Lambda, W^{th}_{\Theta}, W^{th}_{\Lambda}, W^{no}) = \text{AttackIntention}(\Omega(t))$
3: $V = \text{Genproxies}(G, P_{all}, \Theta, W^{th}_{\Theta})$
4: $(R, S) = \text{GenTrafficSteers}(G, P_{all}, \Lambda, W^{th}_{\Lambda}, W^{no})$
5: **return** $\Upsilon = (V, R, S)$
6: **end function**

Algorithm 11.3 Proxy creation generation algorithm

function GENPROXIES($G, P_{\text{all}}, \Theta, W_{\Theta}^{\text{th}}$)
 $Q = \text{Queue}(), \textit{torch} = [-1]_{|N|}$
 $\textit{visited} = [\text{False}]_{|N|}$
 for $w_i^{\text{th}} \in W_{\Theta}^{\text{th}}$ **do**
  $\tilde{L} = \{l_j | l_j \in p_k \wedge p_k[|p_k|] \in W_{\Theta}^{\text{th}}, \forall p_k \in P_{\text{all}}\}$
  $Q.\text{Add}(w_i^{\text{th}})$
  $\textit{torch}[1 \cdots] = -1, \textit{visited}[1 \cdots] = \text{False}$
  while $|Q| > 0$ **do**
   $n_u = Q.\text{Poll}()$
   for $l_j \in \text{Edges}(n_u)$ **do**
    $n_v = \text{PeerNode}(n_u, l_j)$
    if $l_j \notin \tilde{L} \vee \textit{visited}[v]$ **then** continue
    if $l_j \in \Theta$ **then**
     $\textit{torch}[v] = j$
    else
     $\textit{torch}[v] = \textit{torch}[u]$
    end if
    $\textit{torch}[u] = -1$
    $Q.\text{Add}(n_v), \textit{visited}[v] = \text{True}$
   end for
  end while
  $V_{w_i^{\text{th}}} = \{(v(w_i^{\text{th}}), y_j) | \textit{torch}[j] \neq -1, \forall y_j \in Y\}$
 end for
 return $V = \cup_{w_i^{\text{th}} \in W_{\Theta}^{\text{th}}} V_{w_j^{\text{th}}}$
end function

The BFS ensures that only one link ID, closest to the leaf nodes, is propagated for links along the same path. This step is crucial because it reduces the number of proxies that need to be created. By focusing on the link closest to the leaf node, the algorithm ensures that the placement of proxies is both strategic and efficient, maintaining a maximum separation between bots and their corresponding THs. This separation helps prevent attackers from easily moving between compromised nodes and increases the difficulty of executing coordinated attacks.

The overall goal is to minimize the number of proxies created by separating bots from THs as much as possible while maintaining optimal network performance. This process helps achieve resource efficiency, ensuring that the switch and network resources are not overburdened by excessive proxy creation while still providing adequate protection against ongoing threats.

Algorithm 11.4 Traffic steerings generation algorithm

function GENTRAFFICSTEERS($G, P_{\text{all}}, \Lambda, W_{\Lambda}^{\text{th}}, W^{\text{no}}$)
  if $\Lambda = \emptyset$ **then return** $\emptyset$
  $R = \emptyset, S = \emptyset$
  for all $w_i \in W_{\Lambda}^{\text{th}}$ **do**
    $G' = (Y \cup W, L \setminus \Lambda)$
    $P' = \text{ShortestPaths}(G', w_i, \Xi)$
    $W_S^{\text{th}} = \{w_j | p(w_j, w_i) \notin P', \forall w_j \in W_{\Lambda}^{\text{no}}\}$
    $S_{w_i} = \{(y, p, v_{\text{rand}}^{\text{new}}(w_i)) \,|y = \text{Ingress}(l, p) \wedge p(w_j, w_i) \in P_{\text{all}} \wedge (l \in \Lambda, \exists l \in p), \forall w_j \in W_S^{\text{th}}\}$
    $R_{w_i} = \{(p', p, v_{\text{rand}}^{\text{new}}(w_i)) \,|p' \neq p \wedge p'(w_j, w_i) \in P' \wedge p(w_j, w_i) \in P_{\text{all}}, \forall w_j \in (W^{\text{no}} \setminus W_S^{\text{th}})\}$
    $S = S \cup S_{w_i}, R = R \cup R_{w_i}$
  end for
  return (R, S)
end function

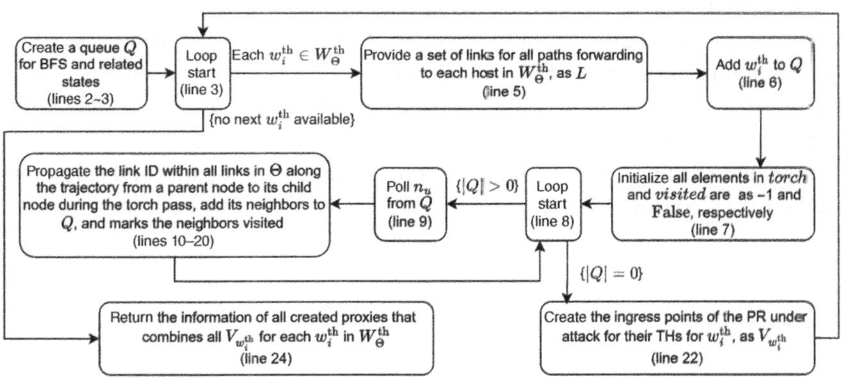

Figure 11.7 Workflow for the GenProxies algorithm.

Proxy implementation: Each proxy is implemented with specific rules on the corresponding switch, including:

1) The rules discard any received ICMP reply sent from the Threatened Host (TH).
2) The rules redirect any ICMP request sent from any Non-Threatened Host (NTH) to the proxy to the controller. Additionally, the rules forward the ICMP reply sent from the controller to its corresponding host.
3) The rules perform bidirectional network address translation for data packets sent between the TH and any NTH. When an NTH sends a packet, the switch replaces its destination field with the proxy address and the real address of its

corresponding TH. Conversely, if the packet is sent from the TH, the switch replaces the source field with the TH address and the proxy address. This replacement is achieved through the OFPAT_SET_FIELD instruction [Open Networking Foundation, 2012].

Moreover, if a packet matches either the first or second rule, the switch will not execute the third rule. To achieve this, the controller inserts these rules in front of the existing rules on the paths between the THs and NTHs of the switch. This implementation incurs only an additional process delay of $O(1)$, which can be neglected.

11.2.3.3 Traffic Steering Countermeasures for Renaissance Attack Stage

During the renaissance attack stage, the network faces specific attack intentions involving Λ, W_{Λ}^{th}, and W^{no}. In response, Algorithm 11.4 presents traffic steering countermeasures to mitigate these attacks by efficiently redirecting and suppressing traffic flows across the network.

This algorithm is designed to minimize the attack's impact by calculating alternative paths for traffic and implementing flow suppression where necessary. The workflow of this algorithm is depicted in Figure 11.8, which provides a step-by-step visualization of the process. In the core computation process (lines 4–6), optional paths for traffic redirection are calculated based on the principles outlined in Chen et al. [2007], ensuring optimized flow divergence while maintaining minimal impact on network performance.

At the end of the workflow, the algorithm outputs a set of paths and rules, represented as (R, S) (line 12):

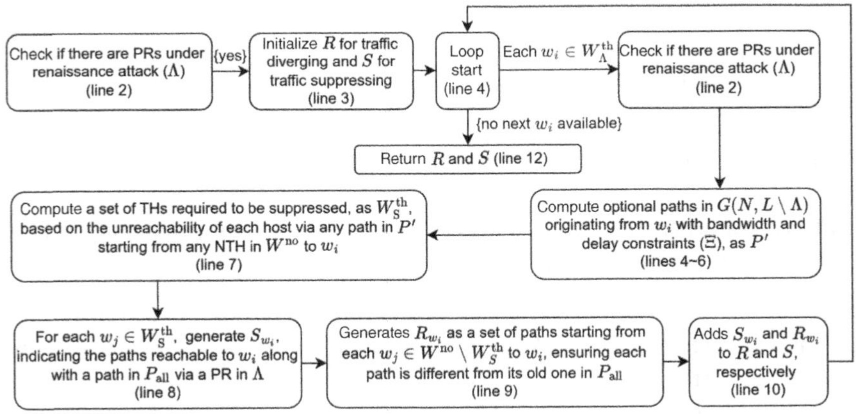

Figure 11.8 Workflow for the GenTrafficSteers algorithm.

- Each item in R, denoted by (p', p, v), indicates that traffic flows currently traversing path p' should be redirected to path p. The algorithm achieves this by generating new forwarding rules for p on each switch along the path and replacing the existing rules for p' to divert the flow. Additionally, the proxies associated with (R, S) are updated to v, reflecting the new path configuration in the network.
- Each item in S, represented by (y_m, p, v), indicates that traffic on path p should be suppressed at node y_m. To implement this, the algorithm generates meter rules for y_m, which are inserted ahead of all existing rules on path p using the OFPIT_METER message [Open Networking Foundation, 2012]. The associated proxies for v are also updated concurrently to reflect these changes.

In practice, this countermeasure ensures that traffic diverging and suppression are efficiently managed without overwhelming network switches or degrading network performance. By dynamically updating the forwarding rules and proxy configurations, the algorithm maintains operational efficiency, preventing attackers from exploiting compromised paths and minimizing the overall attack surface.

11.2.3.4 Execution Complexity of Algorithms

The execution complexity analysis of the GenPolicy algorithm (see Algorithm 11.2) in both time and space unfolds as follows.

For a network G, where $|L|$ is $O(\log|N| \cdot |N|)$, $O(|N|)$, and $O(|N|^2)$ for the expected, best, and worst cases, respectively, assuming $|W| = O(|N|)$.

Concerning time complexity, in Algorithm 11.2, the execution time of line 2 is $O(|W^{th}| \cdot \log|N|)$. Then, in line 3 of Algorithm 11.2, the execution time of Algorithm 11.3 is $O(|W^{th}| \cdot |N|)$, considering the use of BFS for each $w_i^{th} \in W_\Theta^{th}$.

In line 4 of Algorithm 11.2, the execution time of Algorithm 11.4 is $\zeta = O(|W^{th}| \cdot (|N| \cdot \log|N| + |L| + |W| + |W| \cdot \log|N|))$ with the probability $g = Pr\{|\Lambda| \geq 1\}$ (see line 2 of Algorithm 11.4). Here, g is a coefficient linearly correlated with the attack intensity degree, reflecting the number of PRs defended by the proxies countermeasure that are later speculated and flooded in a renaissance attack stage.

The overall execution time of Algorithm 11.2 is $O(|W^{th}| \cdot \log|N| + |W^{th}| \cdot |N| + g \cdot \zeta)$, which can be further simplified in different cases:

a) In both the expected and best cases: $O(|W^{th}| \cdot |N| \cdot (1 + g) + |W^{th}| \cdot \log|N| \cdot (1 + g \cdot |N|))$.
b) In the worst case: $O(|W^{th}| \cdot |N| \cdot (1 + g) + |W^{th}| \cdot \log|N| \cdot (1 + g \cdot |N|) + g \cdot |W^{th}| \cdot |N|^2)$.

Thus, when only the proxy countermeasure is generated, the time complexity is $O(|W^{th}| \cdot (|N| + \log(|N|)))$ for all cases. The time complexity increases based on the value of g in other scenarios. Therefore, the algorithm exhibits scalability in time.

The space complexity analysis for the execution of the algorithms reveals varying memory requirements as outlined below:

a) Lines 2–10 of Algorithm 11.2 is in $O(|W^{th}| \cdot \log(|N|))$.
b) Algorithm 11.3 requires $O(|N|)$ memory (see lines 2–3).
c) Algorithm 11.4 uses $O(|N|)$ memory to execute line 7 and $\vartheta = O(|W^{th}| \cdot (|W| + \log|N| \cdot |W|))$ to store both R and S in lines 8–10.

Thus, for all cases, the total memory usage is in $O(|W^{th}| \cdot \log|N| + |N| + g \cdot \vartheta)$, which simplifies to $O(|W^{th}| \cdot \log|N| \cdot (1 + g \cdot |N|) + |N|)$. This indicates a scalable complexity in terms of memory usage.

11.2.3.5 Overhead Analysis in Defense Policy

i) **Generated proxies count**: In lines 8–21, Algorithm 11.3 explores PRs using BFS on the Shortest Path Tree (SPT) from W^{th}_{Θ}'s source. Concurrently, it verifies PR reachability from W^{th}_{Θ} (see line 12).

   We illustrate the tree as $G_{tree} = (N, L^c)$. Additionally, $|L^c| = |N| - 1$, and each $l'_i \in L^c$ is in L with a probability of $1/\log(|N|)$. Let's consider a subgraph $G'(W^{th} \cup \epsilon(P^{th}_{PR}), L_{PR})$ of $G(N, L)$. Here, ϵ extracts egress nodes from given links, and $l^{PR}_i \in L_{PR}$ is a hypothetical link if there's a subpath in $G_{tree} \backslash P_{PR}$ connecting different $\epsilon(p^{PR}_i \in P^{th}_{PR})$.

   Furthermore, G' is the amalgamation of all traces from torch-passing steps following the SPT starting from each wth in lines 8–21. Consequently, the SPT of G' has a link count of $|\epsilon(P^{th}_{PR})| - 2$ (in $O(|W^{th}|)$), while $|L_{PR}|$ is in $O(|W^{th}| \cdot \log|W^{th}|)$.

   Therefore, the count of generated proxies for w^{th}_i, denoted as $|V_{w^{th}_i}|$ in line 22, is in $O(|W^{th}|/\log|W^{th}|)$. This is because the probability of selecting each $w^{th}_i \in W^{th}$ as a torch-passing node is equivalent to selecting an SPT link in G'. In total, the count of generated proxies is in $O(|W^{th}|^2/\log(|W^{th}|))$. Hence, this complexity is unrelated to the topology size but depends on the number of THs, indicating its suitability for large-scale networks.

ii) **Control message counts**: The messages needed for enforcing proxy creation, diverging, and suppressing countermeasures are in $O(|V|)$, $O(|R| \cdot (\log|N| + 1))$, and $O(|S|)$, respectively.

   Simplifying them, we get $O(|W^{th}|^2/\log(|W^{th}|))$, $O((1-h) \cdot |\Lambda| \cdot (\log|N| + 1))$, and $O(h \cdot |\Lambda|)$, respectively. Here, h is the average probability of $P' = \emptyset$ (see line 6 of Algorithm 11.4), and $\Lambda = O(|W^{th}|)$.

Therefore, the total control message count is in $O(|W^{th}|^2/\log(|W^{th}|) + (1 - h) \cdot \log|N| \cdot |W^{th}| + h \cdot |W^{th}|)$. This term is related to the attacking scale in the number of THs, resulting in $O(|W^{th}|^2/\log(|W^{th}|))$ when $h = 1$. Moreover, it correlates with both the attacking scale ($O(|W^{th}|^2/\log(|W^{th}|))$) and the network size, specifically in a logarithm of $\log|N|$, when $0 < h < 1$. Overall, this indicates scalable message complexity.

11.2.4 Defense Policy Reconfiguration

Defense policy reconfiguration: Initially, the DPE initiates a comprehensive update by replacing outdated rules in all switches with their counterparts from the policy. This replacement method meticulously compares each old rule to its new counterpart, considering real source and destination addresses. Then, it performs dynamic flow adjustment. When a new data flow reaches a switch, the DPE diligently identifies policy-compliant rules whose fields align with the incoming flow. Subsequently, the DPE flexibly and on-demand reconfigures these identified rules within the switches connected to the policy.

Proxy address updates: According to the policy, NTHs receive updated proxy addresses through two methods:

1) **Active method**: The DPE promptly dispatches a Domain Name System (DNS) update message (see IETF RFC 2136 [IETF, 1997]) containing the new addresses to NTHs. Similar to anti-virus programs, each NTH, equipped with an application daemon, actively captures the message and conveys the addresses to relevant applications.
2) **Passive method**: The DNS service in the controller acts as a transparent proxy between all hosts in the current OF data plane and the DNS server, intercepting DNS packets. If a host has a proxy installed, the service dynamically translates the domain name to the corresponding proxy address. Otherwise, it translates the name into the real address of the host. Consequently, NTHs can request new addresses by querying the DNS service. In either method, NTHs then reconnect to their respective THs with the newly provided or requested addresses.

In summary, the PRDD designs facilitate both initial and renaissance attack stage defense processes, offering scalable performance with acceptable overhead.

11.3 Performance Evaluation

We assess the defense effectiveness, performance, and advantages of the proposed PRDD in countering stealthy LFA in a computer network.

11.3.1 Experimental Settings and Parameters

We generated a network topology with 10–2000 nodes, incorporating switches and hosts, using the Erdős–Rényi model [Newman et al., 2001] for diversity. Parameters for the synthesized network are detailed in Table 11.2:

1) The total switch links are twice the number of OF switches, ensuring a proportion of links with optional paths.
2) Each switch connects to a host, representing multiple hosts for abstraction.
3) Link transmission delay ranges randomly between 1ms and 200ms, with link bandwidth and Maximum Transmission Unit (MTU) set at 10MBps and 1500B, respectively.

Each benign host is configured with the standard New Reno TCP/IP stack [Casoni et al., 2015] using the parameters in Table 11.2. Hosts simulate high-throughput applications by sending files of infinite length.

The effectiveness of PRDD was thoroughly assessed using two distinct methods:

- **NS-3 based Discrete Event Simulations** [Riley and Henderson, 2010]: Leveraging NS-3, the test environment incorporates OF switches implemented by

Table 11.2 Experiment setup parameters.

Simulation topology and PRDD			
Parameter	**Value**	**Parameter**	**Value**
Bandwidth(l_i)	10 Mbps	MTU(l_i)	1500 B
Delay(l_i)	[1, 200] ms	$\|L\|$	$2 \cdot \|Y\|$
η for DPG	2	ς_1 for LCLA	0.2
ρ for DPG	0.9	ς_2 for LCLA	0.8
$\tilde{\tau}$ for LCLA	1 s	τ_s for LCLA	100 ms

TCP stack parameter	Value
Maximum receiving buffer size (bytes)	1 MB
Maximum sending buffer size (bytes)	1 MB
Writing buffer size for emulation	1 MB
# packets to wait before sending a TCP ACK	2
Timeout value for TCP delayed ACKs (seconds)	0.2
Nagle algorithm	Disable

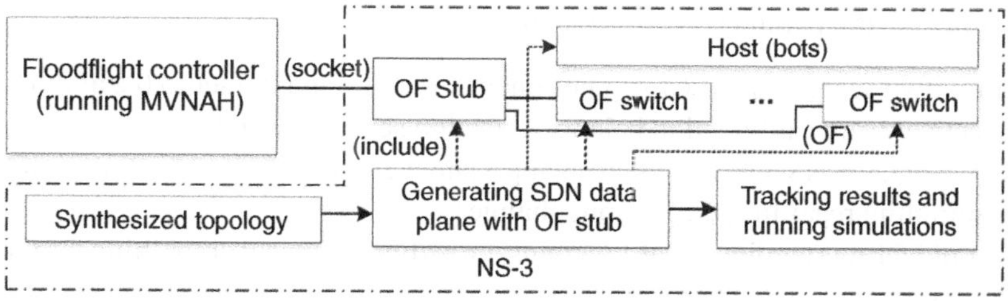

Figure 11.9 Experiment workflow of PRDD in NS-3.

OFSwitch13 [Chaves et al., 2016]. OFSwitch13, supporting OF v1.3, facilitates dynamic interaction between NS-3 and an external OF controller during simulations. The experiment workflow, illustrated in Figure 11.9, begins with NS-3 parsing the topology to generate OF switches, hosts, and bots. The OF controller then communicates with switches through the OF stub module, acting as a transparent network layer proxy for switches via a tap bridge created by OFSwitch13. The simulations offer the advantage of easily tracking detailed testing results.
- **Mininet emulations** [Lantz et al., 2010]: Deploying PRDD on the Floodlight OF controller v1.2 [Floodlight Controller, 2020] – chosen for its simplicity and support for OF v1.3 [Open Networking Foundation, 2012] – the controller interacts with switches in an out-of-band manner.

Furthermore, we evaluated the effectiveness of LCLA (see Section 11.2.2) through simulations in two distinct scenarios:

1) The first scenario involves testing under current high-performance transmission protocols, including: (1) NewReno, LinuxReno, CUBIC, and Illinois, all configured under the same parameters outlined in Table 11.2 [Casoni et al., 2015]; and (2) the default QUIC [De Biasio et al., 2019]. In this setup, each benign host generates traffic using on–off patterns at the same rate as the bot.
2) The second scenario examines realistic internet traffic synthesized with 90.04% TCP flows and 9.96% non-TCP flows, closely resembling a CAIDA real Internet trace found in CAIDA [2016]. Here, the hosts generate packets for each flow based on the Poisson Pareto burst process model [Ammar et al., 2011]. NewReno is chosen for TCP for its generality, configured with recommended parameters of 0.7 for Hurst and 0.2 seconds for mean, as per [Ammar et al., 2011]. The LCLA is tested on the same link in two aspects:
   - Sensitivity to identifying different intensities of flooding, where testing flows comprise Internet flows and flooding from bots.
   - Accuracy in identifying flooding in a heavily congested link, with corresponding heavy flooding from 200 bots sufficient to drain the PR bandwidth.

Meanwhile, the parameters of PRDD are detailed in Table 11.2 as follows:

- For the two scenarios of LCLA mentioned earlier, ς_1 and ς_2 (set to 0.2 and 0.8, respectively) are calibrated to balance detection sensitivity and accuracy.
- The τ_s of LCLA is set to 100ms, ensuring a small enough interval for link statistics sampling regarding throughput changes.
- The $\tilde{\tau}$ of LCLA is configured at 1 seconds, providing sufficient smoothing for the captured throughput changes.
- The steering countermeasure (η) in DPG is set to 2 for optimal performance.
- To ensure victim decoy detection accuracy, the ρ of DPG is configured at 0.9.

11.3.2 Stealthy Link-Flooding Attack Simulations

To assess the defense effectiveness and advantages of PRDD, we employ a synthesized network topology with $|Y| = 2000$ and $|W| = 2000$. In this setup, 30% of the hosts are designated as bots, precisely calibrated to exhaust link bandwidth. The adversary executes the attack in two stages:

i) **Stealthier initial attack stage**: The adversary executes a calculated sequence of moves:
   - Initially, each bot sends irregular ICMP requests to the target with randomized intervals, uniformly distributed within $[1, 10]$ seconds, probing the network paths.
   - Subsequently, bots report their probed paths back to the adversary.
   - Next, the adversary selects the top 40 PRs based on link density toward the target, enough to cause widespread disruption.
   - Afterward, the adversary identifies decoys for the chosen PRs.
   - Finally, bots are assigned to flood the network using their respective decoys on the selected PRs.

ii) **Extreme renaissance attack stage**: The adversary initiates an extreme scenario: successfully speculating all defended PRs and their available decoys within 0 seconds. This extreme test aims to thoroughly evaluate the defense effectiveness of PRDD.

Figure 11.10 illustrates the relationship between the number of bots and their assigned PRs. On average, each PR is flooded by 581.625 bots, with a minimum of 525 and a maximum of 600. These bots engage in a rolling flood, saturating the PR bandwidth with dummy User Datagram Protocol (UDP) packets at a modest rate of 1.719 kB/s in an on–off pattern, as described in Kang et al. [2013] and Zhijun et al. [2020]. Concurrently, regular hosts send ICMP packets with randomized intervals uniformly distributed within $[0, 20]$ seconds, emulating the benign ICMP behavior of typical applications.

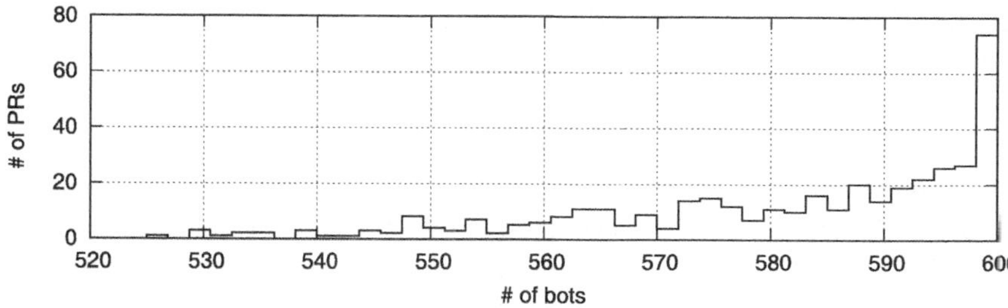

Figure 11.10 The number of bots assigned to PRs in the topology of 4000 nodes.

11.3.3 Evaluation Metrics of Solution

The assessment of PRDD's defense against the two-stage attack (see Section 11.3.2) encompasses the following aspects.

i) **Defense effectiveness**: Our examination includes:
   - Congestion-level detection by LCLA.
   - TCP receiving rate sharpness with decreasing packet losses.
   - CWnd size indicating the attack mitigation effect.

   The detailed results are in Section 11.3.4.
ii) **Defense performance**: We scrutinized the execution time, number of control messages, and generated proxies. Additionally, we compared these values with their theoretical estimations from Sections 11.2.3.4 and 11.2.3.5. The results are presented in Section 11.3.5.
iii) **Defense advantages**: We conducted a comparative analysis of PRDD against existing Moving Target Defense (MTD) and traffic throttling approaches, using the following metrics:
   - **Disappeared PR ratio**: The ratio of disappeared PRs post-defense to congested PRs pre-defense.
   - **Steered PR ratio**: The ratio of diverged or suppressed PRs post-defense to congested PRs pre-defense.

We compared PRDD with MTD approaches such as Proxy-based Moving Target Architecture (PROTAG) [Venkatesan et al., 2016], Aydeger, and EqualNet, as well as throttling approaches like LFADefender [Wang et al., 2019], RADAR [Zheng et al., 2018], and SPIFFY [Kang et al., 2016].

For PROTAG, proxies are set on 1–50% of switches using BIND-SPLIT [Venkatesan et al., 2016]. Aydeger and EqualNet follow configurations from Aydeger et al. [2016] and Kim et al. [2022] respectively. LFADefender sets k to 5 for computing Local Outlier Factor (LOF) in malicious flow ID [Wang et al., 2019]. RADAR and SPIFFY have a sampling period of 10 seconds for data plane statistics collection. Results are presented in Section 11.3.6.

11.3.4 Effectiveness in Defending Attacks

We assessed the defense effectiveness of PRDD in the two-stage attack, focusing on two key aspects:

i) **Congestion level**: The evaluation of LCLA is presented in two scenarios:
- For the transmission protocol scenario, Figure 11.11(a) indicates that the link congestion level, when attacked by 60 bots, increases by 0.211 compared to ς_1. TCP and QUIC benign clients maintain a congestion level of 0 due to their congestion control.
- For the realistic Internet traffic scenario:
  - In the sensitivity test, Figure 11.11(b) shows an increasing congestion level with more bots for flow intensity ranging from 100 to 500.
  - In the accuracy test during flooding, Figure 11.11(c) demonstrates that congestion levels exceed ς_2, gradually decreasing with higher flow intensity as flooding proportions reduce.
  - Without flooding, Figure 11.11(d) displays congestion levels below ς_2, decreasing with higher flow intensity due to reduced CWnd for benign traffic. These results affirm the effectiveness.

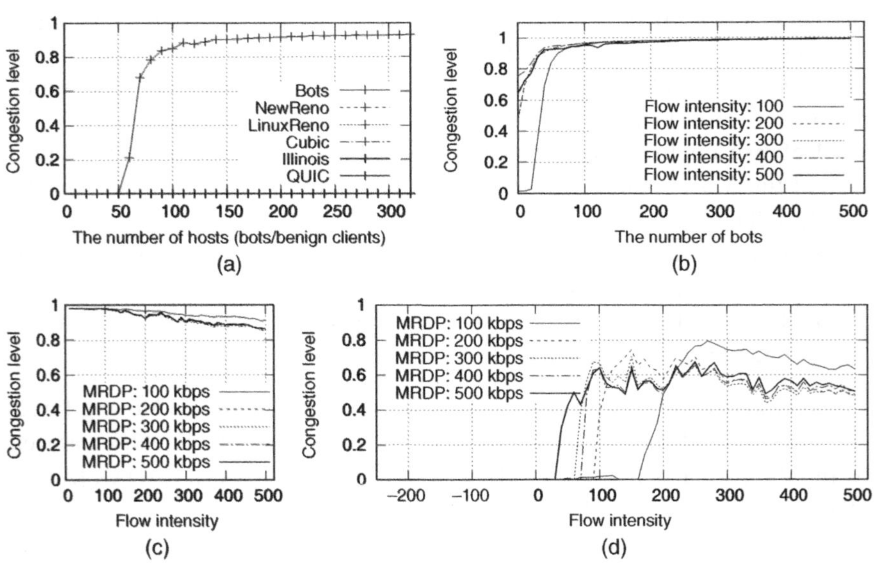

Figure 11.11 Effectiveness for detecting victim PRs under LFA. (a) Congestion level changes for a mixture of the flooding and various transmission protocols, (b) congestion level sensitivity for a mixture of flooding and realistic Internet traffic, (c) congestion level accuracy for a mixture of flooding and realistic Internet traffic, and (d) congestion level accuracy for the realistic Internet traffic.

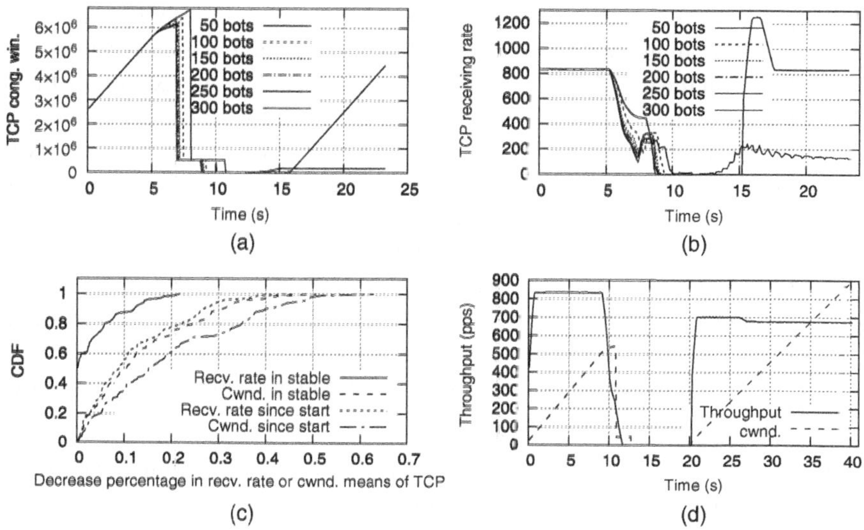

Figure 11.12 Effectiveness for defending LFA in NS-3. (a) TCP Cwnd size at the target for proxy creation, (b) TCP receiving rate at the target for proxy creation, (c) CDF of decreased TCP receiving rate and CWnd for diverging, and (d) TCP receiving rate and CWnd size for suppressing.

ii) **Defense policy**: Post-DPE, the TCP performance between hosts and the target is scrutinized:

a) **Effectiveness of proxy creation countermeasures**: In the initial attack stage, proxy creation countermeasures are tested. Figure 11.12(a) showcases TCP CWnd size changes at the target when an l_i with $e(l_i) = 0$ is flooded by 50 to 300 bots over 5–10 seconds. Simultaneously, Figure 11.12(b) reveals TCP packet receiving rate changes at the target, adjusting based on the congestion state of l_i. From 0 to 5 seconds, they linearly increase and remain stable, respectively, without flooding. Between 5 and 15 seconds, they significantly decrease to nearly zero for 50 bots due to heavy link load, reaching zero for 60–300 bots. These results align with $\kappa(l_i) < \varsigma$ and $\kappa(l_i) \geq \varsigma$ (see Figure 11.11(a)), respectively. Countermeasures are enacted at 15 seconds for bots of 60 or more, with the host receiving DNS updates, reconnecting via the proxy, and TCP CWnd and receiving rate normalization. For 50 bots, DPE is not triggered, as l_i remains available. With this countermeasure, bots fail to relaunch attacks due to the PRs' invisibility, now in a non-congested state. These outcomes underscore the defense effectiveness at the initial attack stage.

b) **Effectiveness of traffic steering countermeasures**: For the renaissance attack stage, traffic steering countermeasures are evaluated. Figure 11.12(c) presents the CDF of the decrease percentage in receiving rate and corresponding CWnd size of TCP when traffic diverging is enforced or not. PRs with the highest link density are attacked by 300 bots. Examining their means, either when TCP stabilizes in 5 seconds or since TCP initiation in 30 seconds, reveals that the TCP packet receiving rate only decreases by 3.32% when its CWnd size stabilizes due to a 14.12% average path delay increase. Additionally, Figure 11.12(d) demonstrates TCP receiving rate and CWnd changes when a critical PR is attacked by 300 bots, with traffic suppressing enforced in 10–20 seconds. The attack suppresses PR bandwidth to 2 Mbps. Post-20 seconds, bots continue flooding at a reduced speed, and the TCP socket reconnects to the target, sending packets at 80% of its previous rate (20% sent by bots). Consequently, PRs between the host and target return to non-congested states.

We evaluate the effectiveness of the defense policy via the Mininet emulation for the topology with 50 switches. Figure 11.13(a)–11.13(c) shows the bandwidth and CWnd statistics measured by iperf for the proxy creation, diverging, and suppressing countermeasures, respectively.

For the initial stage, with the proxies, in Figure 11.13(a), the flooding on a randomly selected PR since 15 seconds is defended at 24.3 seconds due to no probable PR. Later, the renaissance ones are defended as follows: (1) In Figure 11.13(b),

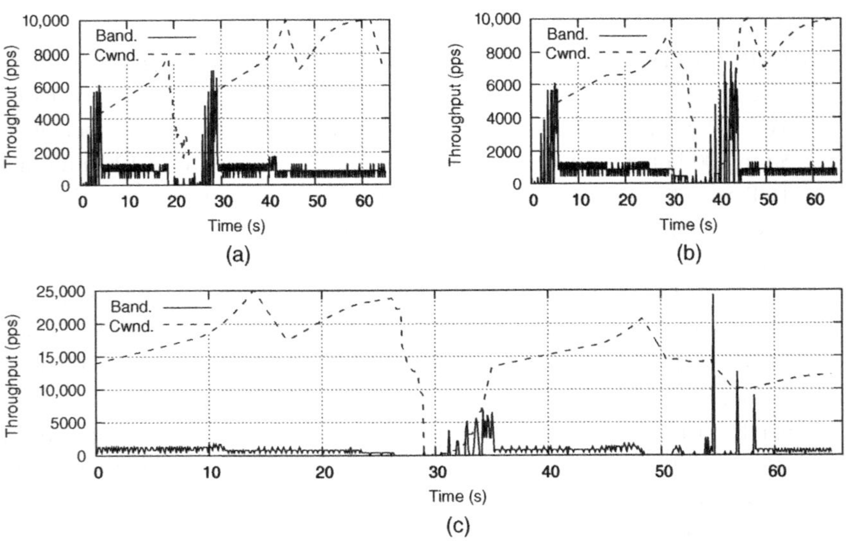

Figure 11.13 Effectiveness for defending LFA in Mininet. (a) TCP bandwidth and CWnd for proxies, (b) TCP bandwidth and CWnd for diverging, and (c) TCP bandwidth and CWnd for suppressing.

the PRDD diverges the flooding on a non-critical PR under attack to an optional bypassed path. (2) In Figure 11.13(c), the PRDD suppresses the flooding on a critical PR at 29 seconds under attack at 20 seconds, due to the suppressing.

Hence, the PRDD mitigates the PRs into noncongestion, and the PRs destructiveness cannot be evaluated. Hence, the PRDD is effective in defending such two-stage attacks.

11.3.5 Performance Results

We delved into assessing the execution time and overhead of the GenPolicy algorithm, considering the following aspects:

i) **Execution time**: Figure 11.14(a) illustrates the execution time of Algorithm 11.2 across topologies ranging from 20 to 4000 nodes for $g = 0$ and

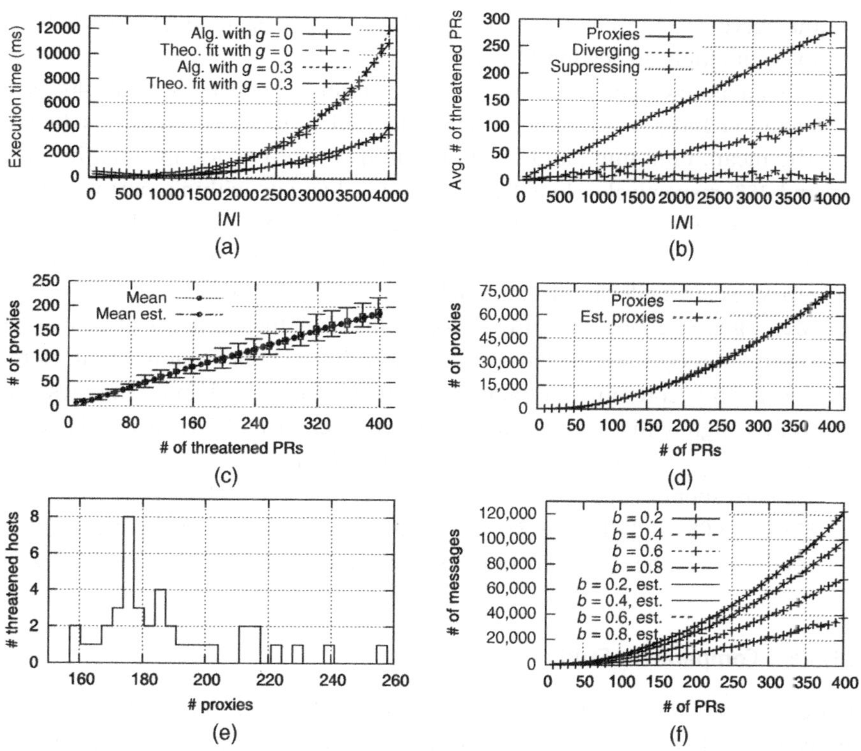

Figure 11.14 Performance in DPG. (a) execution time of GenPolicy, (b) the average number of PRs under attack when $g = 0.3$, (c) the number of proxies of each TH when $|N| = 4000$, (d) the number of total proxies when $|N| = 4000$, (e) the number of THs versus the number of their proxies, $|N| = 4000$, and (f) the number of control messages in enforcing defense policy.

$g = 0.3$. The two curves closely align with their estimations, namely, $0.0015 \cdot |N|^2 - 0.1782 \cdot |N| \cdot \log |N| + 233.7825$ and $O(0.0028 \cdot \log |N| \cdot |N|^2 - 0.0184 \cdot |N|^2 + 440.4651)$, respectively, thus indicating scalability in execution time. When 30% of PRs undergo a renaissance, Figure 11.14(b) demonstrates the average number of PRs under attack across various topology sizes for different countermeasures. Impressively, an average of 75.78% of the PRs is effectively suppressed, underscoring the necessity of the suppressing mechanism.

ii) **Proxies overhead**: Figure 11.14(c) presents a boxplot illustrating the number of generated proxies for 10–400 congested PRs in the synthesized topology with $|N| = 4000$. The mean number of proxies aligns with the curve $3.016 \cdot |W^{\text{th}}| \cdot \log |W^{\text{th}}| - 15.651$. These numbers closely match their estimations from Section 11.2.3.5. Furthermore, in Figure 11.14(d), the total numbers of proxies under varying congested PR scenarios follow the curve $2.834 \cdot |W^{\text{th}}|^2 \cdot \log |W^{\text{th}}| - 1411.8$, which again aligns closely with the estimations in Section 11.2.3.5.

For a more specific scenario, when the number of congested PRs is 400 (see Figure 11.14(e)), the number of THs under different proxy counts is detailed. The mean, minimum, and maximum numbers of proxies are 188.075, 157, and 258, respectively. Notably, these values fluctuate within a narrow range, ensuring they do not exceed the number of leaf nodes in the SPT from any proxy (refer to Section 11.2.3.5). These results underscore the efficiency in proxy resource allocation achieved by the DPG and DPE.

iii) **Messages overhead**: Figure 11.14(f) illustrates the numbers of generated control messages for 10–400 congested PRs during policy enforcement. Here, $b = |\Lambda|/|\tilde{L}|$ represents the ratio of PRs flooded no less than η. The message counts align with their estimation (refer to Section 11.2.3.4) encapsulated by $a_1 \cdot |W^{\text{th}}|^2/a_2 \cdot \log(|W^{\text{th}}|) + a_3$. For different b values of 0.2, 0.4, 0.6, and 0.8, the corresponding (a_1, a_2, a_3) pairs are (5.238, −51.47, 1370), (4.195, −35.19, 878.7), (3.081, −34.15, 999.3), and (1.633, −15.88, 361.6), respectively. These results affirm the scalability of message complexity achieved by the PRDD.

11.3.6 Comparative Study

To demonstrate the advantage, we compared the PRDD with a number of existing benchmark solutions. Here, the evaluation only considers the initial attack stage, as the renaissance one remains unconcerned by the existing solutions (see Section 4.5). The results are detailed as follows:

i) **The MTD-based approach**: Figure 11.15 shows the defense effect of PROTAG [Venkatesan et al., 2016], [Aydeger et al., 2016], and EqualNet [Kim et al., 2022]

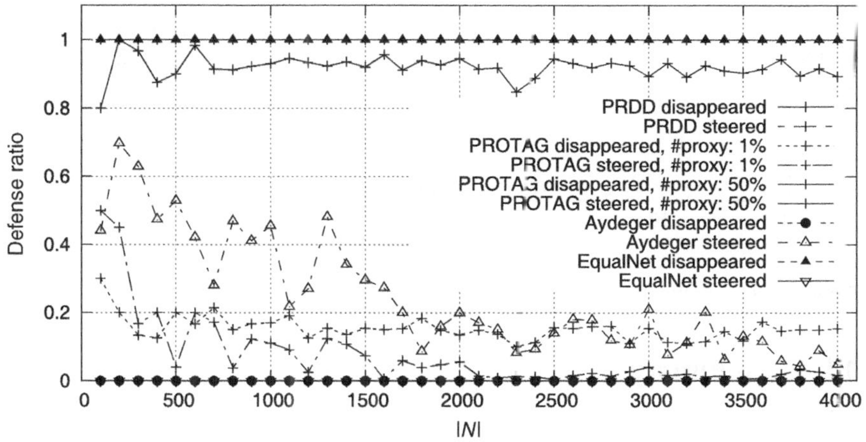

Figure 11.15 Comparative studies on disappeared PRs and steered PRs ratios.

in the ratios of the disappeared PR and the steered PR (see Section 11.3.1) for different network scales. We compared those of PRDD. The results show that the numbers of disappeared PRs and steered PRs are all equal to 100% in PRDD, which are higher than those in PROTAG and Aydeger. It indicates a better attack mitigation effect. Further, for PROTAG in the topology with $|N| = 4000$, when the proportion is 1% and 50%, 1.5% and 0% of PRs are diverged, respectively. It is because the target is occasionally moved to the proxies and not passed via the PRs. For Aydeger, 95.3% of PRs are unable to be diverged due to their criticality. However, PRDD can steer all PRs with created proxies to avoid their congestion.

ii) **The traffic throttling-based approach**: Figure 11.16 illustrates the defense effect of LFADefender [Wang et al., 2019], RADAR [Zheng et al., 2018], and SPIFFY [Kang et al., 2016] in terms of their positive ratios (see Section 11.3.3). Here, the curve of LFADefender fluctuates for different numbers of bots due to the LOF deviating from accuracy when ICMP packets of normal hosts arrive at switches in bursts, thanks to the on–off model. Additionally, the RADAR curve shows an average positive ratio of 12.63% since the sampling period is too long to capture throughput changes of bots. Moreover, the positive ratio values in SPIFFY are 0% because the sketch-based measurement fails with the stable rate of bot flooding. However, the positive ratio values in PRDD are 100% since detecting congested PRs in PRDD is easier than detecting malicious flows or bots in those three solutions. Meanwhile, DPE can mitigate each PR.

Hence, PRDD holds the defensive advantage in mitigating more congested PRs and reacting to more flooded bots compared to the solutions mentioned above.

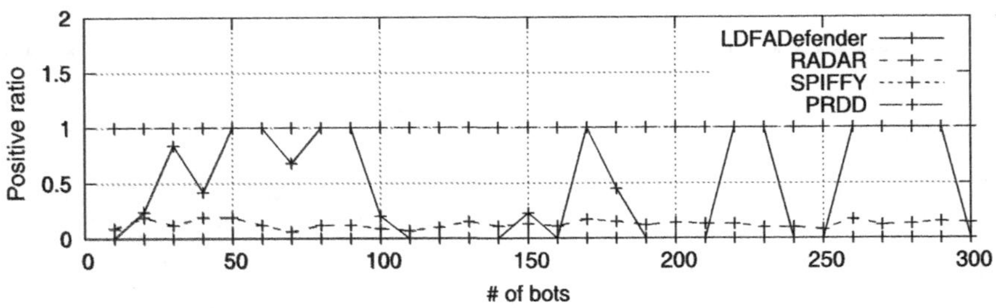

Figure 11.16 Comparative studies on positive ratios.

11.4 Summary

The proposed PRDD solution proficiently guards against the two-stage stealthy LFA, targeting the PRs in the network. PRDD adopts proxy creation and traffic steering countermeasures to diversify the identified PRs during its initial and renaissance attack stages, respectively. These measures not only halt the ongoing flooding but also make it challenging for adversaries to locate the PRs, effectively thwarting the two-stage attacks.

We conduct a comprehensive evaluation of PRDD performance, employing both NS-3 simulation and MiniNet emulation. The numerical results affirm that PRDD can adeptly defend against attacks with a scalable computation cost and an acceptable overhead. Moreover, PRDD exhibits a superior mitigation effect compared to existing approaches.

Questions

11.1 How does the PRDD solution defend against the two-stage stealthy LFA attacks targeting PRs in a network?

11.2 How does the PRDD solution leverage proxy creation and traffic steering countermeasures to thwart attacks during both the initial and renaissance stages?

11.3 What are the specific steps involved in the initial attack stage of a two-stage LFA targeting PRs, as described in the chapter?

11.4 How does the adversary execute a calculated sequence of moves during the stealthier initial attack stage, and what is the significance of probing the network paths with irregular ICMP requests?

11.5 Can you explain the key aspects evaluated in the defense effectiveness of PRDD, particularly focusing on congestion levels and defense policies?

11.6 How does PRDD perform in defending against attacks compared to existing approaches, specifically in mitigating congested PRs and reacting to flooded bots?

11.7 How does the proposed PRDD solution contribute to the field of network security, and what are the key takeaways from the comprehensive evaluation conducted in this chapter?

Bibliography

Alexander Afanasyev, Neil Tilley, Peter Reiher, and Leonard Kleinrock. Host-to-host congestion control for TCP. *IEEE Communications Surveys & Tutorials*, 12(3):304–342, 2010.

Doreid Ammar, Thomas Begin, and Isabelle Guerin-Lassous. A new tool for generating realistic Internet traffic in NS-3. In *SIMUTools*, pages 81–83, 2011.

A. Aydeger, N. Saputro, K. Akkaya, and M. Rahman. Mitigating crossfire attacks using SDN-based moving target defense. In *IEEE LCN*, pages 627–630, 2016.

CAIDA. The CAIDA UCSD anonymized Internet traces - equinix-Chicago.dira.20160406-130000.UTC.anon, 2016. URL https://www.caida .org/catalog/datasets/passive_dataset.

Maurizio Casoni, Carlo Augusto Grazia, Martin Klapez, and Natale Patriciello. Implementation and validation of TCP options and congestion control algorithms for NS-3. In *Workshop on NS-3*, pages 112–119, 2015.

Luciano Jerez Chaves, Islene Calciolari Garcia, and Edmundo Roberto Mauro Madeira. OFSwitch13: Enhancing NS-3 with OpenFlow 1.3 support. In *Workshop on NS-3*, pages 33–40, 2016.

Mo Chen, Rezaul A. Chowdhury, and Vijaya Ramachandran. Priority queues and Dijkstra's algorithm. Technical report, University of Texas, 2007.

Alvise De Biasio, Federico Chiariotti, Michele Polese, Andrea Zanella, and Michele Zorzi. A QUIC implementation for NS-3. In *Workshop on NS-3*, pages 1–8, 2019.

Floodlight Controller, 2020. URL http://projectfloodlight.org/.

IETF. IETF RFC 2136: Dynamic updates in the domain name system (DNS update). Standard, IETF, 1997.

M. S. Kang, S. B. Lee, and V. D. Gligor. The crossfire attack. In *IEEE S & P*, pages 127–141, 2013.

Min Suk Kang, Virgil D. Gligor, and Vyas Sekar. SPIFFY: Inducing cost-detectability tradeoffs for persistent link-flooding attacks. In *USENIX NDSS*, 2016.

Jinwoo Kim, Eduard Marin, Mauro Conti, and Seungwon Shin. EqualNet: A secure and practical defense for long-term network topology obfuscation. In *USENIX NDSS*, 2022.

Adam Langley, Alistair Riddoch, Alyssa Wilk, Antonio Vicente, Charles Krasic, Dan Zhang, Fan Yang, Fedor Kouranov, Ian Swett, Janardhan Iyengar, Jeff Bailey, Jeremy Dorfman, Jim Roskind, Joanna Kulik, Patrik Westin, Raman Tenneti, Robbie Shade, Ryan Hamilton, Victor Vasiliev, Wan-Teh Chang, Zhongyi Shi. The QUIC transport protocol: Design and Internet-scale deployment. In *ACM SIGCOMM*, 2017.

Bob Lantz, Brandon Heller, and Nick Mckeown. A network in a laptop: Rapid prototyping for software-defined networks. In *ACM HotNets*, 2010.

M. E. Newman, S. H. Strogatz, and D. J. Watts. Random graphs with arbitrary degree distributions and their applications. *Physical Review E Statistical Nonlinear & Soft Matter Physics*, 64(2):026118, 2001.

Open Networking Foundation. OpenFlow specification v1.3, 2012.

George F. Riley and Thomas R. Henderson. The NS-3 network simulator. *Modeling & Tools for Network Simulation*, pages 15–34. Springer-Verlag, Berlin, Heidelberg, 2010.

Sridhar Venkatesan, Massimiliano Albanese, Kareem Amin, Sushil Jajodia, and Mason Wright. A moving target defense approach to mitigate DDoS attacks against proxy-based architectures. In *IEEE CNS*, pages 198–206, 2016.

Juan Wang, Ru Wen, Jiangqi Li, Fei Yan, Bo Zhao, and Fajiang Yu. Detecting and mitigating target link-flooding attacks using SDN. *IEEE Transactions on Dependable and Secure Computing*, 16(6):944–956, 2019.

Minlan Yu, Lavanya Jose, and Rui Miao. Software defined traffic measurement with OpenSketch. In *USENIX NSDI*, pages 29–42, 2013.

Jing Zheng, Qi Li, Guofei Gu, Jiahao Cao, David K. Y. Yau, and Jianping Wu. Realtime DDoS defense using COTS SDN switches via adaptive correlation analysis. *IEEE Transactions on Information Forensics and Security*, 13(7): 1838–1853, 2018.

Wu Zhijun, Li Wenjing, Liu Liang, and Yue Meng. Low-rate DoS attacks, detection, defense, and challenges: A survey. *IEEE Access*, 8:43920–43943, 2020.

12

Practice Guidelines for High-Available Networks

As discussed in Chapters 3–11, both classical and cutting-edge solutions, including the IEC 62439 protocols, are essential for enhancing network availability. These solutions encompass OpenFlow-based Non-Disjoint Path Aggregation Mechanism (OPAM), Traffic-Aware Fast Reroute (TA-FRR), Disruption Resilient Transport Protocol (DRTP), as well as its related methods such as Resilient Sensor Data Dissemination (RSDD), HeuMPSG, Adaptive RaptorQ Encoder and Decoder (ARED), and Persistent Route Diversification Defense (PRDD). To effectively implement these solutions, operators, researchers, and developers must establish an environment that supports their deployment, enabling performance testing and integration into existing platforms. Section 12.1 of this chapter will present concrete guidelines for designing such an environment. Using this setup, we will select a typical example of DRTP to demonstrate its practicality and effectiveness in significantly improving transport reliability, thereby enhancing power grid stability. Lastly, we will provide guidelines for implementing these solutions.

12.1 Development Environments

The development environment is critical for assessing how the availability of various networking solutions can influence power grid stability. As previously mentioned, the interactions among diverse terminals, data concentrators, and Supervisory Control and Data Acquisition (SCADA) systems [Thomas, 2020] – the secondary systems – play a pivotal role in determining the performance of the electric physical components of power grids, known as the primary systems. These SCADA-based systems rely heavily on measurement and control processes to maintain stable grid operations.

Communication Networks in Smart Power Grids, First Edition. Boyang Zhou.

Companion Website: www.wiley.com/go/SmartPowerGrids1e

Initially, the system measures electrical data from generators, transformers, power lines, and other equipment. This statistical data typically encompasses voltage and power levels. Based on the collected measurements, State Estimation (SE) [Huang et al., 2012] is conducted, leading to informed control decisions. Current research on cyber–physical systems primarily focuses on real-time digital simulations. A typical simulation framework comprises three layers: the power supply layer, the network layer, and the control center layer. In this context, a hardware-in-the-loop approach integrates real-time power system simulators with control and protection devices. However, these studies often overlook the impact of packet loss in measurement data or control errors on SE during network operations. This gap has motivated us to design a test platform that merges the decision-making process with SE and the physical power system.

12.1.1 Overall Design

As depicted in Figure 12.1, the architecture of the power grid simulation test environment is designed to replicate the communication networks that interact with the SCADA systems [Thomas, 2020], terminals, and data concentrators interfacing with the primary systems of power grids (see Section 1.2). The outlook of our testbed is shown in Figure 12.2. To achieve this, the architecture comprises four main components: (1) the physical power grid and process layer, which is

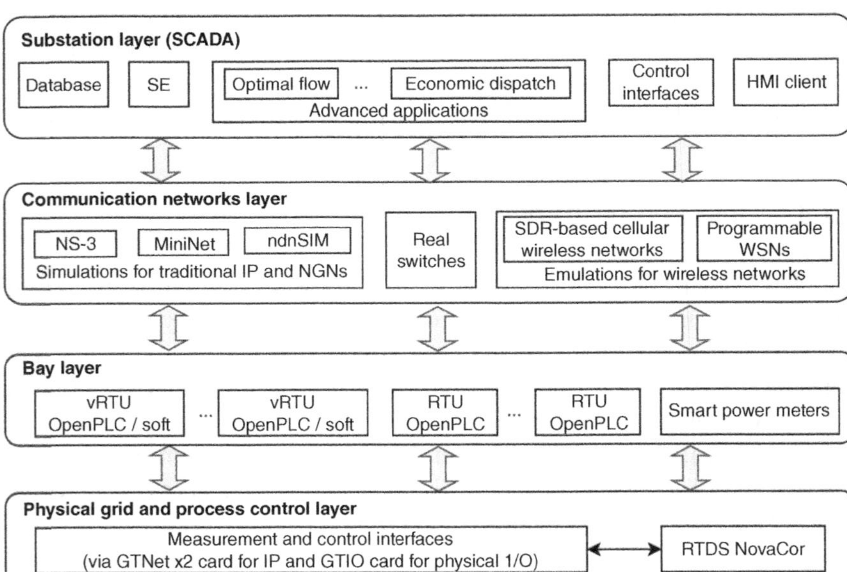

Figure 12.1 Development environments used for solution verification.

Figure 12.2 Outlook of our testbed.

simulated using Real-Time Digital Simulator (RTDS) [Kuffel et al., 1995; RTDS Technologies Inc., 2023] to represent the primary systems; (2) the bay layer, which employs real Programmable Logic Controllers (PLCs) as Remote Terminal Units (RTUs) to simulate the measurement and control functions within the power grid environment; (3) the communication network layer, which uses simulations and emulations to replicate the entire optical–wireless communication process, integrated with the bay layer and substation control layer; and (4) the substation control layer, which provides essential SCADA functions [Thomas, 2020], primarily encompassing SE [Huang et al., 2012] and related control decision applications.

12.1.2 Communication Networks Layer

The communication networks layer provides an integration of virtual and physical communication simulation resources based on NS-3 and the access of physical terminal devices, as well as the construction of communication network scenarios. This subsystem is designed with the following three layers:

- Communication resource layer, which includes the NS-3 simulator [Riley and Henderson, 2010] based on discrete time simulation and access to real network terminals and LXC virtual machines [Linuxcontainers.org, 2024], interconnected through low-latency data-exchange pathways.

- Network topology construction and operational state collection, where automated scene construction is realized based on XML language descriptions, allowing users to customize the network topology and facilitate resource destruction. Operational state collection is achieved via discrete event reporting, network measurements, and data collection from switch ports.
- Communication network scenario construction and status visualization user interface, which enables the editing and management of scene topology and the dynamic display of network operational status.

12.1.2.1 Network Simulation Scene Compiler

The experimental communication network simulation scene compiler (hereinafter referred to as the compiler) provides automated compilation from XML-based communication network scenes-to-scene execution instances. Figure 12.3 shows the functional structure of the compiler, which consists of two parts: (1) Main implements a command-line-based user interface, processing based on the XML topology file and parameters provided by the user; (2) SimInstanceGenerator implements the compilation and generation of simulation instances. The generated scene execution instances include: (1) C++ executable files based on NS-3 [Riley and Henderson, 2010]; (2) scene initialization scripts; (3) scene destruction scripts; (4) LXC virtual machine [Linuxcontainers.org, 2024] configuration files. The operating system environment on which this instance is based includes the following components: NS-3 runtime environment, LXC virtual machine, Virtual Local Area Network (VLAN) physical devices, etc.

The topology description language uses XML syntax, allowing users to flexibly define nodes in the topology and their operating applications, the connections

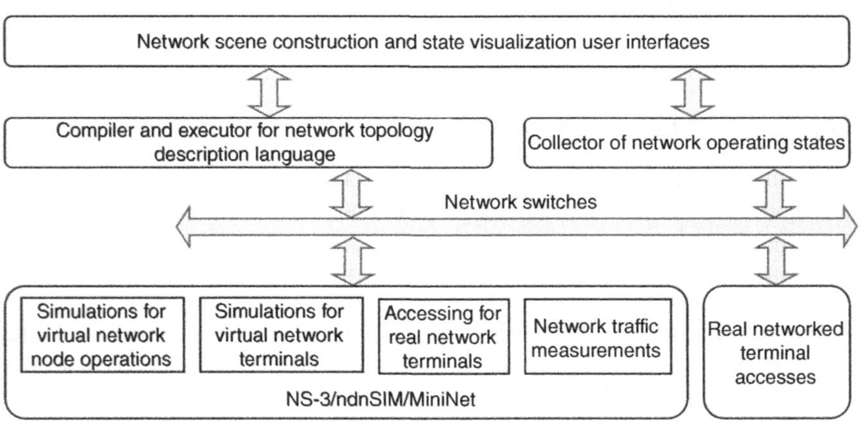

Figure 12.3 Communication network simulation.

between nodes, routing generation methods, trackers, external listening addresses, and other components, thereby achieving flexible definitions for IP networks. The specific components are described as follows:

```
<?xml version="1.0" encoding="utf-8"?>
<topo xmlns:xsi="http://www.w3.org/2001/XMLSchema-instance"
 xsi:schemaLocation="topo_schema.xsd">
    (Node Definition)
    (Link Definition)
    (Routing Generation Method)
    (Simulation Instance Parameter Configuration)
    (Tracker Settings)
</topo>
```

Node types include the following four:

(1) `vir_router`, virtual router, simulated using discrete mathematical models; (2) `vir_host`, virtual host, simulated using discrete mathematical models, which includes an application defined by the <app> ...</app> tags; (3) `lxc_router`, LXC virtual machine, where users can define the operating system, IP, and gateway of the virtual machine, which will be automatically generated during simulation instance execution via the initialization script; (4) `vlan_accessor`, VLAN physical host, where the initialization script automatically creates VLAN bridging. Using this VLAN, the NS-3 host can interact with external physical hosts, for example, enabling communication with external hosts through a layer 3 switch.

The specific configuration of the four node types is illustrated in the following example:

```
<node name="[Node Name]" type="[Node Type]" />
<node name="[Node Name]" type="[Node Type]">
    <app type="[TCP or UDP]" role="source' name="[Application Name]">
        <start_time_secs>[Start Time]</start_time_secs>
        <end_time_secs>[End Time]</end_time_secs>
        <local_host>[Local Host Name]</local_host>
        <local_port>[Local Port]</local_port>
        <remote_host>[Remote Host Name]</remote_host>
        <remote_port>[Remote Port]</remote_port>
        <traffic type="onoff">
            <on_time_us>[Traffic Sending Duration]</on_time_us>
            <off_time_us>[Traffic Stop Duration]</off_time_us>
            <data_rate_bps>[Sending Rate]</data_rate_bps>
            <packet_size>[Packet Size]</packet_size>
            <max_bytes>[Total Bytes to Send]</max_bytes>
       </traffic>
        <!--
         <traffic type="bulk">
            <send_size>[Single Packet Size]</send_size>
            <max_bytes>[Total Bytes to Send]</max_bytes>
        </traffic>
         -->
   </app>
</node>
```

Links are divided into the following two types: (1) Point-to-point links, such as simulating optical channels and xDSL, which bind two network cards; (2) Carrier Sense Multiple Access (CSMA) links, such as simulating switches, which bind at least one network card. Each link defines delay, transmission rate, network mask, and the network card settings related to the link. Each network card interfaces with the above nodes or network cards through node name (`node_name`) and interface name (`int_name`), and configures the IP address and maximum transmission unit through `ipv4_address` and `MTU`. An example of link definition is shown below.

```
<link type="[p2p or csma]">
   <delay_us>[Link Delay]</delay_us>
   <data_rate_bps>[Maximum Rate]</data_rate_bps>
   <net_mask>[Subnet Mask]</net_mask>
   <interface node_name="[Node Name]" int_name="[Network Card Name]"
    ipv4_addr="[IPv4 Address]" mtu="[MTU Size]" />
   <interface... />
    ...
</link>
```

Routing Generation Method Users can specify global static routing. When defining this reason, the system will automatically calculate the shortest path based on Dijkstra's algorithm and configure the routing table for each node. The configuration command is as follows:

```
<routing type="global" />
```

Users can specify the IP and port number of external event listeners through configuration parameters, as well as the simulation duration and simulation type. There are two simulation types: (1) real-time simulation (realtime), which simulates according to real system time, allowing real-time interaction with external VLAN devices and LXC virtual machines; (2) non-real-time (default), which simulates at the fastest simulation time.

```
<sim>
    <server_ip>127.0.0.1</server_ip>
    <server_port>668</server_port>
    <stop_time_secs>600</stop_time_secs>
    <sim_type>realtime</sim_type>   <!-- default or realtime -->
</sim>
```

The tracker is used to monitor the data and state transmission in the topology and sends the monitored data to the specified `server_ip` and `server_port`. An example of the settings is as follows; for specific syntax, refer to Sections 2.5.1–2.5.2.

```
<tracers_settings>
    <counters>
        <counter .../>  (Counter Syntax Specification)
```

```
        <counter .../>
        ...
    </counters>
    <points>
        <point .../> (Monitoring Point Syntax Specification)
        <point .../>
        ...
    </points>
</tracers_settings>
```

12.1.2.2 Event Tracking

i) **Cumulative period counter:** This counter accumulates variable statistics and reports the variable value to the client specified by the user at the `server_ip` and `server_port` tags at a specified period. Users can define the counter variable name and statistical period through the following tags:

```
<counter type="cumulative_period" name="[counter_name]"
 interval_us="[interval_us]" />
```

`[counter_name]`: Counter name, format is `[A-Za-z0-9_]+`
`[interval_us]`: Period (microseconds)

ii) **Cumulative immediate counter:** This counter accumulates variable statistics and reports the variable value immediately to the client specified by the user at the `server_ip` and `server_port` tags when the variable value changes. Users can define the counter name through the following tags:

```
<counter type="cumulative_imm" name="[counter_name]" />
```

`[counter_name]`: Counter name, format is `[A-Za-z0-9_]+`

iii) **Single immediate counter:** This counter records the current value of the counter and reports the variable value immediately to the client specified by the user at the `server_ip` and `server_port` tags when the variable value changes. Users can define the counter name through the following tags:

```
<counter type="one_imm" name="[counter_name]" />
```

`[counter_name]`: Counter name, format is `[A-Za-z0-9_]+`

iv) **Single period counter:** This counter records the current value of the counter and reports the variable value to the client specified by the user at the `server_ip` and `server_port` tags at a specified period. Users can define the counter variable name and statistical period through the following tags:

```
<counter type="one_period" name="[counter_name]"
 interval_us="[interval_us]" counter_class="Time"/>
```

`[counter_name]`: Counter name, format is `[A-Za-z0-9_]+`
`[interval_us]`: Period (microseconds)

12.1.2.3 Software-Defined Radio (SDR)-Based Cellular Wireless Networks

In modern power grid communications, SDR has played an important role in enabling flexible and cost-effective experiment environment for cellular networks requiring high-speed signal processing [Ettus, 2025]. The following content presents a detailed overview of the SDR-based cellular wireless network architecture, providing insight into its capabilities for testing and improving communication reliability in the power grid.

i) **Experimental setup:** The test environment for SDR-based cellular networks is established in a secure, electromagnetically shielded chamber (see Figure 12.4), capable of operating in the 20 MHz to 6 GHz frequency range. This chamber is effectively isolated from external electromagnetic interference by utilizing a steel-enclosed Faraday cage. Within this controlled environment, a range of 2G to 4G wireless communication simulations can be conducted, offering a robust platform for performance testing.

Concretely, the 2G environment reproduces a system based on Global System for Mobile Communications (GSM) operating in the 850 and 900 MHz bands and is deployed to simulate 2G power grid wireless terminals (see Figure 12.5). This setup enables various 2G-based devices, such as smart meters and distribution network terminals, to be tested for security vulnerabilities, including rogue base stations and data interception.

Additionally, the 4G simulation environment (see Figure 12.6) consists of two components: the radio frequency (RF) section and the core network section. The RF section, built on the OpenAirInterface platform [OpenAirInterface.org, 2024], supports Long Term Evolution (LTE) base stations operating at 1.8 GHz, 2.6 GHz, and 230 MHz, offering comprehensive testing for signal interference, transmission performance, and security. The core network employs Software-Defined Networking (SDN) technology, which is aligned with 5G architecture as defined by the 3GPP standards [3GPP, 2024],

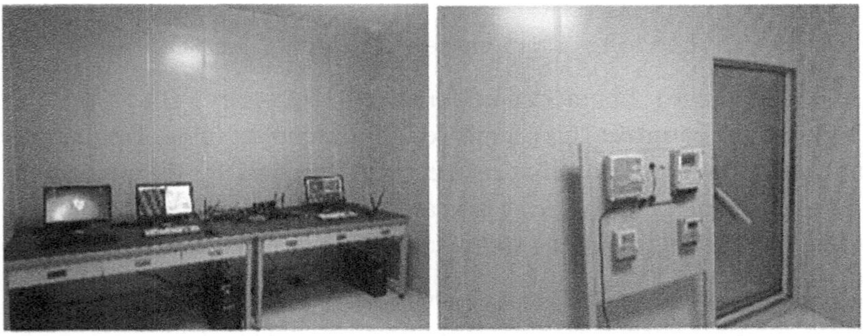

Figure 12.4 Electromagnetically shielded chamber for wireless networks.

Figure 12.5 2G emulation environment with power meters.

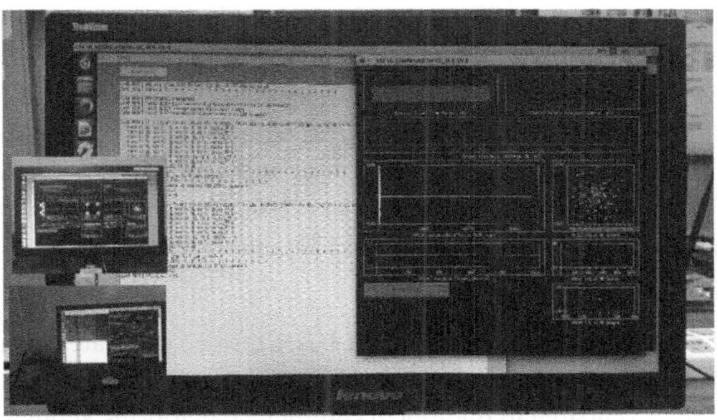

Figure 12.6 4G emulation environment with power meters.

making it suitable for structured network simulations in commercial power grid environments.

1) **Physical network simulation:** The simulation is executed using four programmable ONetSwitch network switches, connecting the power grid control center, industrial control systems, virtual simulation network, and wireless base stations, allowing for a realistic testing environment.
2) **Virtual network simulation:** A virtual network topology simulates the power grid's internal communication infrastructure, providing a flexible and adaptable platform for testing network performance under various conditions.

ii) **Spectrum analysis for public 2G networks:** The Electric Wireless Security Situational Awareness System (EWSSA) is a platform designed to monitor the resource usage of wireless communication in the power grid (see Figure 12.7).

It provides a comprehensive platform for analyzing the security of 2G terminals in the distribution and metering networks. This system enables the detection of security vulnerabilities such as rogue base stations, terminal-side attacks, and unauthorized data access.

- **2G monitoring units:** EWSSA is composed of multiple units, including the spectrum monitoring, Site-Control (for controlling monitoring points), Analyzer (for data analysis), and WebSys (for web-based monitoring).
- **Security feature analysis:** EWSSA can assess current base station configurations by analyzing the system information of Absolute Radio Frequency Channel Number (ARFCN) and Broadcast Control Channel (BCCH) [3GPP, 1999], providing insights into base station authenticity and performance characteristics.
- **Radio resource and spectrum load analysis:** The system evaluates the allocation of uplink and downlink channels, identifying potential anomalies in time slot allocation and detecting issues such as Distributed Denial-of-Service (DDoS) attacks targeting frequency resources.
- **Wireless base station power and terminal access analysis:** The power levels of base station pulses are monitored in real time to identify inconsistencies, which could indicate signal jamming or unauthorized interference. Similarly, the system tracks all terminal access behaviors, enabling the detection of unauthorized devices attempting to gain access to network resources.

Key innovations and platform flexibility. This SDR-based testbed integrates both SDR and SDN technologies to create a customizable, transparent, and low-cost simulation environment. Compared to traditional commercial platforms, it offers superior flexibility in terms of code-level customization, module functionality, and radio channel configurations. As a result, it enables thorough testing of network security and performance, supporting improvements in wireless communication systems for power grids.

Additionally, EWSSA excels in identifying base station spoofing, intelligent interference, and DDoS attacks at the resource level, filling existing gaps in the security of wireless communication in power grids.

12.1.2.4 Programmable Wireless Sensor Networks (WSNs)

WSNs play a pivotal role in modern power grids, facilitating the monitoring and control of a wide array of devices across vast geographical regions. A programmable WSN framework offers a versatile platform for testing the communication performance of various power grid terminals, providing critical insights into the reliability and scalability of communication systems under different scenarios.

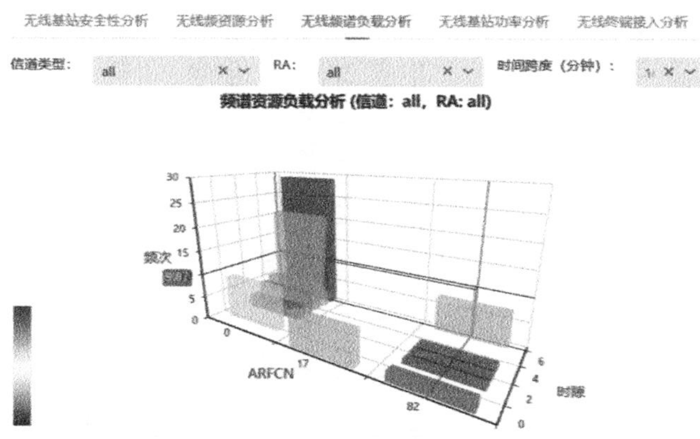

Spectrum resource load

Spectrum resource analysis

Terminal access analysis

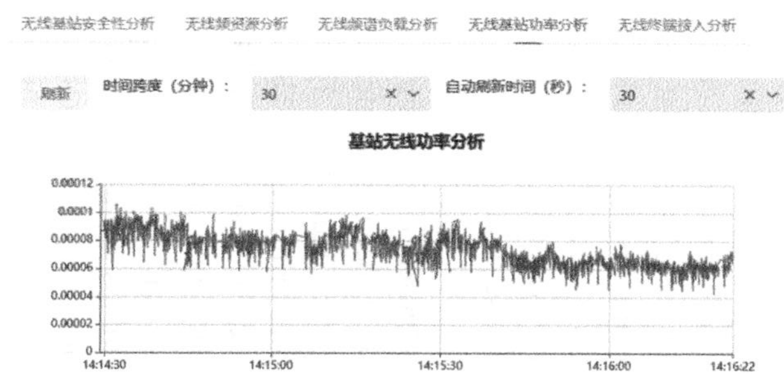

Base station power analysis

Figure 12.7 Analysis of 2G wireless resources.

One of the common technologies used in WSNs for power grid applications is ZigBee, which is built on the IEEE 802.15.4 standard [IEEE, 2020]. ZigBee provides low-power, low-data-rate wireless communication, making it an ideal choice for sensor networks that require efficient energy consumption. In a typical power grid environment, ZigBee-enabled WSNs can monitor and collect data such as temperature, humidity, and light levels, as well as signal metrics like Received Signal Strength Indicator (RSSI) and Link Quality Indicator (LQI). These metrics are essential for evaluating the communication link quality between different sensor nodes and base stations.

Simulation Environment To simulate a realistic power grid communication environment, a ZigBee-based Internet-of-things (IoT) simulation platform has been developed (see Figure 12.8). This environment incorporates TelosB sensor nodes [TelosB, 2025] running the TinyOS operating system [TinyOS, 2025], which are widely used for academic and industrial research due to their flexibility and compatibility with low-power wireless communication protocols like ZigBee. The simulation environment also includes TelosB-based base stations and ZigBee wireless communication software to emulate the interaction between power-monitoring systems and ZigBee devices.

In this setup, the ZigBee network enables seamless communication between different IoT devices deployed in a power grid, such as smart meters, temperature sensors, and humidity monitors. The power ZigBee wireless communication software facilitates the transmission and reception of data between sensor nodes and the central monitoring system. This allows researchers and engineers to simulate typical ZigBee wireless communication access scenarios encountered in power grid monitoring, helping to identify potential challenges and optimize performance.

Communication Testing The ZigBee-based simulation environment supports a variety of testing scenarios for different power grid terminals, allowing for the collection of critical data on communication performance. By measuring key parameters such as RSSI and LQI, it is possible to assess the signal strength and the overall quality of communication links in the network. These parameters are particularly important in power grid environments, where interference from physical structures, weather conditions, and other wireless devices can significantly impact communication reliability.

In addition to environmental factors, testing also includes performance evaluations under different traffic loads and network topologies, helping to determine

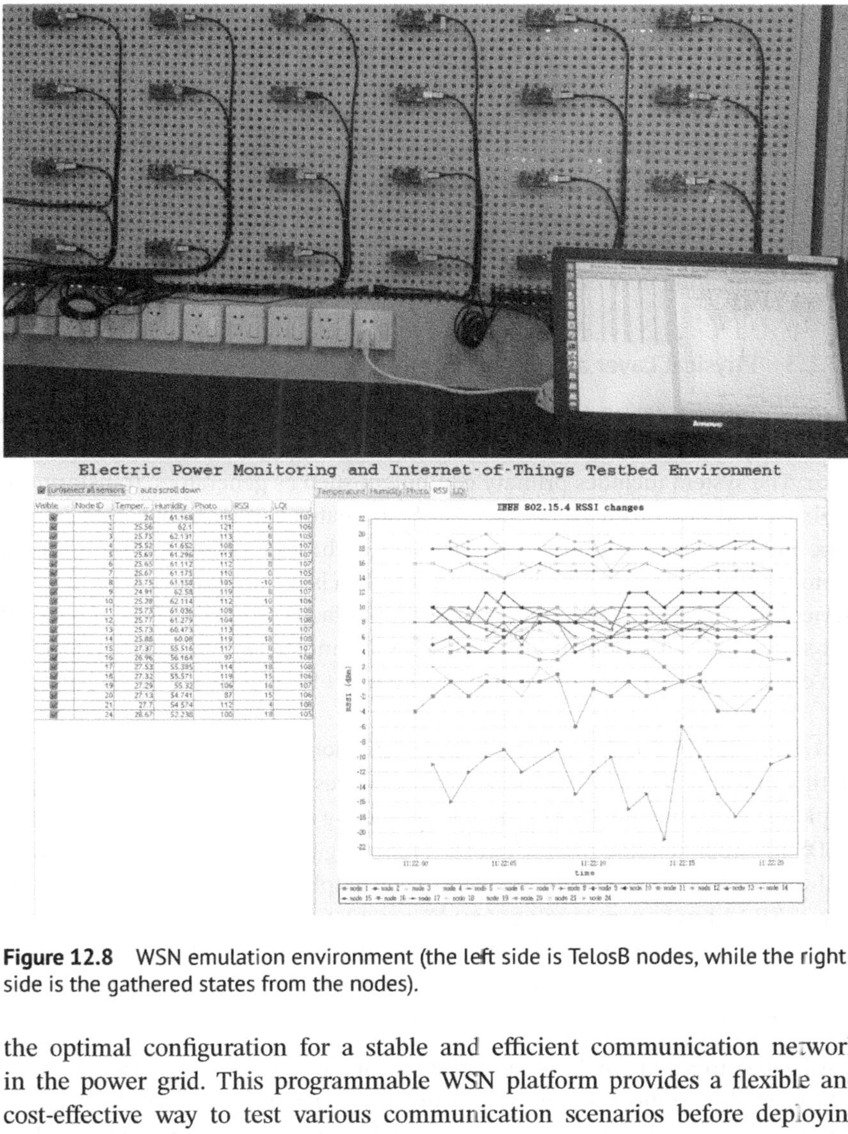

Figure 12.8 WSN emulation environment (the left side is TelosB nodes, while the right side is the gathered states from the nodes).

the optimal configuration for a stable and efficient communication network in the power grid. This programmable WSN platform provides a flexible and cost-effective way to test various communication scenarios before deploying large-scale systems in the field.

Those programmable WSNs, particularly those based on ZigBee and the IEEE 802.15.4 standard, offer significant advantages for testing the communication performance of power grid terminals. By simulating realistic network environments and collecting valuable data on communication metrics, these systems provide critical insights that help ensure the reliability and scalability of communication infrastructures in smart power grids.

In conclusion, the functional implementation of the experimental communication network simulation scene compiler enables users to flexibly define the communication network simulation environment, including node definitions, link definitions, routing generation methods, and event-tracking mechanisms. This provides a foundation for building complex simulation scenarios in the field of communication networks for power grids.

12.1.3 Physical Layer and Process Layer

In this study, an RTDS [Kuffel et al., 1995; RTDS Technologies Inc., 2023] is used for power grid simulation modeling of the physical layer and process layer. The system includes primary equipment such as generators, transformers, busbars, and circuit breakers. The RTDS simulator can exchange data in real time with external hardware or software devices through various communication protocols. Power grid simulation operational data is communicated with external devices via the RTDS simulator's network interface card GTNETx2 and gigabit transceiver input/output card GTIO [RTDS Technologies Inc., 2023]. GTNETx2 can easily connect to external devices via Ethernet for real-time data exchange.

12.1.3.1 IEEE 14-Bus Power System Implementation

In this study, the IEEE 14-bus power system serves as the primary case example, implemented using the RTDS system for modeling and real-time simulation. The RTDS incorporates RSCAD FX [RTDS Technologies Inc., 2023], a user-friendly real-time simulation software that provides a graphical interface for ease of use. The IEEE 14-bus system topology (see Figure 12.9) is derived from data collected from a power system in the Midwestern United States, consisting of 14 buses, 5 generators, and 11 loads, making it a widely recognized benchmark in power system studies. The key features of this system are listed below:

- All generators and synchronous capacitors in the system are equipped with the "IEEET1" excitation system, which utilizes a Direct Current (DC) exciter to provide excitation voltage and current. Additionally, each generator employs the "TGOV1" type thermal regulator, which plays a crucial role in regulating generator output.
- The primary-side voltage connecting nodes 5 and 6 is set at 18 kV, while the secondary-side voltage measures 16.78 kV. The maximum allowable voltage

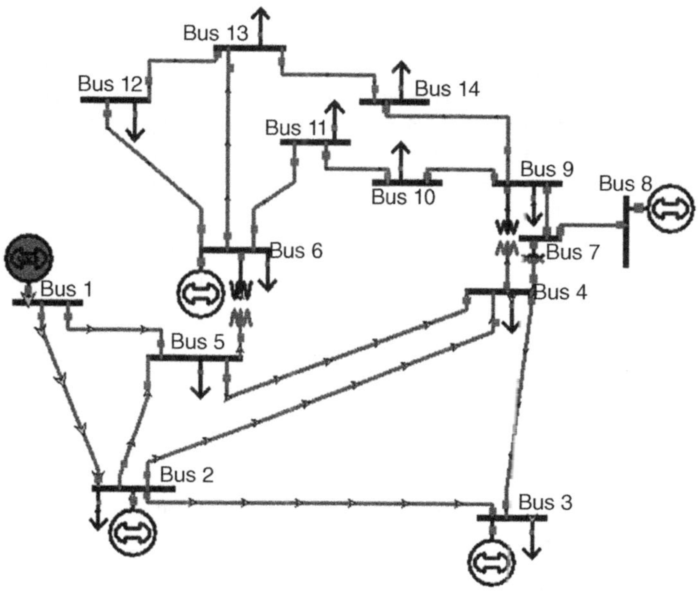

Figure 12.9 IEEE 14-bus power systems (This figure is copied from: https://icseg.iti.illinois.edu/files/2013/10/WSCC14.png).

at each bus is regulated to be 1.06 p.u., with a minimum allowable voltage of 0.94 p.u. These voltage limits are essential for maintaining system stability and reliability; operating outside these limits can lead to system disturbances and affect overall performance.

- Details regarding the types of buses, as well as active and reactive power demands, can be found in Table 12.1. The IEEE 14-node power grid system [IEEE, 2025] consists of 20 branches, with information about the starting and ending node numbers, resistance, reactance, admittance, and branch ratios detailed in Table 12.2.
- The system includes 5 generators, whose data is summarized in Table 12.3. The relationship between the cost y of each generator and its active power output x is expressed in Table 12.4. These tables collectively provide a comprehensive overview of the system's operational characteristics.

This detailed setup facilitates accurate simulation of the IEEE 14-bus power system, allowing for in-depth analysis of power flow, stability, and performance under various operational conditions. The simulation aims to investigate specific scenarios, such as fault conditions and load variations, to assess the

Table 12.1 Node information for the IEEE 14-bus system.

Node	Node type	Active power demand (MW)	Reactive power demand (Mvar)	Parallel conduct. (S)	Parallel susceptance (S)	Initial voltage mag. (p.u.)	Initial voltage angle (degrees)
1	Slack	0	0	0	0	1.06	0
2	PV	21.7	12.7	0	0	1.045	−4.98
3	PV	94.2	19	0	0	1.01	−12.72
4	PQ	47.8	−3.9	0	0	1.019	−10.33
5	PQ	7.6	1.6	0	0	1.02	−8.78
6	PV	11.2	7.5	0	0	1.07	−14.22
7	PQ	0	0	0	0	1.062	−13.37
8	PV	0	0	0	0	1.09	−13.36
9	PQ	29.5	16.6	0	19	1.056	−14.94
10	PQ	9	5.8	0	0	1.051	−15.1
11	PQ	3.5	1.8	0	0	1.057	−14.79
12	PQ	6.1	1.6	0	0	1.055	−15.07
13	PQ	13.5	5.8	0	0	1.05	−15.16
14	PQ	14.9	5	0	0	1.036	−16.04

system's resilience and efficiency. The comprehensive implementation of the IEEE 14-bus system not only serves as a critical testbed for real-time simulation but also provides valuable insights into the operational dynamics of power systems.

12.1.3.2 500 kV Hub Substation Implementation

The 500 kV hub substation plays a pivotal role in modern power systems, serving as a central aggregation point for multiple power sources and facilitating interconnections between various grid segments. This subsection outlines the implementation of the hub substation in the RSCAD environment [RTDS Technologies Inc., 2023], detailing its electrical main wiring and operational characteristics.

Key Functions The hub substation fulfills several critical functions within the power grid:

- **Aggregation of transmission lines:** It consolidates transmission trunk lines from various power plants, providing essential connections to key points in the grid and linking to the next lower-voltage level.

Table 12.2 Branch data for the IEEE 14-bus standard power system.

Branch	From node	To node	Resistance (p.u.)	Reactance (p.u.)	Susceptance (p.u.)	Ratio
1	1	2	0.01938	0.05917	0.0528	0
2	1	5	0.05403	0.22304	0.0492	0
3	2	3	0.04699	0.19797	0.0438	0
4	2	4	0.05811	0.17632	0.034	0
5	2	5	0.05695	0.17388	0.0346	0
6	3	4	0.06701	0.17103	0.0128	0
7	4	5	0.01335	0.04211	0	0
8	4	7	0	0.20912	0	0.978
9	4	9	0	0.55618	0	0.969
10	5	6	0	0.25202	0	0.932
11	6	11	0.09498	0.1989	0	0
12	6	12	0.12291	0.25581	0	0
13	6	13	0.06615	0.13027	0	0
14	7	8	0	0.17615	0	0
15	7	9	0	0.11001	0	0
16	9	10	0.03181	0.0845	0	0
17	9	14	0.12711	0.27038	0	0
18	10	11	0.08205	0.19207	0	0
19	12	13	0.22092	0.19912	0	0
20	12	14	0.17093	0.34865	0	0

- **Connection point for generating plants:** The substation serves as the primary interface for large and medium-sized power plants, connecting them to the high-voltage grid, thereby enhancing system reliability and efficiency.
- **Framework of the main grid:** Multiple hub substations, in conjunction with transmission trunk lines, form the backbone of the main grid, ensuring robust power distribution and facilitating effective load management.
- **Interconnection between adjacent power systems:** The substation acts as a vital connection point for interconnecting neighboring power systems, enabling resource sharing and enhancing overall grid resilience.
- **Primary source for lower-voltage grids:** It serves as a major power source for the next lower-voltage grid, ensuring stable and reliable energy supply to downstream systems.

Table 12.3 Generator data for the IEEE 14-bus system.

Generator	Type	Node	Rated power (MW)	Maximum power (MW)
1	synch.	1	0	40
2	synch.	2	40	60
3	synch.	3	50	90
4	synch.	6	70	100
5	synch.	5	40	60

Table 12.4 Generator costs for IEEE 14-bus system.

Generator	Power generation costs
1	$y = 0.04303x^2 + 20x$
2	$y = 0.25x^2 + 20x$
3	$y = 0.01x^2 + 40x$
6	$y = 0.01x^2 + 40x$
8	$y = 0.01x^2 + 40x$

Implementation The implementation of the 500 kV hub substation in the RSCAD environment is illustrated in Figure 12.10. This substation is connected to buses 8 and 6 within the IEEE 9-node grid, effectively aggregating transmission trunk lines from three power plants. The internal electrical main wiring of the substation, constructed in RSCAD, is depicted in Figure 12.11.

Configurations The electrical main wiring adopts a single-half breaker configuration, which offers several advantages:

- **High reliability:** Each of the four circuits is powered by two circuit breakers, ensuring that when one is under maintenance, the circuit remains energized. In the event of a fault on any bus, only the circuit breakers connected to that bus will trip, allowing other circuits to remain operational. Even when maintenance coincides with faults, the number of circuits experiencing outages will not exceed two, significantly enhancing overall reliability.
- **Flexible operational scheduling:** Under normal operational conditions, both sets of buses and all circuit breakers are in service, creating a multi-loop power supply configuration that allows for flexible operational scheduling and optimal resource utilization.

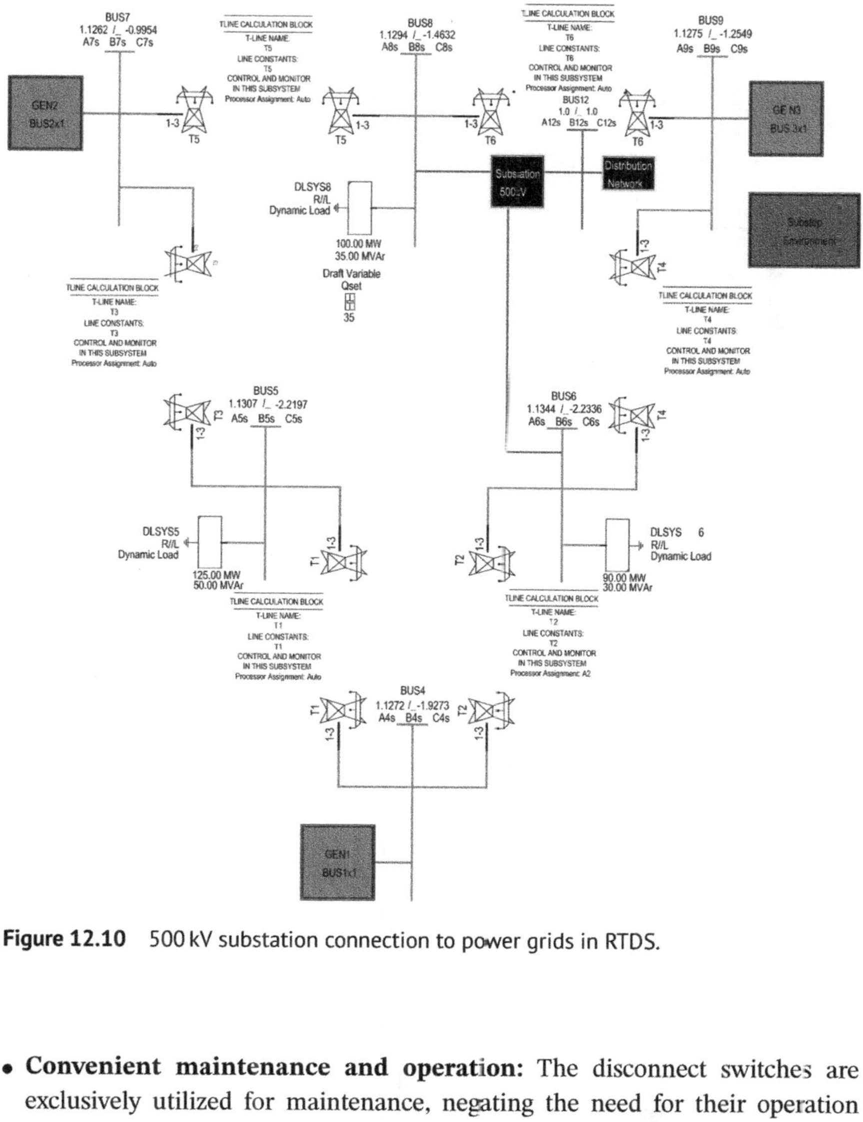

Figure 12.10 500 kV substation connection to power grids in RTDS.

- **Convenient maintenance and operation:** The disconnect switches are exclusively utilized for maintenance, negating the need for their operation during regular activities. When maintaining a circuit breaker, there is no requirement for bypass operations, and similarly, the circuit remains energized during bus maintenance.

This comprehensive setup allows for accurate simulation and analysis of the 500 kV hub substation, enabling assessments of power flow, stability, and performance under various operational conditions.

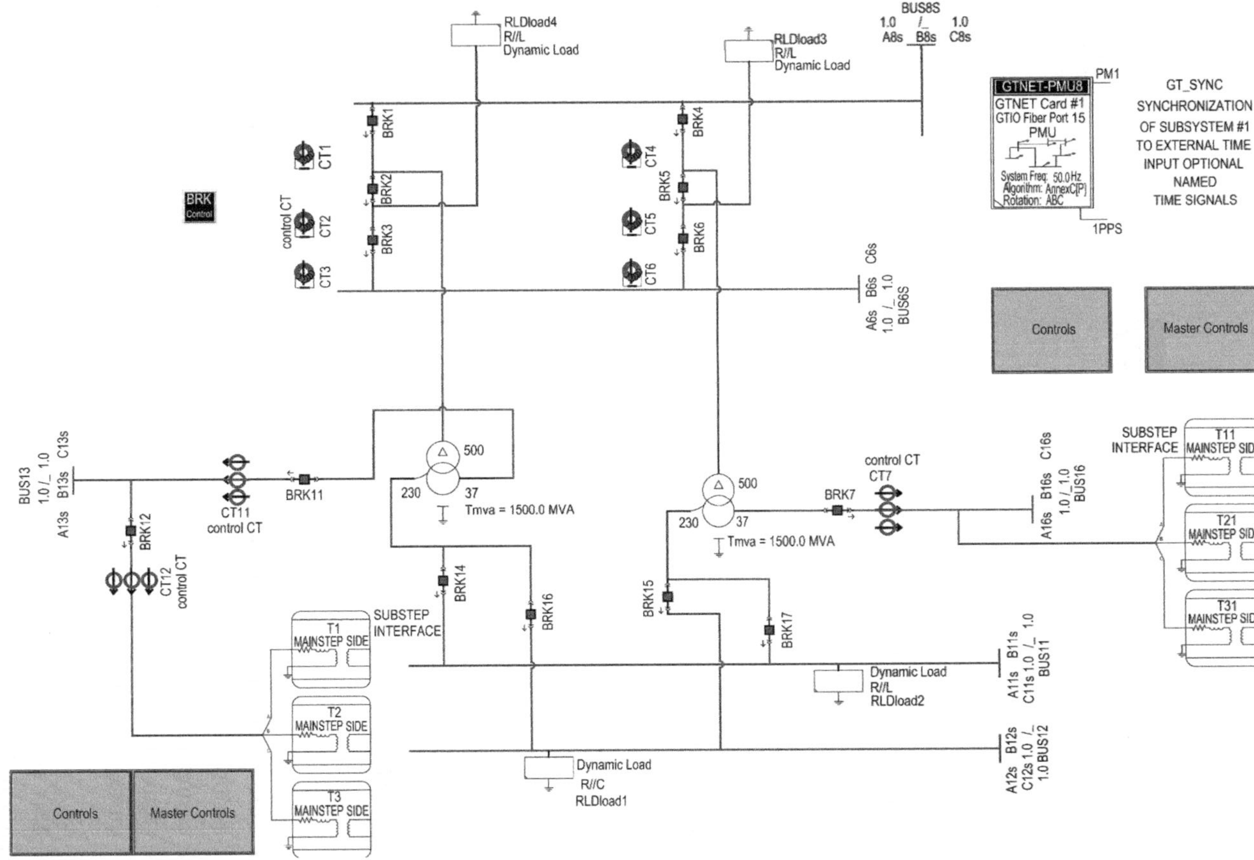

Figure 12.11 500 kV substation implementation in RTDS.

12.1.3.3 10 kV Power Distribution Grid Implementation

This subsection describes the implementation of a 10 kV power distribution grid based on the IEEE 14-node network topology [IEEE, 2025] within the RTDS simulation environment [RTDS Technologies Inc., 2023]. The single-line diagram of the IEEE 14-node system is illustrated in Figure 12.9. This topology serves as a critical component in the overall grid architecture, facilitating efficient power distribution from higher-voltage levels to end-users.

The distribution network system model, constructed in the RTDS simulation environment, is depicted in Figure 12.12. This model enables the analysis and simulation of various operational scenarios, allowing for a detailed understanding of the distribution network's performance.

Power Flows of the System The operational parameters of the 10 kV distribution grid have been established based on standard practices and data derived from the IEEE 14-node topology. Specific values for active and reactive power flows, as well as voltage levels at each bus, should be provided here to give a complete picture of the system's performance. These parameters are crucial for evaluating the grid's operational capabilities and efficiency.

Operating Status Following the successful compilation of the model in the Drft interface of the RTDS software [RTDS Technologies Inc., 2023], waveform diagrams for monitoring voltage, current, and power were constructed in the Runtime interface. The setup includes control elements, such as switches and sliders, which facilitate interactive monitoring and control of the system's operational status, as shown in Figure 12.13.

These control elements allow operators to adjust parameters dynamically, simulating various conditions and observing the resulting impacts on voltage and current flows within the distribution network. This functionality is essential for training purposes and for analyzing the system's response under different operational scenarios.

The simulation of the 10 kV distribution grid provides valuable insights into the grid's performance, enabling the identification of potential issues and facilitating the optimization of power delivery to end-users. This comprehensive approach ensures that the distribution network can reliably serve its critical function within the broader power system.

12.1.4 Bay Layer

The bay layer primarily includes secondary equipment such as protection, measurement, and control, and metering devices. In this study, real PLC devices are used as RTUs to simulate the measurement and control of the power grid

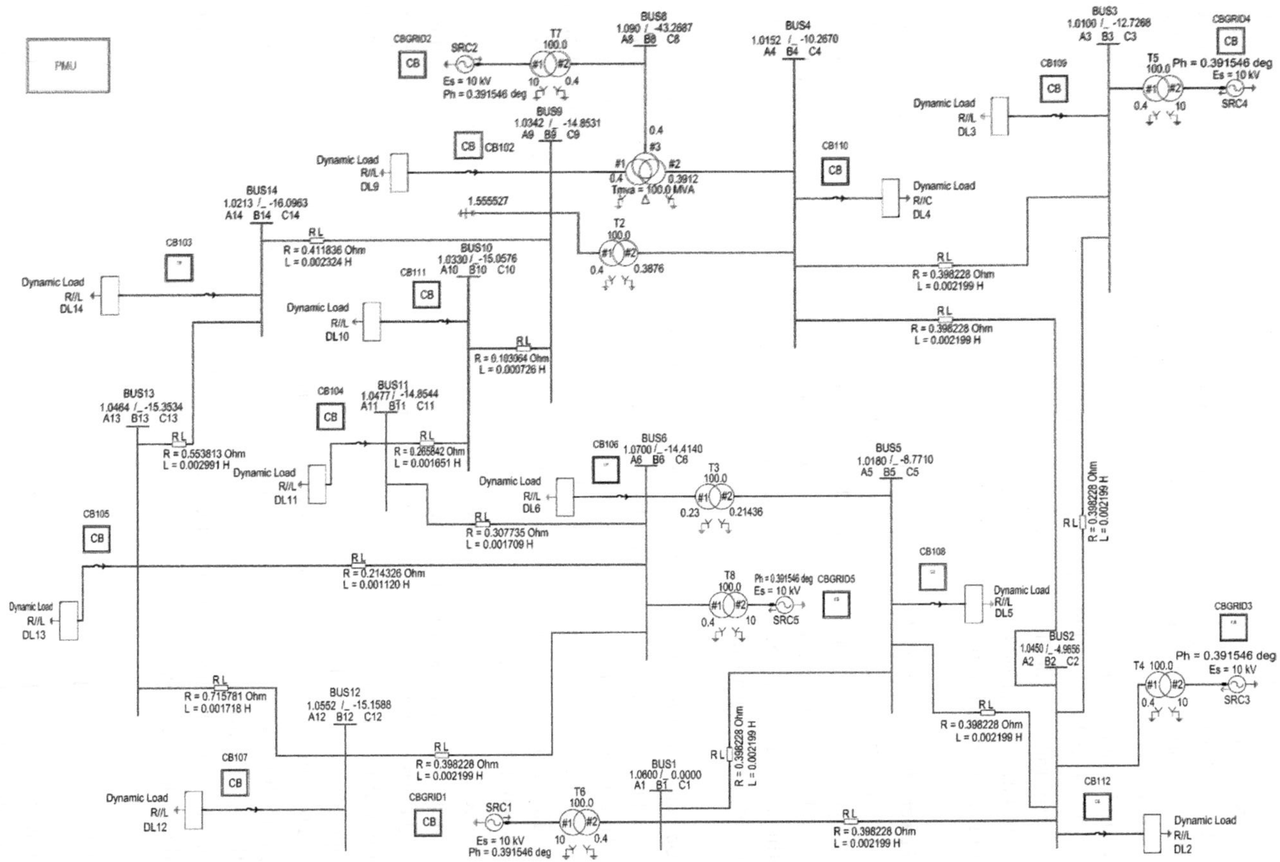

Figure 12.12 10 kV power distribution grid implementation in RTDS.

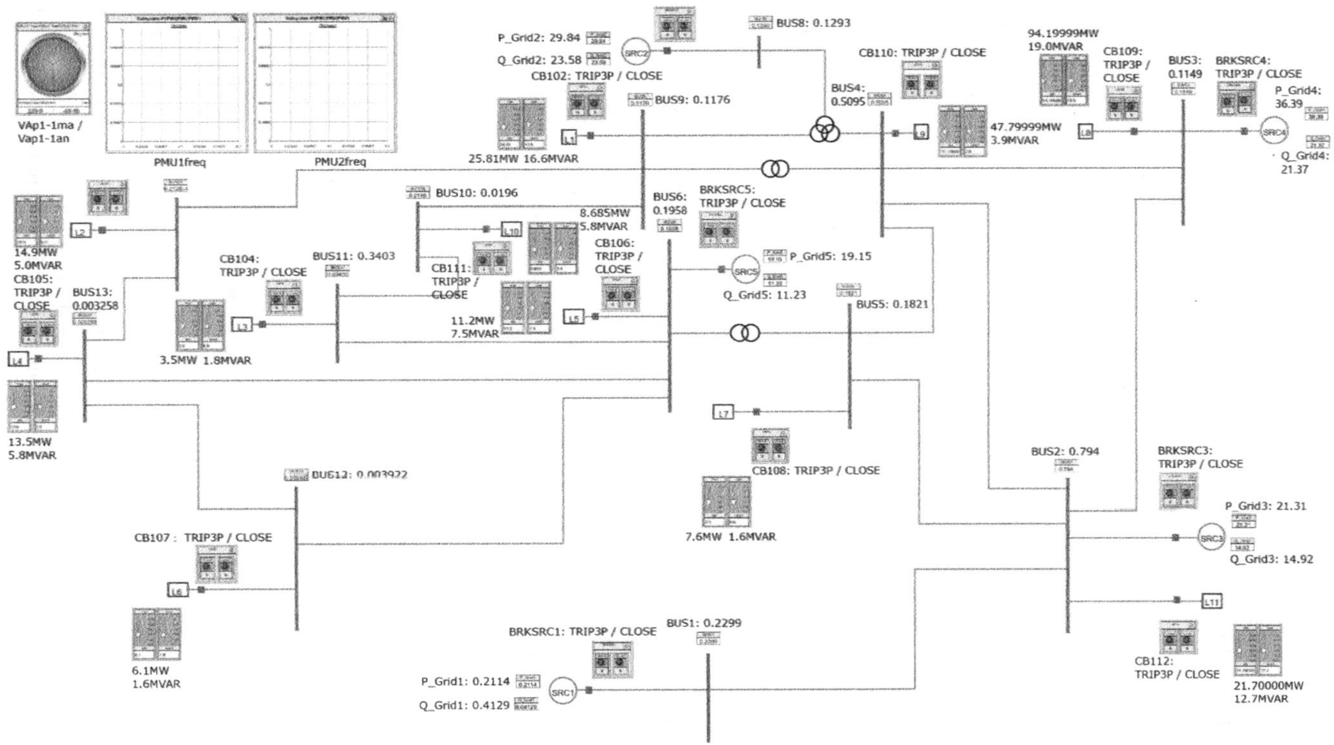

Figure 12.13 Operating status of power grids in RTDS.

environment. The PLC devices communicate with the circuit breaker switches in the power grid system via the GTIO card of the RTDS.

12.1.5 Station Control Layer

The power-grid-monitoring system is built based on the open-source Spire SCADA system that incorporates with Matpower for SE-based power control and dispatches.

12.1.5.1 SCADA

The Spire SCADA system [Babay et al., 2018] provides Byzantine fault-tolerant semantics, ensuring the timeliness of data transmission even under attack. The system can continue to operate normally even if one site becomes disconnected from the rest of the network. Additionally, the Human Machine Interface (HMI) interface is designed using PvBrowser [PvBrowser, 2025], which is an open-source HMI software used by the Romanian power grid. SCADA can interact with diverse PLC/RTU agents through various protocols chosen according to emulation needs.

12.1.5.2 Matpower

Matpower [Zimmerman et al., 2011] is used to implement the monitoring and control of the power grid, divided into three main steps: (1) Based on the power grid topology and the available measurements obtained from the physical grid, state estimation (SE) is performed using Matpower's classic weighted least squares method. (2) Based on the SE results, advanced applications such as optimal power flow and economic dispatch are employed to calculate control strategies that maintain system stability and minimize operational costs. (3) According to the strategy, the control center operates the bay layer's remote terminal units.

SE [Huang et al., 2012] is a crucial component of modern energy management systems in power grids. In the estimation process, the state vector is indirectly determined using measurement equations, which express the functional relationship between the system state vector and the measurement vector. The mathematical model of state estimation is based on measurement equations that reflect the relationships between the network structure, line parameters, state variables, and real-time measurements, as shown in Eq. (12.1).

$$\mathbf{z} = h(\mathbf{x}) + \mathbf{v} \tag{12.1}$$

The vector $\mathbf{z}$ is an m-dimensional measurement vector, which includes measurements such as branch power, nodal injection power, and nodal voltage magnitudes. The system state vector to be determined is $\mathbf{x}$, which consists of the voltage magnitudes and voltage angles at each node. $h(\mathbf{x})$ represents the state variable vector, that is $h(\mathbf{x}) = [h_1(\mathbf{x}), h_2(\mathbf{x}), \dots, h_m(\mathbf{x})]^{\mathrm{T}}$, and $\mathbf{v}$ is the measurement error vector, that is $\mathbf{v} = [\mathbf{v}_1, \mathbf{v}_2, \dots, \mathbf{v}_m]^{\mathrm{T}}$. The covariance matrix of the measurement error, $\mathbf{R}$, can be expressed as a diagonal matrix of the variance of each measurement error: $\mathbf{R} = \mathrm{diag}(\sigma_1^2, \sigma_2^2, \dots, \sigma_m^2)$, where σ_i is the standard deviation of the error. A larger variance indicates a higher probability of error.

The least squares method is one of the most widely used approaches in power system state estimation. The main advantage of this method is that it does not require any statistical characteristics of random variables. It is an estimation method that aims to minimize the sum of the squares of the differences between the measured values $\hat{\mathbf{z}}$ and the estimated values.

12.2 Example for Testing DRTP

Based on the above development environment, we investigate how the DRTP (see Chapters 7 and 8) enhances network reliability within power grid communication systems, focusing on its effect on state estimation (SE) [Huang et al., 2012] accuracy. SE is pivotal for determining the real-time operating conditions of power grids, which in turn supports efficient control and reliable grid operations. Our analysis is grounded in simulations using a practical model of a power transmission grid.

We utilize an IEEE 14-bus transmission grid model, integrating data from Phasor Measurement Units (PMUs) strategically deployed at ten buses to ensure topological observability. Observability is vital for accurate SE [Huang et al., 2012]. To assess how transport protocol reliability, particularly DRTP, impacts SE, we employ two core metrics: the Observable Grid Ratio (OGR) and estimation accuracy.

12.2.1 Key Evaluation Metrics

12.2.1.1 OGR

OGR quantifies the proportion of data that remains observable and intact relative to the total data received. Ensuring a high OGR is essential to maintaining the integrity of the SE process.

12.2.1.2 Accuracy

This metric evaluates the precision of SE outputs by comparing estimated phase angle and voltage magnitude values to actual measurements. Errors in non-observable values are considered infinite ($+\infty$), highlighting a complete loss of observability.

12.2.2 Grid Dynamics and DRTP's Role

The grid's dynamic behavior is captured through periodic fluctuations, recorded at a 1 Hz frequency, as illustrated in Figure 12.14(a) and 12.14(b). These fluctuations, driven by load changes, present challenges for SE in maintaining reliable and accurate outputs.

By simulating disruptions like Primary Forwarding Path (PP) and Multipath Subgraph (MPSG) failures, we analyze how well different transport protocols, including DRTP, sustain SE accuracy and observability. The key disruption metric used is Link Loss Rate in Packets (LLR), representing data loss caused by network interference or failures.

12.2.3 Performance of DRTP

Under PP disruption, DRTP demonstrates robust reliability, measured by End-to-End Failure Rate in Packet Delivery (EEFR). Even under 50% LLR, DRTP maintains an EEFR of 2.68%, a stark contrast to the higher error rates of conventional protocols like IEEE C37.118.2 and MFTP.

Similarly, in MPSG scenarios, DRTP exhibits strong resilience, maintaining an EEFR of 43.74% at 50% LLR, outperforming other protocols like TCP-OPAM and UDP-OPAM.

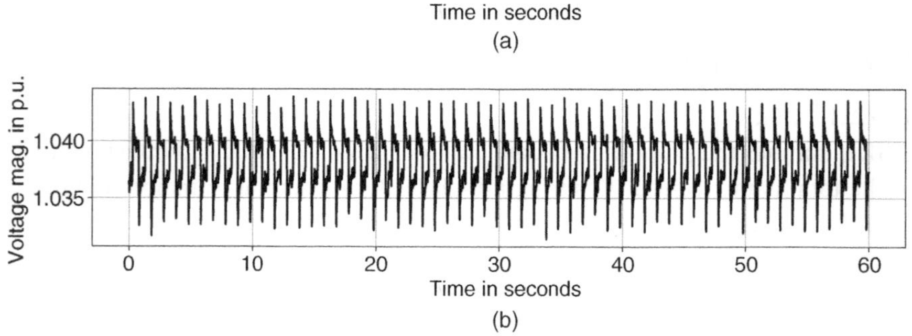

Figure 12.14 Periodic electric variations at bus #1 in a power transmission grid built based on the IEEE 14-bus system. (a) Phase angle changes over time and (b) voltage magnitude changes over time.

12.2.4 Observability and Accuracy Impacts

The reliability of DRTP translates directly into superior grid observability and accurate SE results. Under PP disruptions, DRTP consistently achieves near-complete observability, even with LLRs as high as 100%. Competing protocols, however, experience a significant drop in observability as LLR increases.

Regarding accuracy, DRTP consistently produces lower errors in both phase angle and voltage magnitude estimates compared to traditional protocols (see Figure 12.15(c)). Under 50% LLR, for example, DRTP maintains a phase angle error of only 0.79 milliradians, while other protocols often fail to maintain observability, resulting in infinite errors (see Figure 12.16).

In conclusion, the above results of our analysis underscore the importance of choosing transport protocols capable of withstanding the types of network disruptions commonly encountered in power grids. DRTP not only outperforms established protocols like IEEE C37.118.2 and MFTP in terms of data reliability but also ensures better observability and accuracy in the SE process. This is critical for maintaining grid stability under adverse network conditions.

Incorporating DRTP into power grid communication frameworks can significantly bolster the resilience of SE operations, supporting the real-time accuracy and reliability necessary to manage modern, dynamic grid environments effectively.

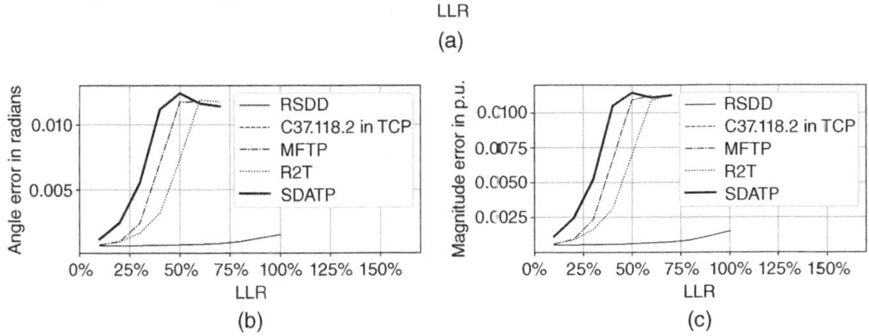

Figure 12.15 Comparative analysis in protocol reliability impacts on SE under full PP disruptions in the IEEE 14-bus system. (a) OGR versus LLR, (b) phase angle errors versus LLR, and (c) voltage magnitude errors versus LLR.

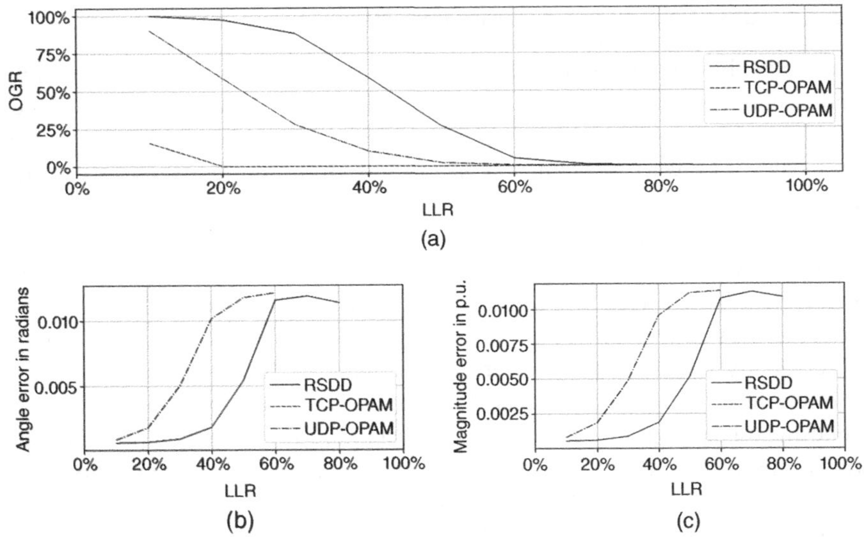

Figure 12.16 Comparative analysis in protocol reliability impacts on SE under full MPSG disruptions in the IEEE 14-bus system. (a) OGR versus LLR, (b) phase angle errors versus LLR, and (c) voltage magnitude errors versus LLR.

12.3 Solution Guidelines

For practical applications, we offer detailed guidelines based on the classical IEC 62439 standard and cutting-edge solutions.

12.3.1 Guidelines of IEC 62439 Protocols

The IEC 62439 protocols [IEC 62439, 2008] (see Chapter 3) offer various approaches to increase network redundancy and resilience, but their limitations in scalability, overhead, and adaptability must be addressed to meet the evolving demands of power grid communication systems. This set of guidelines aims to bridge the gap between theoretical protocol capabilities and practical implementations that optimize performance, reduce unnecessary complexity, and enhance network availability.

12.3.1.1 Strategic Applications of Parallel Redundancy Protocol (PRP) and High-availability Seamless Redundancy (HSR)

While both Parallel Redundancy Protocol (PRP) [IEC, 2021] and High-availability Seamless Redundancy (HSR) [IEC, 2021] provide instantaneous failover capabilities by duplicating traffic across independent paths for wired networks, their use can lead to significant bandwidth overhead and network congestion, especially when applied universally.

PRP and HSR should not be applied across all data streams, as doing so unnecessarily duplicates traffic that does not require strict zero-loss and zero-delay guarantees. Instead, the protocols should be reserved for mission-critical data such as protection commands, grid control signals, and phasor measurement data. These data streams must be delivered without interruption to ensure grid stability and safety.

For example, in SCADA systems [Thomas, 2020], only the control signals that affect the grid's operational state (e.g., relay-tripping signals) should use PRP or HSR. This focused approach ensures the available bandwidth is preserved for other functions, while critical messages are guaranteed fast, uninterrupted delivery.

12.3.1.2 Topology-Tailored Fast Reroute (FRR) Usage

Cross-Network Redundancy Protocol (CRP), Distributed Redundancy Protocol (DRP), and Media Redundancy Protocol (MRP) must be aligned with the network's topology due to their design purposes.

- CRP [IEC, 2010a] is suitable for hierarchical or tree-like topologies, ensuring that control traffic is rerouted with minimal delay.
- DRP [IEC, 2010b] targets ring topologies, where it can leverage two-direction paths on the ring to bypass failed links quickly.
- MRP [IEC, 2023] is highly effective in ensuring availability in ring topologies, but it can be optimized further by segmenting traffic. Traffic segmentation ensures that high-priority traffic (such as grid control signals) receives immediate attention during failover, while lower-priority data (e.g., historical logs or maintenance data) can tolerate slightly longer recovery times. This way maximizes MRP's failover efficiency, minimizing downtime for critical systems while preventing the network from being overwhelmed, reducing the likelihood of congestion during network failures.

12.3.1.3 Hybrid Routing for Flexibility and Scalability

Static routing provides the predictability needed for certain control paths, but as power grid networks scale and diversify, relying solely on static routes can limit flexibility and responsiveness to failures. Introducing hybrid routing – combining static and dynamic routing – provides the necessary adaptability without sacrificing control.

Hybrid routing models can allow operators to maintain static routes for critical control paths, ensuring low latency and high reliability. Dynamic routing, such as Open Shortest Path First (OSPF) [Moy, 1998] or Border Gateway Protocol (BGP)

[Rekhter et al., 2006], can be introduced for less critical traffic or for handling situations where static routes may fail, such as during dynamic load changes or grid reconfigurations.

This approach allows networks to automatically adapt to topology changes or link failures without compromising the performance of essential traffic. Static routing for core network paths (e.g., between control centers) ensures consistency, while dynamic routing in less critical areas or edge networks (e.g., IoT devices) provides scalability, resilience and adaptable without manual intervention.

In conclusion, by addressing the inherent limitations of IEC 62439 protocols [IEC 62439, 2008] with these engineering guidelines, operators can enhance the availability of power grid communication networks, ensuring they are scalable, efficient, and resilient. Targeted application of redundancy, topology-aware protocols, adaptive FRR mechanisms, and hybrid routing models can significantly reduce operational overhead while maximizing fault tolerance, aligning communication networks with the increasing demands of smart grid infrastructures.

12.3.2 Practice Guidelines for OPAM

OPAM (see Chapter 5) offers several technical advantages, especially in smart grid communication networks that demand high availability and data reliability. OPAM ensures that data packets can traverse multiple non-disjoint paths, significantly enhancing network resilience against link or path failures. Those non-disjoint paths are much more diverse than the disjoint paths employed by [IEC 62439 [2008], according to our studies in Section 4.1, which can significantly enhance reliability (see experiment results presented in Section 5.3). By duplicating data across multiple non-disjoint paths, OPAM minimizes the risk of single-point failures, ensuring continuous operation of critical grid applications like SCADA systems, fault detection, and real-time monitoring.

In addition to redundancy, OPAM supports real-time path management, leveraging centralized SDN controllers for dynamic path planning, Wherein, SDN has widely recognized as popular architecture in today's power grids (see Section 1.3.1). This real-time management capability reduces failover times when disruptions occur, helping maintain minimal latency in critical grid communications. Furthermore, the protocol's ability to reorder packets at the destination ensures that data transmission remains reliable, even when packets arrive out of order due to path variations.

Moreover, OPAM's centralized control structure enables enhanced visibility and flexibility for network operators. By monitoring traffic in real time and adjusting paths dynamically, operators can optimize network performance and ensure load balancing, leading to better resource utilization across the communication infrastructure.

12.3.2.1 Applied Scenarios

OPAM is ideally suited for several smart grid applications that require high availability, low latency, and reliable data delivery. Given the importance of timely data in SCADA systems, OPAM's path redundancy ensures that control signals and measurement data are reliably transmitted, even during link failures. Additionally, OPAM can be integrated with many existing Transmission Control Protocol (TCP)/User Datagram Protocol (UDP)-based protocols such as IEC 61850 and IEC 60870-5-104 in power grids. This integration allows for even greater resilience by providing redundancy at the network layer of the network.

12.3.2.2 Deployment Practices

While OPAM offers better data delivery reliability, there are deployment practices:

- **Balancing redundancy and complexity:** When deploying OPAM, it is important to balance the need for redundancy with network complexity. Excessive use of non-disjoint paths can increase overhead and lead to performance bottlenecks, particularly in large-scale grid networks. A recommended practice is to limit the number of paths per flow based on application criticality and network capacity.
- **Managing scalability:** In larger grid environments, scalability becomes a key concern. To manage this, operators should deploy distributed SDN controllers to ensure that path computations and traffic management are decentralized, reducing the risk of congestion and single points of failure. Additionally, traffic engineering techniques such as Multiprotocol Label Switching (MPLS) can be used alongside OPAM to efficiently manage large-scale data flows.
- **Security measures:** While OPAM provides reliability in data delivery, ensuring the security of the SDN controller is paramount, as it is the core component in managing the network. Implementing security protocols like encryption for control messages, firewalling the controller, and integrating intrusion detection systems can mitigate risks. Ensuring controller redundancy can also protect against single points of failure.
- **Fault tolerance optimization:** To fully leverage OPAM's benefits, operators should configure fast failover mechanisms and maintain hot standby paths. This reduces downtime by immediately switching to backup paths when a failure is detected. Network simulations and testing in controlled environments can help fine-tune the configuration before live deployment.

12.3.2.3 Open Concerns

- **Legacy system compatibility:** Integrating OPAM with existing legacy systems in the grid infrastructure may pose compatibility challenges. Operators may need to consider hybrid approaches that allow phased integration of OPAM while maintaining legacy communication paths for critical systems.

- **Performance bottlenecks:** While OPAM improves redundancy, care must be taken to avoid performance bottlenecks caused by increased path management overhead. Regular monitoring and adjusting network parameters to balance load across paths is essential.

12.3.3 Enhancing Path Reliability with TA-FRR

TA-FRR (see Chapter 6) stands out from other fast reroute mechanisms due to its unique design, which combines several powerful features that are particularly beneficial in the demanding environment of smart grid networks. Below are the key advantages that make TA-FRR a superior solution for ensuring path reliability:

- **Proactive path recovery:** One of the most critical aspects of TA-FRR is its proactive nature. Traditional reroute mechanisms often react to network failures by triggering global recalculations of routing paths, leading to delays in restoring connectivity. TA-FRR eliminates this delay by precomputing alternative paths, allowing for immediate redirection of traffic when a failure is detected. This proactive path recovery significantly reduces Mean Time to Repair Network (MTTRN) and enhances overall network availability, ensuring that communication is not disrupted even momentarily.
- **Near-zero latency in rerouting:** In smart grid communications, particularly in critical applications like SCADA, any delay in communication can lead to operational inefficiencies or, in worst cases, grid instability. TA-FRR addresses this by achieving near-zero latency in rerouting. By maintaining a precomputed set of backup paths and immediately switching to these paths upon failure, TA-FRR ensures rerouting latency as low as 0.4 ms. This level of performance is critical for maintaining the real-time data exchange necessary for accurate state estimation, grid dispatching, and fault isolation.
- **High reliability:** The ability to instantaneously reroute traffic also contributes to reducing packet loss, which is a critical factor in maintaining reliable communication. During evaluations, TA-FRR has been shown to reduce End-to-End Packet Loss Rate (EELR) to just 2.03%, far lower than traditional rerouting mechanisms. This low packet loss is crucial for maintaining the integrity of time-sensitive data, such as measurements from PMUs or commands from control centers, which are essential for the stable operation of the grid.
- **Scalability and flexibility:** Smart grid networks are highly diverse, ranging from densely connected urban grids to more sparsely populated rural transmission networks. TA-FRR is designed to be computationally efficient and scalable, making it applicable across a wide range of network sizes and topologies without imposing an unsustainable computational burden on the system.

These advantages collectively make TA-FRR a critical component of any strategy aimed at enhancing the resilience and reliability of communication networks in smart grids.

12.3.3.1 Applied Scenarios

The versatility of TA-FRR allows it to be deployed in a range of smart grid environments, from highly connected urban networks to isolated rural grids. The following scenarios highlight how TA-FRR can be applied to different network architectures to ensure continuous and reliable communication:

- **Urban smart grids:** In metropolitan areas, smart grid networks are densely interconnected to support the high demand for electricity and the critical need for real-time monitoring and control. In these environments, the failure of a single link can cause significant disruptions due to the high volume of data traffic. TA-FRR ensures that, even in the event of link or node failures, data can be rerouted seamlessly, maintaining the necessary low latency and reliability that urban grid operations require. For example, during peak demand periods or emergencies like natural disasters, TA-FRR enables uninterrupted communication between substations, control centers, and distributed energy resources.
- **Rural transmission networks:** In contrast to urban environments, rural transmission networks often have fewer redundant paths and may rely on a smaller number of communication links to cover vast geographic areas. In these cases, the failure of a single link can have a disproportionately large impact on grid operations. TA-FRR provides an effective solution by utilizing the few available connections to reroute traffic, ensuring that critical data can still be delivered even in the face of infrastructure limitations. This is particularly important in areas prone to natural disasters like hurricanes or floods, where physical damage to communication lines is a constant risk.
- **Mixed-technology networks:** Many smart grids feature a mix of legacy and modern communication technologies, where legacy systems rely on older FRR protocols such CRP, DRP, and MRP. TA-FRR can be deployed in these mixed environments to bridge the gap between old and new technologies. By providing an independent FRR mechanism, TA-FRR ensures that even legacy systems benefit from enhanced reliability and resilience.

These scenarios demonstrate the wide applicability of TA-FRR in ensuring reliable communication across diverse network environments, making it an invaluable tool for grid operators.

12.3.3.2 Deployment Practices

To achieve the best results from deploying TA-FRR, smart grid operators should follow a set of best practices that ensure optimal performance and integration with existing systems:

- **Comprehensive network assessment:** Before implementing TA-FRR, it is critical to perform a thorough assessment of the network topology. This includes identifying key communication nodes, determining potential failure points, and analyzing traffic patterns. A comprehensive assessment allows operators to strategically deploy TA-FRR at critical points in the network where it can provide the most benefit in terms of reroute efficiency and minimizing downtime.
- **Interoperability with legacy and modern systems:** Many smart grids still rely on legacy communication systems that may not support modern fast reroute protocols. It is important to ensure that TA-FRR can interoperate with these systems. This may involve implementing additional protocols or middleware that enable seamless communication between legacy and modern components of the grid. Testing for compatibility and potential bottlenecks should be performed prior to deployment.

Following these practices ensures that TA-FRR is deployed effectively, maximizing the reliability and resilience of smart grid communication networks.

12.3.3.3 Open Concerns

While TA-FRR offers many benefits, there are also open questions:

- **Increased computational overhead:** Maintaining the Resilient Subgraph (RSG) and precomputing alternative paths can impose additional computational demands. Although TA-FRR is designed to be computationally efficient, larger networks may require more advanced hardware or cloud-based solutions to handle the computational load. Operators must ensure that their infrastructure can support these demands without causing delays in path computation or rerouting.
- **Legacy system integration:** Integrating TA-FRR with legacy systems can be a challenge, particularly if these systems were not designed with redundancy or fast rerouting in mind. In some cases, operators may need to retrofit legacy systems with additional hardware or software to enable compatibility with modern rerouting mechanisms. This process requires careful planning and execution to minimize disruptions.
- **Cost implications:** The implementation of TA-FRR may involve significant upfront costs, including investments in new hardware, software licenses, and potential training for network operators. While the long-term benefits of improved network reliability and efficiency often outweigh these initial costs, budget constraints can pose a barrier to adoption. Operators must conduct a thorough cost–benefit analysis to justify the investment and seek funding or partnership opportunities to alleviate financial burdens.

- **Security concerns:** As TA-FRR relies on real-time traffic data to make routing decisions, operators must ensure that their systems are secure from potential cyber threats. Implementing robust security measures, such as encryption and intrusion detection systems, is crucial to safeguarding data integrity and maintaining user confidence.
- **Change management and training:** The deployment of TA-FRR often requires significant changes to existing operational processes. Operators must invest time and resources in change management initiatives to facilitate a smooth transition.
- **Network monitoring and maintenance:** Effective monitoring and maintenance are vital to ensuring that TA-FRR functions as intended. Operators need to implement comprehensive monitoring tools to assess the performance of the rerouting algorithms continuously.

12.3.4 Practice Guidelines for DRTP

DRTP (see Chapters 7, 8, and 9) is a purpose-built solution designed to enhance data transport reliability in smart grids, where continuous, reliable communication is vital for grid stability and control. DRTP achieves this through its core mechanism, RSDD, and is further reinforced by the HeuMPSG algorithm, which fortifies route reliability. Together, these mechanisms address the challenges of data transport in environments prone to disruptions, such as sensor network failures or sudden network congestion, ensuring the resilience and integrity of grid communications. This solution achieves superior performance in maintaining reliable data transport over the existing transport protocols tailored to power grids under dynamic and sometimes harsh conditions. This DRTP solution leverages intermediate redundant subpaths on primary forwarding paths for performing reliable retransmissions, minimizing much lower data loss rate, and ensuring a low delay in large-scale communication networks in power grids.

12.3.4.1 Applied Scenarios

DRTP is well suited for a variety of critical smart grid applications where data transport reliability is paramount. These scenarios include:

- **Wide area monitoring and control:** In such systems, synchrophasor data is continuously transmitted from PMUs to Phasor Data Concentrators (PDCs). DRTP's ability to maintain data delivery despite disruptions ensures that the power grid's real-time state is accurately monitored, enabling operators to take quick corrective actions during disturbances.
- **Substation automation:** Substations require reliable data transmission to control circuit breakers, transformers, and other critical components. With

DRTP's robust data transport and routing mechanisms, it ensures that control signals and monitoring data can be transmitted without interruption, even during network instability.
- **Distributed energy resource management:** In a smart grid, resources such as solar panels and wind turbines are distributed across the network. DRTP's ability to dynamically adapt to changes in network conditions ensures that data about energy production and grid conditions reaches control centers reliably, facilitating effective load balancing and grid stability.

12.3.4.2 Deployment Practices

For effective deployment of DRTP in smart grids, certain practices should be followed to maximize its performance:

- **Optimizing network routes:** To fully utilize DRTP's multipath routing, the network topology should be designed to offer sufficient redundancy in communication paths. This involves deploying additional routers or communication nodes at key points to provide alternate routes, ensuring the HeuMPSG algorithm can operate effectively.
- **Real-time network monitoring:** Continuous monitoring of network performance is crucial for DRTP's real-time adaptation capabilities. Implementing network management systems that collect and analyze sensor data can help preemptively detect and resolve issues, allowing DRTP to respond to potential failures more quickly.
- **Load balancing:** DRTP's multipath capability should be leveraged to balance data loads across multiple paths, preventing congestion and ensuring that no single route becomes a bottleneck.
- **Training and deployment testing:** Operators should be trained to understand how DRTP functions in their specific environment, and comprehensive testing should be conducted prior to full deployment. This includes simulating network failures and congestion to ensure that DRTP's RSDD and HeuMPSG mechanisms perform as expected.

12.3.4.3 Open Concerns

While DRTP offers substantial data transport reliability for smart grid communications, there are still some open questions:

- **Resource constraints:** DRTP's use of multiple routes and dynamic routing adjustments can be bandwidth-required, particularly in networks with limited bandwidth or processing power. Deploying DRTP in such environments may require careful planning to optimize performance without overwhelming network resources.

- **Integration with legacy systems:** DRTP provides a new data transport protocol that can seamlessly support application-layer subscription-based protocols such as IEEE C37.118.2. DRTP can be coupled with existing legacy transport protocols such as TCP and UDP through an additional protocol gateway that can be easily done by using programmable network devices, so that they can interact with each other seamlessly.

12.3.5 Guidelines for Practicing ARED

ARED (see Chapter 10) is a robust solution designed to enhance data transmission reliability in smart distribution networks, particularly in environments requiring high availability and fault tolerance. Leveraging the RaptorQ coding scheme, ARED efficiently handles varying data packet sizes and loss rates, making it particularly suitable for critical applications such as smart grid communications, IoT devices, and SCADA systems. ARED offers several significant advantages:

1) **Robust error correction:** Utilizing RaptorQ's fountain coding, ARED provides strong error correction capabilities, ensuring data integrity even in the presence of high packet loss. This is crucial for applications in unreliable network environments.
2) **Adaptability:** ARED dynamically adjusts its encoding parameters based on real-time network conditions. This adaptability allows it to optimize performance, maximizing throughput and minimizing latency during periods of fluctuating network quality.
3) **Scalability:** The architecture of ARED supports seamless integration into existing infrastructures, accommodating an expanding number of devices without degradation in performance. This makes it an ideal choice for growing smart grid applications.
4) **Efficiency:** ARED's encoding process is designed for efficiency, reducing the overhead associated with data transmission. This leads to lower bandwidth consumption and cost savings, particularly beneficial for resource-constrained environments.

12.3.5.1 Applied Scenarios

ARED's versatility enables its application across various scenarios:

1) **Wireless networks in power grids:** ARED provides robust error correction which enhances the reliability of critical operational data, supporting grid stability and efficiency. Additionally, ARED provides the necessary reliability for transmitting telemetry and control signals across potentially lossy networks. This is essential for maintaining operational integrity and timely responses in industrial environments.

2) **IoT device communication:** ARED is well suited for IoT ecosystems where devices often operate under variable conditions. Its adaptability allows for efficient data transfer, enhancing the performance of connected devices in applications ranging from smart metering to environmental monitoring.

12.3.5.2 Deployment Practices

To maximize the effectiveness of ARED, consider the following best practices:

1) **Network assessment:** Before implementation, conduct a thorough assessment of the existing network infrastructure. Understanding the network's characteristics, including typical packet loss rates and latency, will inform the configuration of ARED.
2) **Parameter optimization:** Fine-tune the encoding parameters based on real-time network conditions to achieve optimal performance. Regular monitoring and adjustments can help maintain efficiency and reliability.
3) **Integration testing:** Implement ARED in a controlled environment before full-scale deployment. This allows for testing under various scenarios, ensuring that the system operates as intended in real-world conditions.
4) **Training:** Provide training for personnel involved in the deployment and operation of ARED to facilitate understanding and troubleshooting.

12.3.5.3 Open Concerns

While ARED presents significant benefits, there are open questions to consider:

1) **Integration optimization:** Integrating ARED into existing systems requires careful planning and testing to balance between reliability and redundant packet overhead under constrained bandwidth resources. It is because that ARED's adaptability and efficiency depend on adequate computational resources.
2) **Continuous monitoring:** The effectiveness of ARED hinges on real-time monitoring of network conditions. Establishing a robust monitoring system is essential to ensure that ARED can adapt dynamically and maintain optimal performance.

In conclusion, ARED represents a forward-thinking solution for enhancing data transmission reliability in smart distribution networks. By understanding these, practitioners can effectively leverage ARED to improve the performance of critical communication systems.

12.3.6 Practice Guidelines for PRDD

To combat the Link Flooding Attack (LFA) threat under the evolving integration of Operational Technology (OT) and Information Technology (IT), PRDD (see Chapter 11) emerges as a robust solution designed to enhance the resilience of smart grid communications. PRDD offers several key advantages that make it a compelling choice for smart grid operators:

- **Enhanced resilience against LFAs:** PRDD actively mitigates the risks associated with LFAs by diversifying routes and deploying decoys. This approach makes it significantly more challenging for attackers to target specific paths, thereby preserving the integrity of data transmission.
- **Dynamic route management:** By constantly monitoring network conditions and dynamically adjusting routes, PRDD ensures optimal performance even under attack scenarios. This adaptability allows for continuous service delivery, minimizing disruptions.
- **Scalability:** The architecture of PRDD is designed for scalability, making it suitable for large and complex networks. Its efficient algorithms maintain its effectiveness without excessive overhead.

12.3.6.1 Applied Scenarios

PRDD can be deployed in various scenarios within smart grid communications:

- **Real-time monitoring and control systems:** In environments where data from PMUs, SCADA, and distributed sensors is critical, PRDD can safeguard against LFAs that seek to disrupt real-time telemetry.
- **Cloud managements:** As more utilities shift to cloud computing for data management and analytics, PRDD provides a necessary layer of security. By protecting cloud communication channels from LFAs, utilities can ensure uninterrupted service delivery and accurate data analysis.
- **Integrated networks:** In hybrid environments where OT and IT networks converge, PRDD facilitates seamless communication while enhancing the security posture against multifaceted threats, ensuring that relevant assets are protected.

12.3.6.2 Deployment Practices

To maximize the effectiveness of PRDD in smart grid environments, consider the following best practices:

- **Network assessment:** Conduct a thorough assessment of existing network infrastructure to identify vulnerable Persistent Routes (PRs) and establish baseline performance metrics, which guides the implementation of PRDD.
- **Training programs:** Implement training programs for IT and OT staff to familiarize them with PRDD mechanisms and best practices, to ensure that personnel can effectively respond to evolving threats.
- **Integration with existing security measures:** PRDD should be deployed as part of a broader security framework that includes anomaly detection systems and traffic analysis tools. This multi-layered approach enhances overall network security.

12.3.6.3 Open Concerns

While PRDD presents numerous advantages, several open questions may arise during its implementation:

- **Integration:** PRDD requires an SDN environment, as a popularized architecture in today's power grids (see Section 1.3.1). Those SDNs employ an IP architecture that can seamlessly integrated with legacy environments which possibly use static or dynamic distributed routing. An integration of SDN controllers to monitoring the border switches to connect legacy environment is beneficial for allow PRDD to more broadly deal with LFA threats.
- **Evolving LFA threats:** As attackers develop more sophisticated strategies, PRDD must continually evolve. This requires ongoing research and development efforts on configurations and strategies to enhance the system's effectiveness in countering emerging threats.

In conclusion, PRDD represents a powerful defense mechanism in the ongoing battle against LFAs in smart grids. By enhancing route resilience, smart grid operators can significantly bolster their defenses, which can safeguard their infrastructure and ensure a reliable integration of IT and OT in future.

12.4 Summary

This chapter provides practical guidelines for achieving high-availability networks in smart power grids. It begins by introducing a development environment and offering an example to guide testing approaches. The chapter then outlines solution guidelines for enhancing communication network availability, incorporating both classical IEC 62439 methods and advanced solutions. These cover areas such as data delivery reliability, routing path stability, data transport protocols, application-layer data integrity, and defenses against link-flooding attacks (LFAs). IEC 62439 protocols excel in scenarios tailored to specific ring, tree, and mesh topologies. Meanwhile, cutting-edge solutions push availability goals further with distinct technical advantages, though their proper implementation is essential. Some open questions remain, offering potential for future research to further improve network availability in power grids.

Questions

12.1 Evaluate the effectiveness of DRTP in managing distributed energy resources within a smart grid. How does it facilitate load balancing and grid stability?

12.2 Describe the importance of a development environment in testing communication protocols like DRTP in smart grid applications. What factors should be considered when establishing such an environment?

12.3 Discuss how the deployment of TA-FRR can enhance the reliability of wide-area monitoring and control systems in smart grids. What specific challenges does it address in this context?

12.4 Identify and discuss the implications of increased computational overhead when deploying TA-FRR in large networks. What considerations should operators take into account?

12.5 Outline the key deployment practices necessary for the successful implementation of ARED in a smart distribution network. Why is each practice critical?

12.6 Evaluate the role of XML-based network topology descriptions in enhancing the simulation capabilities of communication networks in smart grids. How do they contribute to better design and testing?

Bibliography

3GPP. 3rd generation partnership project; technical specification group GSM/EDGE radio access network; multiplexing and multiple access on the radio path, 1999.

3GPP. 5G system overview, 2024. URL https://www.3gpp.org/technologies/5g-system-overview.

Amy Babay, Thomas Tantillo, Trevor Aron, Marco Platania, and Yair Amir. Network-attack-resilient intrusion-tolerant SCADA for the power grid. In *IEEE/IFIP DSN*, pages 255–266, 2018.

Ettus. Ettus universal software radio peripheral (USRP), 2025. URL https://www.ettus.com/.

Yih-Fang Huang, Stefan Werner, Jing Huang, Neelabh Kashyap, and Vijay Gupta. State estimation in electric power grids: Meeting new challenges presented by the requirements of the future grid. *IEEE Signal Processing Magazine*, 29(5):33–43, 2012.

IEC. Industrial communication networks - high availability automation networks - Part 4: Cross-network redundancy protocol (CRP), 2010a.

IEC. Industrial communication networks - high availability automation networks - Part 6: Distributed redundancy protocol (DRP), 2010b.

IEC. Industrial communication networks - high availability automation networks - Part 3: Parallel redundancy protocol (PRP) and high-availability seamless redundancy (HSR), 2021.

IEC. Industrial communication networks - high availability automation networks - Part 2: Media redundancy protocol (MRP), 2023.

IEC 62439. High-availability automation networks, 2008.

IEEE. IEEE standard for low-rate wireless networks, 2020.

IEEE. IEEE standard bus network dataset, 2025. URL http://www.ee.washington.edu/research/pstca/.

R. Kuffel, J. Giesbrecht, T. Maguire, R. P. Wierckx, and P. McLaren. RTDS-a fully digital power system simulator operating in real time. In *IEEE EMPD*, volume 2, pages 498–503, 1995.

Linuxcontainers.org. Container and virtualization tools, 2024. URL https://linuxcontainers.org/.

J. Moy. OSPF version 2, 1998.

OpenAirInterface.org. OpenAirInterface –The fastest growing community and software assets in 5G wireless, 2024. URL https://openairinterface.org/.

PvBrowser. PvBrowser, 2025. URL https://github.com/pvbrowser/pvb.

Y. Rekhter, T. Li, and S. Hares. A border gateway protocol 4 (BGP-4), 2006.

George F. Riley and Thomas R. Henderson. The NS-3 network simulator. *Modeling and Tools for Network Simulation*, pages 15–34. Berlin, Heidelberg, Springer, 2010.

RTDS Technologies Inc. Real-time digital simulator (RTDS), 2023. URL https://www.rtds.com/.

TelosB. TelosB, 2025. URL http://www.cmt-gmbh.de/Produkte/WirelessSensorNetworks/TelosB.html.

Mini S. Thomas. *Power System SCADA and Smart Grids*. CRC Press, 2020.

TinyOS. TinyOS, 2025. URL http://www.tinyos.net/.

Ray Daniel Zimmerman, Carlos Edmundo Murillo-Sánchez, and Robert John Thomas. MATPOWER: Steady-state operations, planning, and analysis tools for power systems research and education. *IEEE Transactions on Power Systems*, 26(1):12–19, 2011.

13

Conclusive Remarks and Future Prospects

13.1 Concluding Remarks

This book provides a comprehensive overview of the pioneering technologies in communication networks for smart power grids, covering four key areas: vision, traditional approaches, challenges and current landscape, advanced solutions, and practical applications.

13.1.1 Vision

The book begins by introducing the communication networks of smart power grids, focusing on the background, optical–wireless hybrid network architecture, the complex structure of highly interconnected components, various network protocols, and the emerging threats from Link Flooding Attacks (LFAs) in the context of the increasing convergence of Information Technology (IT) and Operational Technology (OT). Within these networks, we examine the vision for network availability, which is crucial for the stability of power grids. This includes a discussion of technical standards, particularly the IEC 62439 protocols, and the five major aspects of threats and demands. These include the reliability of data delivery, path and data transport, the completeness of application data, and defenses against LFAs. Addressing network availability concerns requires a comprehensive technical approach that involves every layer of the network stack. While optimizing the physical and link layers can be challenging, as they are dependent on the capabilities of network equipment providers, the network, transport, and application layers offer more flexibility and customization, making them especially important for improving network availability.

Communication Networks in Smart Power Grids, First Edition. Boyang Zhou.

Companion Website: www.wiley.com/go/SmartPowerGrids1e

13.1.2 Classical Approaches

Within this context, we present both classical and advanced solutions aimed at optimizing network availability by improving failure recovery effectiveness and efficiency. The classical solutions leverage the IEC 62439 protocols, which include 12 distinct approaches. These approaches encompass end-node-based protocols such as Parallel Redundancy Protocol (PRP), High-availability Seamless Redundancy (HSR), Cross-Network Redundancy Protocol (CRP), and Beacon Redundancy Protocol (BRP), as well as network-based protocols like IP routing, Spanning Tree Protocol (STP), Rapid Spanning Tree Protocol (RSTP), Distributed Redundancy Protocol (DRP), and Media Redundancy Protocol (MRP). Among these, PRP and HSR offer packet aggregation over disjoint paths, while BRP provides failure detection between two end nodes. The remaining protocols deliver Fast Reroute (FRR) capabilities. We detail the principles and interaction processes of these solutions, along with key considerations for their implementation.

These classical approaches perform effectively in wired networks with specific topologies or predefined routes that incorporate multiple-node connectivity. For example, PRP requires independent or disjoint routes and doubly attached nodes to function optimally. However, protocols like IP routing, STP, and RSTP often suffer from prolonged routing convergence times, resulting in inefficiency. Other classical approaches, such as HSR, CRP, DRP, and MRP, necessitate specialized topologies like ring configurations, while BRP is designed for doubly meshed networks. We discuss the limitations of these protocols in the context of modern smart power grid networks, which feature more complex topologies and structures. These limitations underscore the need for enhanced, cutting-edge solutions to ensure higher levels of network availability.

13.1.3 Challenges and Landscape

In addition, we present the specific challenges and corresponding solutions for achieving high availability in the communication networks of smart power grids. Based on the five key demands for network availability, we conduct a thorough analysis of existing solutions, highlighting their strengths and weaknesses. This analysis leads to the identification of five key technical trends for pursuing high-quality solutions: robust non-disjoint path aggregation, efficient and resilient FRR, a robust data transport protocol, and effective defenses against LFAs. Through detailed discussions, we explore the intrinsic reasons why these trends are critical to ensuring the network availability of smart power grids.

13.1.4 Cutting-edge Approaches

From Chapters 5 to 11, we detail the cutting-edge technologies that significantly enhance the communication network availability of smart power grids. These technologies make key contributions to the following related fields.

i) **OpenFlow-based Non-Disjoint Path Aggregation Mechanism** (**OPAM**): To achieve a robust network, we introduce OPAM to enhance the IEC 61850 MMS packet delivery reliability, considering the presence of communication link failures. This solution outperforms many existing solutions, including widely adopted IEC 62439 PRP and iPRP standards, in terms of reliability and data packet delivery time. It efficiently handles both route generation and packet aggregation, making it suitable for large-scale power networks.
ii) **Traffic-Aware Fast Reroute** (**TA-FRR**): To provide a resilient network, we introduce the traffic-aware fast reroute (TA-FRR) mechanism that achieves very efficient communication link failovers in power grids in comparison to today's rerouting techniques. This TA-FRR is complementary to the classical IEC 62439 approaches in FRR. The specific mechanism involves an efficient Resilient Subgraph (RSG) construction algorithm that significantly reduces the Reroute Failure Rate (RFR) with a lower maximum reroute stretch. The Disjoint Subpaths (DSPs) involved in the RSG have rarely been exploited to achieve a significantly lower RFR. Additionally, it includes an efficient link fault discovery method that decreases Link Recovery Time (LRT) during a reroute with cross-layer optimization via dynamically detecting packet arrival timeout. It is beneficial in lowering LRT with lower message cost. The performance of TA-FRR was evaluated, demonstrating strong resilience and efficiency under serious random link faults, compared to today's widely adopted arborescences and disjoint paths-based approaches. Additionally, we have released the source code of TA-FRR to facilitate further developments (GitHub, Inc., 2022).
iii) **Disruption Resilient Transport Protocol** (**DRTP**): To satisfy resilient data transport demand, we delve into DRTP, enabling hop-by-hop retransmission with Redundant Sub-Paths (RSPs) based on Named Data Network (NDN). Simultaneously, we identify a non-trivial distributed collaboration issue during retransmission under multiple link failures, potentially leading to cache mismatching and deterioration of End-to-End Failure Rate in Packet Delivery (EEFR). This issue has not been considered in existing hop-by-hop approaches. To address it, the primary idea is to selectively choose retransmission opportunities, avoiding cache mismatching in a best-effort manner by exploiting fixed frequency. We validate this idea in

experiments conducted on real transmission grid topologies. The results clearly demonstrate that DRTP excels in EEFR with a bounded End-to-End Payload Packet Delivery Time (EEDT) under recursively controlled collaboration, compared to existing hop-by-hop approaches. Additionally, its reliability advantage ensures observability and accuracy, guaranteeing the quality of power grid operations. In the grand showdown of transport protocols, DRTP emerged as the hero, adept at minimizing EEFR while gracefully dancing with low EEDT. The study wasn't just about numbers; it uncovered the profound impact of transport protocol reliability on the power transmission grid. The story was crystal clear – DRTP took the lead, ensuring the grid's observability and accuracy during the storms of full Primary Forwarding Path (PP) and Multipath Subgraph (MPSG) disruptions, outshining its protocol peers.

iv) **Resilient Sensor Data Dissemination (RSDD)**: To implement DRTP, we present the concrete design of RSDD. RSDD consists of eight State Transition Machines (STMs) that efficiently handle incoming DRTP messages while adhering to the recursive collaboration steps of DRTP, thereby reducing EEFR. We confirm RSDD's correctness as a DRTP implementation, adhering to its recursive nature with constant message processing costs, through rigorous verification. Experiment results demonstrate RSDD's substantial advantages in enhancing sensor data dissemination reliability. The source code of RSDD, along with its relevant test environment, is available online (GitHub, Inc., 2023).

In a comparative analysis against various existing transport protocols, RSDD showcased a significant advantage in minimizing EEFR while maintaining a low EEDT. Our study underscores the substantial impact of transport protocol reliability on the power transmission grid. The results unequivocally indicate that RSDD excels in maintaining the grid's topological observability and accuracy for State Estimation (SE) during full PP and MPSG disruptions, outperforming other protocols.

RSDD presents a promising approach to reduce EEFR under link faults in optical Internet-of-things (IoT) networks of the power transmission grid. Researchers interested in replicating our findings can access the RSDD prototype with its algorithms online (GitHub, Inc., 2023).

v) **HeuMPSG – Heuristic MPSG generation algorithm**: To enhance the resilience of DRTP, we introduce HeuMPSG. HeuMPSG efficiently generates Multipath Subgraphs (MPSGs) with reduced fragility under the maximum EEDT constraint, where each generated MPSG features more RSPs per hop in PP. Compared with GenMPSG, the MPSGs generated by HeuMPSG significantly render DRTP more resilient by reducing the retransmission failure probabilities of DRTP under harsh fault links. We comprehensively evaluated the performance of HeuMPSG against GenMPSG by integrating them with DRTP in several IoT networks built according to real power transmission grids. The results show significant improvements of HeuMPSG in EEFR and EEDT under harsh faults, and the execution time of HeuMPSG is much lower than that of GenMPSG. We have implemented and released the source code of HeuMPSG (Zhou, 2023) online, which is fully compliant with DRTP. The code is helpful for further development.

vi) **Adaptive RaptorQ Encoder and Decoder (ARED)**: To enhance the reliability of control message exchanges within the wireless cellular networks of power distribution grids, we introduce ARED, designed to mitigate packet losses in adversarial interference environments. ARED employs RaptorQ coding with adaptive parameter tuning based on channel quality feedback, allowing for dynamic adjustments that optimize performance under varying conditions. Comprehensive evaluations of ARED demonstrate its effectiveness, showing a significant reduction in packet loss rates amid channel noise interference while satisfying the one-way transmission delay requirements. Specifically, ARED achieves a remarkable decrease in uplink and downlink loss rates, reducing them by 64.35% and 89.92%, respectively, compared to traditional methods used in the Long Term Evolution (LTE) domain. These improvements underscore the ARED as a robust solution for reliable communication in challenging wireless environments.

vii) **Persistent Route Diversification Defense (PRDD) against LFAs**: To meet the safeguarding demand of LFAs, we introduce the PRDD defense mechanism against two-stage attacks, leaving the adversary with no choice in launching a disruptive attack on the target. Implementing the defense processes involves designing PRDD to efficiently generate and enforce the defense policy with scalable complexity. PRDD effectively halts flooding on the Persistent Routes (PRs) to mitigate congestion. Simultaneously, it prevents the adversary from probing or speculating the PRs under their corresponding attack stages, disabling every flooding choice for the adversary and avoiding attacks. A comprehensive evaluation of PRDD is provided, assessing its defense effectiveness and performance against two-stage attacks through comparative studies. The results demonstrate that PRDD is more effective in mitigating attacked PRs compared to existing solutions and is efficient in defense.

Those cutting-edge solutions mitigate or improve the classical IEC 62439 standards with strengthened network availability of smart power grids. The reliability of the network layer can be greatly enhanced through the adoption of OPAM and TA-FRR. Additionally, the transport layer's dependability sees a significant uplift with the implementation of RSDD utilizing DRTP, assisted by HeuMPSG. Moreover, the effectiveness of PRDD ensures the efficient mitigation of LFAs in the network layer. These introduced solutions can guide a more stable power grid in operations and dispatches.

13.1.5 Practices

To illustrate this, Chapter 12 provides practical guidelines for readers to develop and apply both classical and cutting-edge solutions in real-world power grid environments. We outline the methods, principles, and a detailed example to guide the creation of high-fidelity development environments for verifying these solutions within power grids, enabling readers to apply them in actual production systems. Additionally, we offer comprehensive solution guidelines for both classical and cutting-edge approaches, covering their features, implementation practices, and unresolved concerns, helping readers better understand how to build high-availability networks for modern power grids.

In conclusion, this book delivers valuable insights into solutions that leverage recent technological advancements to enhance the availability of smart grid communications. Researchers, developers, and operators can utilize the contributions made in this book to improve communication network availability in smart power grids.

13.2 Future Prospects

The communication networks within smart power grids are poised to remain critical infrastructure, driven by advancing technologies that enhance their availability. This availability is not only essential but also a consistent factor that enables secure interactions among various terminals, data concentrators, and Supervisory Control and Data Acquisition (SCADA) systems, all of which are vital for maintaining the stability of power grids.

Over the past few decades, development trends indicate that these networks are evolving into a hybrid form characterized by the tight integration of optical and wireless technologies. The traditional Internet Protocol (IP) architecture increasingly faces challenges in realizing a vision of highly dynamic networks that support heterogeneous services and ubiquitous connectivity. These challenges are spurring the adoption of advanced networks, exemplified by Software-Defined Networkings (SDNs) and NDNs.

To achieve the vision of enhanced availability, it is crucial to apply classical solutions in core power grid networks while simultaneously embracing cutting-edge solutions to improve availability across a broader networking landscape. Addressing these challenges will necessitate various ongoing efforts to advance the field and ensure the resilience of future smart grid communication networks. Future prospects may involve several critical dimensions, particularly those related to operators, researchers, and educators.

13.2.1 Power Grid Operators

Today's power grids are continuously evolving to align with humanity's vision of achieving carbon peak and carbon neutrality. As this vision progresses, smart power grids are becoming increasingly complex in both structure and function. Power transmission grids serve as the backbone and hubs of these systems, while power distribution and consumption grids integrate a blend of traditional and renewable energy sources, often characterized by dynamic and variable loads. The rising demand for reliable communication networks compels operators to adopt a proactive and strategic approach, utilizing both classical IEC 62439 solutions and cutting-edge technologies to enhance network resilience. In the following, we outline key prospects for operators to consider in the coming years to address these evolving challenges.

i) **Leverage packet aggregation**: Operators should maximize the use of packet aggregation protocols such as the cutting-edge OPAM, along with classical solutions like PRP and HSR, all of which provide near-zero recovery times. These protocols are essential in environments where downtime can lead to significant power outages and cascading failures. By ensuring data is transmitted simultaneously along multiple paths, these protocols prevent packet loss during failures. While OPAM exploits non-disjoint paths, PRP requires fully disjoint paths, and HSR operates in a ring topology. Operators must assess the physical layout of transmission substations to implement these protocols effectively, avoiding bottlenecks or single points of failure. Regular performance monitoring, particularly regarding latency from routing convergence, is crucial for ensuring seamless transitions between paths during failures.

ii) **Implement FRR mechanisms**: Operators should adopt both state-of-the-art solutions like TA-FRR and classical IEC 62439 protocols such as MRP (for ring topologies) and DRP, as well as IP routing for arbitrary topologies. These mechanisms provide robust solutions for various network configurations and offer fast recovery times, which are particularly important in urban power distribution grids where minimizing downtime is critical. The choice of technology should be tailored to the specific network topology, ensuring optimal performance.

iii) **Enhancing data transport reliability**: As the demand for smarter and more efficient power grids continues to escalate, the significance of reliable data transport becomes increasingly critical, especially in applications such as wide-area phase measurement. While traditional protocols like TCP and UDP have historically served their purpose, they often struggle to meet the high reliability and low-latency requirements of modern power grid operations. The state-of-the-art DRTP employs intermediate RSPs on PP to achieve significantly more reliable data transport. This precision in data transmission is essential for the accurate functioning of power grid operations and dispatches, particularly during natural disasters or device failures, where maintaining overall grid stability is paramount.

iv) **Enhancing application data completeness**: With the increasing reliance on wireless networks within power grids, the implementation of application data redundancy has become essential for improving the reliability of control and monitoring systems in power distribution and consumption networks. The advanced ARED, which is based on the principles of TCP or UDP, greatly enhances data transmission reliability in smart distribution networks through the innovative use of the RaptorQ coding scheme. By moving beyond traditional protocols, operators can significantly bolster the resilience and efficiency of application protocols, ultimately contributing to the stability and sustainability of power grids in a rapidly evolving energy landscape.

v) **Addressing defenses against LFAs**: Despite the deployment of firewalls, gatekeepers, and intrusion detection systems within power grid networks, threats posed by LFAs present unique challenges. These threats are particularly insidious, as the victim targets – bots, decoys, and other vulnerable devices – can be distributed throughout the network. The consequences of LFAs can lead to network disruptions, jeopardizing essential power grid services. The state-of-the-art PRDD framework offers an effective strategy for mitigating these threats within the network infrastructure. By employing this method, operators can enhance their network security protocols, ensuring greater availability and reliability in their power grid operations.

vi) **Enhance multi-node failure detection**: The multi-node failure detection capabilities of BRP in the IEC 62439 standard should be adopted to effectively detect and isolate failures between nodes, particularly in mesh networks. Operators should prioritize deploying BRP in key sections of the network and ensure robust failure detection mechanisms are in place to prevent fault propagation. Additionally, implementing real-time network monitoring systems can provide insights into node communication performance, enabling quicker fault identification and resolution.

vii) **Prepare to support network evolution**: The future of smart power grids lies in supporting heterogeneous services and ubiquitous connectivity, requiring networks that are flexible, efficient, and highly reliable. To meet these demands, operators should adopt advanced networks such as SDNs and NDNs, which offer scalability and adaptability for evolving grid architectures. The integration of diverse communication technologies, including enhanced optical networks, 5G, and low-power wide-area networks, will be critical across large geographical areas. Traditional IEC 62439 protocols, currently tailored to IP architectures, must evolve accordingly. Operators should look to deploy those advanced networks for enabling better integration with the cutting-edge solutions to address the challenges associated with realizing this vision.

In conclusion, ensuring the availability of communication networks in smart power grids will remain a priority as the complexity and scale of these networks continue to grow. By strategically deploying appropriate solutions across transmission, distribution, and consumption grids, operators can maintain and enhance network availability in the face of emerging challenges. However, achieving this requires ongoing research, careful planning, and collaboration with industry stakeholders. The protocols discussed, when applied correctly, provide a robust foundation for building highly available and resilient communication networks that future power grids will depend on.

13.2.2 Researchers and Developers

The demands outlined in this book highlight the need for innovative solutions in smart grid communications, offering valuable guidance for researchers and developers aiming to advance the field.

For *data delivery reliability* led by OPAM, particularly in critical scenarios like wide-area phase measurement, researchers are encouraged to explore hop-by-hop packet retransmission methods derived from wireless sensor networks. This approach can significantly improve reliability by preventing the retransmission of lost TCP packets at intermediate nodes during failures, thereby enhancing the resilience of smart grids, especially under environmental disruptions (Wang et al., 2006). Additionally, investigating non-disjoint path aggregation in a distributed architecture for large-scale transmission grids can facilitate more robust data transport, crucial for redundancy and continuous communication. Leveraging autonomic communications and nature-based computation of Cognitive Packet Network (CPN) for both the data plane and control plane of OPAM can promote self-awareness of network conditions, which is essential for adapting to dynamic environments and meeting differentiated QoS objectives.

In the realm of *robust and efficient fault recovery* pioneered by TA-FRR, two key areas of research are vital. First, integrating traffic steering policies can optimize data flows and improve overall system performance by dynamically adjusting routing based on real-time network conditions. Second, enhancing link fault detection by increasing awareness of data caching within NDN can yield better strategies for detecting and recovering from link faults. Furthermore, integrating TA-FRR with high-performance data planes presents additional optimization opportunities. By employing fast reroute primitives with low memory overhead

(Chiesa et al., 2019) and exploring forwarding table compression algorithms (Stephens et al., 2016), researchers can significantly enhance performance during reroute processes.

For *data transport reliability* led by DRTP, crafting MPSG with an EEDT constraint may fine-tune DRTP's performance. Moreover, exploring emerging communication technologies, such as 5G, can provide insights into enhancing smart grid communications (Esenogho et al., 2022). Researchers should consider integrating RSDD prototypes into existing high-performance data planes with streamlined pipelines and efficient congestion control mechanisms, leveraging RSDD's consistent message processing costs. Additionally, adapting RSDD for data dissemination applications in diverse machine-to-machine communications could further enhance data transport efficiency through the incorporation of path delay prediction techniques.

The application data completeness represented by ARED warrants research into adaptive fountain or RaptorQ coding, which holds promise for improving terminal communications in wireless networks, particularly for IEC 60870-5-104-enabled terminal devices within cellular networks of power distribution grids. Investigating broadcasting and multicasting features in these networks may lead to further performance optimization. Furthermore, utilizing mature System-on-Chip (SoC) hardware platforms can reduce RaptorQ (RQ) encoding and decoding latency, which is essential for meeting the high-latency demands of power distribution wireless communication services like IEC 61850.

Finally, in the context of *defenses against LFAs* provided by PRDD, its potential to counter various network bandwidth consumption attacks is significant. PRDD effectively neutralizes attempts to consume bandwidth on specific links or hosts, making it a valuable asset in network security. To optimize its defense performance, researchers must carefully consider adversarial attack strategies, especially in the selection of victim decoys, to enhance resilience against exploitation attempts.

13.2.3 Educators

This book delves into the interdisciplinary field of smart grid communications, which exists at the intersection of computer networks, electrical engineering, and telecommunications. Given the advancements and insights presented in "Communication Networks in Smart Power Grids," it is critical that educators in universities and technical institutes adopt a proactive approach to prepare future engineers and technologists. Below are suggestions for how educators can integrate this book into their curricula.

i) **Integrating smart grid communication networks**: It is increasingly vital to incorporate smart grid communication networks into graduate courses for students majoring in electrical engineering with a focus on electric power systems. Prominent institutions, including the University of California, Berkeley, University of California, Los Angeles, Carnegie Mellon University, Tsinghua University, Zhejiang University, Chongqing University, and North China Electric Power University, have already made strides in integrating this field into their curricula. In China, this discipline has been formalized into specialized programs, such as the major in information engineering for power grids.

This book serves as a valuable resource for structuring such courses, offering systematic coverage of both classical and cutting-edge approaches to power grid networks. To enhance this point, educators could develop a course that incorporates specific case studies or project-based learning to illustrate real-world applications. Through these courses, universities have successfully prepared students for roles in the operation, research, and development of the information systems powering today's electrical grids. With a blend of theoretical knowledge and practical application, these programs equip graduates with the skills needed to advance smart grid communications and meet the demands of modern power grid corporations.

ii) **Curriculum development for advanced networking**: Incorporating smart grid communications as part of the curriculum in graduate courses on advanced networking topics, especially for the industrial Internet, is crucial. For over a decade, power grid communication networks have led the way in industrial networking, influencing sectors like oil and railway operations. These networks leverage advanced optical–wireless communication technologies and highly available architectures, and the lessons learned from power grids can provide valuable insights for other industries.

This book thoroughly covers various technologies aimed at enhancing network availability and offers a distinctive perspective on the challenges and cutting-edge solutions within the field. Educators should integrate discussions on topics such as packet aggregation, data transport protocols, and application data coding into the curriculum to provide students with comprehensive knowledge. The foundational concepts presented in the book can serve as technical guidelines for students, helping them grasp the principles and key technologies that underpin network availability. This, in turn, facilitates deeper knowledge acquisition and stimulates academic research in these areas.

iii) **Theoretical and practical integration**: Utilizing this book, courses can be designed to encompass both theoretical and practical aspects of high-availability communication networks in smart grids. By incorporating modules on IEC 62439 standards and their applications, educators can equip students with a solid foundation in industry-relevant knowledge. Additionally, educators should encourage students to engage in research initiatives that focus on high-availability networks, exploring unresolved challenges, such as optimizing fault recovery processes and developing transport protocols.

Students should also have access to state-of-the-art laboratories that simulate real-world scenarios in smart power grid communications, leveraging the environments, methods, and codes introduced in this book. Engaging in practical projects will provide students with invaluable hands-on experience, enhancing their problem-solving skills. Furthermore, instructors can utilize the exercise questions at the end of each chapter to reinforce understanding of the technical content. Establishing research centers dedicated to smart grid technologies will facilitate collaboration with industry partners, ultimately bolstering the institution's reputation in this critical field.

In conclusion, this book serves as a valuable resource for educators aiming to cultivate expertise in the field of smart grid communications by thoroughly examining the integration of computer networks within complex cyber-physical power grid systems. This holistic approach equips students to address the multifaceted challenges inherent in smart grid communications more effectively. By adopting this proactive strategy, educational institutions can enhance the learning experience, ensuring that graduates are well prepared to meet the demands of tomorrow's energy systems.

Bibliography

Marco Chiesa, Roshan Sedar, Gianni Antichi, Michael Borokhovich, Andrzej Kamisiński, Georgios Nikolaidis, and Stefan Schmid. PURR: A primitive for reconfigurable fast reroute: Hope for the best and program for the worst. In *ACM CoNEXT*, pages 1–14, 2019.

Ebenezer Esenogho, Karim Djouani, and Anish M. Kurien. Integrating artificial intelligence Internet of things and 5G for next-generation smartgrid: A survey of trends challenges and prospect. *IEEE Access*, 10:4794–4831, 2022.

GitHub, Inc. Traffic aware fast reroute mechanism (TA-FRR), 2022. URL https://github.com/zhouby-zjl/tafrr.

GitHub, Inc. The source code of the resilient sensor data dissemination mechanism (RSDD) implementing the disruption resilient transport protocol (DRTP) for power transmission grids, 2023. URL https://github.com/zhouby-zjl/drtp.

Brent Stephens, Alan L. Cox, and Scott Rixner. Scalable multi-failure fast failover via forwarding table compression. In *ACM Symposium on SDN Research*, pages 1–12, 2016.

Chonggang Wang, K. Sohraby, Bo Li, M. Daneshmand, and Yueming Hu. A survey of transport protocols for wireless sensor networks. *IEEE Network*, 20(3):34–40, 2006.

Boyang Zhou. The source code of the HeuMPSG algorithm, 2023. URL https://github.com/zhouby-zjl/heumpsg.

Copyright Page

Communication Networks in Smart Power Grids

Reused Content Acknowledgments:

Communication Networks in Smart Power Grids, First Edition. Boyang Zhou.

Companion Website: www.wiley.com/go/SmartPowerGrids1e

Bibliography

Boyang Zhou. Adaptive RaptorQ coding to improve reliability of LTE communication in distribution power grid. *Computer Engineering and Applications*, 55(3):76–82, 2019.

Boyang Zhou, Qiang Yang, Yansong Wang, and Chunming Wu. Reliable communication in transmission grids based on nondisjoint path aggregation using software-defined networking. *IEEE Transactions on Industrial Informatics*, 15(2):842–855, 2019.

Boyang Zhou, Chunming Wu, Qiang Yang, and Xiang Chen. DRTP: A disruption resilient hop-by-hop transport protocol for synchrophasors measurement in electric transmission grids. *IEEE Access*, 10:133898–133914, 2022a.

Boyang Zhou, Chunming Wu, Qiang Yang, Xiang Chen, and Dong Zhang. A persistent route diversification mechanism for defending against stealthy crossfire attack. *Computer Engineering and Applications*, 2022:2566681, 1–20, 2022b.

Boyang Zhou, Chunming Wu, Yang Qiang, Qian Yaguan, and Nie Yinghui. Resilient sensor data dissemination to mitigate link faults in IoT networks with long-haul optical wires for power transmission grids. *IEEE Internet of Things Journal*, 11(9):15919–15939, 2023a.

Boyang Zhou, Chunming Wu, Qiang Yang, Zhen Zhu, Yinghui Nie, and Wenjie Yu. Traffic-aware fast reroute mechanism exploiting disjoint subpaths for named data networks. In *IEEE ICC*, pages 4013–4019, 2023b.

Boyang Zhou, Tong Ye, Yinghui Nie, Wenjie Yu, and Chunming Wu. Generating routes less fragile to improve disruption-resilient transport protocol for phasor data in optical IoT networks of power transmission grids. In *IEEE ICPADS*, 2023c.

Boyang Zhou, Zhen Zhu, Yinghui Nie, Xiang Chen, Bingjing Yan, Qiang Yang, and Chunming Wu. Fast proactive re-route with detours on joint paths in named-data networks for communications of smart electric power transmission grids. In *COMSNETS*, pages 640–648, 2023d.

Index

Communication Networks in Smart Power Grids, First Edition. Boyang Zhou.

Companion Website: www.wiley.com/go/SmartPowerGrids1e

Printed and bound by CPI Group (UK) Ltd, Croydon, CR0 4YY
17/09/2026
14964333-0005